PHYSICS IN COLLISION 19

PHYSICS IN COLLISION 19

Edited by

M. Campbell
J.M. Qian
University of Michigan

World Scientific
Singapore • New Jersey • London • Hong Kong

Published by

World Scientific Publishing Co. Pte. Ltd.

P O Box 128, Farrer Road, Singapore 912805

USA office: Suite 1B, 1060 Main Street, River Edge, NJ 07661

UK office: 57 Shelton Street, Covent Garden, London WC2H 9HE

British Library Cataloguing-in-Publication Data
A catalogue record for this book is available from the British Library.

PHYSICS IN COLLISION XIX

ISBN 981-02-4354-5

Printed in Singapore.

PREFACE

We are privileged to host the 1999 Physics in Collision conference in Ann Arbor. The three days of plenary sessions were held in our newly renovated physics colloquium room on the central campus of the University of Michigan. Michigan has a long tradition of excellence in particle physics. Coincidentally Professor Emeritus Martinus J.G. Veltman shared 1999 Nobel prize in physics for his contribution to the quantum structure of electroweak interactions.

This year's program featured 27 invited speakers reviewing and updating key topics in elementary particle physics. Twenty-four of those presentations are summarized in this edition of the proceedings. Thanks to the articulate speakers and enthusiastic participants, the conference provided a great forum for informal discussions on experimental results and their implications.

The conference opened with reports on electroweak physics. A decade of precision experiments in laboratories around the world failed to uncover any significant deviations from standard model predictions. Precise W boson and top quark mass measurements suggest a low mass Higgs boson in the standard model, possibly within the reach of the LEP II and the upgraded Tevatron colliders. These presentations were naturally followed by a summary of the latest results on searches for Higgs and supersymmetry. Though no positive search result was reported, our hope is high for discovering the Higgs boson at the LEP II or at the Tevatron.

A number of speakers reviewed the current status of Quantum Chromodynamics, covering topics such as two photon physics, hard and soft diffraction, parton distribution functions and spin structure of nuclei, as well as jet physics. Tremendous progress has been made in recent years in understanding the source of nucleon spins and in refining parton distribution functions for precision physics.

As in the past few series of the conference, the mystery in neutrino sector continued to generate excitement at this conference. We heard three reports on neutrino physics: atmospheric, solar and reactor/accelerator based neutrino experiments and results. Impressive Super-K results strongly suggest $\nu_\mu \to \nu_\tau$ oscillation, a scenario confirmed by less precise Soudan2 and MACRO data. A number of new experiments scheduled to come online in the next few years will provide valuable additional information to solve the neutrino puzzle.

On the last day of the conference, speakers summarized the latest results on strange and heavy quark physics. High statistics from CLEO, LEP I and Tevatron has enabled experimenters to study many rare charm and bottom quark decays for the first time. As the conference was going on, we learnt that both SLC and KEK B-factories were coming online. A flood of new results is expected from these factories in coming years. Among many other interesting results presented, the first measurements of $\sin 2\beta$ by CDF and ε'/ε by KTeV provide us a preview what to expect in the near future.

Astrophysics has slowly crawled its way into the scientific program of the conference. In this conference, we heard two presentations, one on the first observation of the optical counterpart of a gamma ray burst and the other on the issue of cosmological constant and dark matter. The speakers provided detailed account of and in depth analyses of the latest observations. Needless to say, these two talks were among the best attended and generated lots of coffee-hour discussions.

Finally, we are grateful to the financial support provided by the US Department of Energy, Exabyte, ADCO Circuits Inc., Hewlett-Parkard, Tektronix, Rittal Corporation and Micron Semiconductor, as well as the Department of Physics at the University of Michigan. We thank the International Advisory Committee for the exciting and comprehensive scientific program and the Local Organizing Committee for the successful hosting of the conference. In particular, special thanks goes to Mrs. Tina Wells. Without her dedication, the conference would not be possible. Lastly we'd like to express our appreciation to all speakers and participants for their contributions to the conference.

Myron K. Campbell
Jianming Qian

International Advisory Committee

J.A. Appel	Fermilab, USA
B. Aubert	LAPP Annecy, France
G. Bellini	Milano University, Italy
A. Bettini	Padova University, Italy
W. Blum	CERN, Switzerland
J-M. Brom	IRes Strasbourg, France
M. Campbell	University of Michigan, USA (Chair)
H. Castilla	CINVESTAV, Mexico
F.L. Fabbri	INFN Frascati, Italy
B. Foster	Bristol University, United Kingdom
R.S. Galik	Cornell University, USA
A. Goshaw	Duke University, USA
J. Grunhaus	Tel Aviv University, Israel
E-E. Kluge	Heidelberg University, Germany
L. Littenberg	BNL, USA
A. Maki	KEK, Japan
B. Naroska	University of Hamburg, Germany
K. Rybicki	INP Krakow, Poland
P. Strolin	Napoli University, Italy
D. Su	SLAC, USA
H.D. Wahl	Florida State University, USA

Local Organizing Committee

D. Amidei	D. Gerdes	J. Qian
B. Ball	W. Lorenzon	K. Riles
M. Campbell	G. Kane	R. Thun

Scientific Secretaries
B. Bonde, A. Carter, E. Lang, K. Logan, E. Luplow

Secretariat
Tina Wells

Table of Contents

THE MASS AND WIDTH OF THE W BOSON

MARK LANCASTER

*Department of Physics and Astronomy, University College London, Gower Street,
London, WC1E 6BT, UK.
E-mail: markl@hep.ucl.ac.uk*

The Tevatron and LEP2 experiments presently provide the most precise direct
determinations of the mass and width of the W boson. The combined results are
: M_W = 80.394 ± 0.042 GeV and Γ_W = 2.095 ± 0.106 GeV. The results are in
excellent agreement with the predictions of the Standard Model. In this article
the latest results are described and the systematic errors which could limit further
significant improvements in the precision of these measurements are reviewed.

1 Introduction

A precise W mass measurement allows a stringent test of the Standard Model (SM)
beyond tree level where radiative corrections lead to a dependence of the W mass on
both the top quark mass and the mass of the, as yet unobserved, Higgs boson. The
dependence of the radiative corrections on the Higgs mass is only logarithmic whilst
the dependence on the top mass is quadratic. Simultaneous measurements of the
W and top masses can thus ultimately serve to further constrain the Higgs mass
beyond the LEP1/SLD limits and potentially indicate the existence of particles
beyond the SM. Similarly, non SM decays of the W would change the width of the
W boson. A precise measurement of the W width can therefore be used to place
constraints on physics beyond the SM. The latest results on the W mass from the
LEP2 and Tevatron experiments are now of such a precision that the uncertainty
on the top mass [2] is beginning to become the limiting factor in predicting the mass
of the Higgs boson.

The results described in this article are from the LEP experiments with data
taken above the WW production threshold in e^+e^- collisions at $\sqrt{s} = 161-189$ GeV
and from the two Tevatron experiments with data taken at $\sqrt{s} = 1.8$ TeV from $p\bar{p}$
collisions. Significant improvements in the statistical error of these measurements
is anticipated since the LEP2 experiments are continuing to take data in the range
$192 < \sqrt{s} < 202$ GeV and the Tevatron experiments resume data taking in the
year 2000 at $\sqrt{s} = 2.0$ TeV. However, the precision of both sets of measurements
are now becoming limited by systematic uncertainties. Further progress will need
to be made in the understanding of various issues for a significant improvement in
the precision of these results to be realised.

2 Data Samples and Event Selection

The first observation and measurements of the W boson were made at the CERN
SpS by the UA1 and UA2 experiments [1]. These measurements were based on modest event samples ($\sim 4k$ events) and integrated luminosity (12 pb^{-1}). Since that
time the Tevatron and LEP2 experiments have recorded over 1 fb^{-1} of W data.
The Tevatron experiments have the largest sample of W events : over 180,000 from

a combined integrated luminosity of ~ 220 pb^{-1}. The LEP experiments, despite a very large integrated luminosity (~ 15000 pb^{-1} total across all experiments), have event samples substantially smaller than the Tevatron experiments. The LEP2 results presented here are based on event samples of $\sim 6k$ events per experiment. However, despite the smaller statistics of the event sample in comparison to the Tevatron experiments, the LEP2 experiments ultimately achieve a comparable precision. On an event by event basis, the LEP2 events have more information; in particular the LEP2 experiments can impose energy and momentum constraints because they have a precise knowledge of the initial state through the beam energy measurement. The NuTeV experiment [3] at FNAL also has a large sample ($\sim 10^6$) of charged current events mediated by the exchange of a W-boson. This allows an indirect determination of the W mass through a measurement of $\sin^2\theta_w$. This is done by comparing the neutral and charged current cross sections in νFe and $\bar{\nu}$Fe collisions. In this article only the direct determinations of the W mass and width from LEP2 and the Tevatron will be discussed.

3 Mass Measurement : Method

At LEP2, W bosons are produced in pairs though s-channel Z or γ exchange or by t-channel neutrino exchange. These three production channels are referred to as the CC03 channels. The other small non-CC03 contributions to the measured cross section are corrected for by Monte-Carlo such that the data can be directly compared to the CC03 predictions. The W pairs decay 46% of the time to a purely hadronic final-state, 44% of the time to a semi-leptonic final state, i.e. where one W decays to qq' and the other to $l\nu$ and 10% of the time to a purely leptonic final state. Leptons are detected in all three flavours : electron, muon and tau. The W mass at LEP2 has been measured by two methods :

- Through a measurement of the WW cross section at threshold i.e. $\sqrt{s} = 161$ GeV.

- By explicit reconstruction of the invariant masses of the two W bosons.

The first measurement has a small systematic error and its error is entirely dominated by the statistical uncertainty. Ultimately, given a sufficient amount of integrated luminosity, this is the most precise method for determining the W mass.

At the Tevatron W bosons are predominantly produced singly by quark anti-quark annihilation. The quarks involved are mostly valence quarks becauase the Tevatron is a $p\bar{p}$ machine and the x values involved in W production ($0.01 \lesssim x \lesssim 0.1$) are relatively high. The W bosons are only detected in their decays to $e\nu$ (CDF and D$\emptyset$) and $\mu\nu$ (CDF only) since the decay to qq' is swamped by the QCD dijet background whose cross section is over an order of magnitude higher in the mass range of interest. At the Tevatron one does not know the event $\hat{s}$ and one cannot determine the longitudinal neutrino momentum since a significant fraction of the products from the $p\bar{p}$ interaction are emitted at large rapidity where there is no instrumentation. Consequently, one must determine the W mass from transverse quantities [4] namely : the transverse mass ($\mathrm{M_T}$), the charged lepton $\mathrm{P_T}$ ($\mathrm{P_T^l}$) or the missing transverse energy ($\not{E}_T$). $\not{E}_T$ is inferred from a measurement of P_T^l and the remaining P_T in the detector, denoted by $\vec{U}$ i.e.

$$\vec{\not{E}_T} = -(\vec{U} + \vec{P_T^l}) \quad \text{and } M_T \text{ is defined as}$$

$$M_T = \sqrt{2P_T^l \not{E}_T (1 - \cos\phi)} \quad \text{where } \phi \text{ is the angle between } \vec{\not{E}_T} \text{ and } \vec{P_T^l}$$

$\vec{U}$ receives contributions from two sources. Firstly, the so-called W recoil i.e. the particles arising from initial state QCD radiation from the $q\bar{q}$ legs producing the hard-scatter and secondly contributions from the spectator quarks ($p\bar{p}$ remnants) and additional minimum bias events which occur in the same crossing as the hard scatter. This second contribution is generally referred to as the underlying-event contribution. Experimentally these two contributions cannot be distinguished. Owing to the contribution from the underlying-event, the missing transverse energy resolution has a significant dependence on the instantaneous $p\bar{p}$ luminosity. M_T is to first order independent of the transverse momentum of the W (P_T^W) whereas P_T^l is linearly dependent on P_T^W. For this reason, and at the current luminosities where the effect of the $\not{E}_T$ resolution is not too severe, the transverse mass is the preferred quantity to determine the W mass. However, the W masses determined from the P_T^l and $\not{E}_T$ distributions provide important cross-checks on the integrity of the M_T result since the three measurements have different systematic uncertainties.

The systematics of the LEP2 and Tevatron measurements are very different and thus provide welcome complementary determinations of the W mass. The systematics at LEP2 are dominated by the uncertainty in the beam energy (which is used as a constraint in the mass fits) and by the modeling of the hadronic final state, particularly for the events where both W bosons decay hadronically. At the Tevatron, the systematics are dominated by the determination of the charged lepton energy scale and the Monte-Carlo modeling of the W production, in particular its P_T and rapidity distribution. At the Tevatron, one cannot use a beam energy constrain to reduce the sensitivity of the W mass to the absolute energy (E) and momentum (p) calibration of the detector. Any uncertainty in the detector E, p scales thus enters directly as an uncertainty in the Tevatron W mass. This means that the absolute energy and momentum calibration of the detectors must be known to better than 0.01%. By contrast at LEP, an absolute calibration of 0.5 % is sufficient.

4 Tevatron W Mass

The W mass at the Tevatron is determined through a precise simulation of the transverse mass line-shape, which exhibits a Jacobian edge at $M_T \sim M_W$. The simulation of the line-shape relies on a detailed understanding of the detector response and resolution to both the charged lepton and the recoil particles. This in turn requires a precise simulation of the W production and decay. The similarity in the production mechanism and mass of the W and Z bosons is exploited in the analysis to constrain many of the systematic uncertainties in the W mass analysis. The lepton momentum and energy scales are determined by a comparison of the measured Z mass from $Z \to e^+e^-$ and $Z \to \mu^+\mu^-$ decays with the value measured at LEP. The simulation of the W P_T and the detector response to it are determined by a measurement of the Z P_T which is determined precisely from the decay leptons and by a comparison of the leptonic (from the Z decay) and non-leptonic E_T quantities in Z events. The reliance on the Z data means that many of the systematic uncertainties in the W mass analyses are determined by the statistics of the Z sample.

The W and Z events in these analyses are selected by demanding a single isolated high P_T charged lepton in conjunction with missing transverse energy (W events) or a second high P_T lepton (Z events). Depending on the analyses, the $\not{E}_T$ cuts are either 25 or 30 GeV and the lepton P_T cuts are similarly 25 or 30 GeV. CDF only uses W $\to$ eν and W $\to$ $\mu\nu$ events [5] in the rapidity region : $|\eta| < 1$, whereas DØ [6] uses W $\to$ eν events out to a rapidity of $\sim$ 2.5. In total $\sim$ 84k events are used in the W mass fits and $\sim$ 9k Z events are used for calibration.

4.1 Lepton Scale Determination

The lepton scales for the analyses are determined by comparing the measured Z masses with the LEP values. The mean lepton P_T in Z events ($P_T \sim$ 42 GeV) is $\sim$ 5 GeV higher than in W events, consequently in addition to setting the scale one also needs to determine the non-linearity in the scale determination i.e. to determine whether the scale has any P_T dependence. DØ does this by comparing the Z mass measured with high P_T electrons with J/ψ and π^0 masses measured using low P_T electrons as well as by measuring the Z mass in bins of lepton P_T. In the determination of CDF's momentum scale the non-linearity is constrained using the very large sample of $J/\psi \to \mu\mu$ and $\Upsilon \to \mu\mu$ events which span the P_T region : $2 < P_T < 10$ GeV. The non-linearity in the CDF transverse momentum scale is consistent with zero (see Fig. 1). This fact in turn can be exploited to determine the non-linearity in the electron transverse energy scale through a comparison of the measured E/p with a MC simulation of E/p where no E_T non-linearity is included. The lepton scale uncertainties form the largest contribution to the W mass systematic error. The non-linearity contribution to the scale uncertainty is typically $\sim$ 10% or less.

The Z lineshape is also used by both experiments to determine the charged lepton resolution functions i.e. the non-stochastic contribution to the calorimeter resolution and the curvature tracking resolution in the case of the CDF muon analysis.

4.2 W Production Model

The lepton P_T and $\not{E}_T$ distributions are boosted by the non zero P_T^W and the $\not{E}_T$ vector is determined in part from the W-recoil products. As such a detailed simulation of the P_T^W spectrum and the detector response and resolution functions is a necessary ingredient in the W mass analysis. The W P_T distribution is determined by a measurement of the Z P_T distribution (measured from the decay leptons) and a theoretical prediction of the W to Z P_T ratio. This ratio is known with a small uncertainty and thus the determination of the W P_T is dominated by the uncertainty arising from the limited size of the Z data sample. The P_T^Z distribution

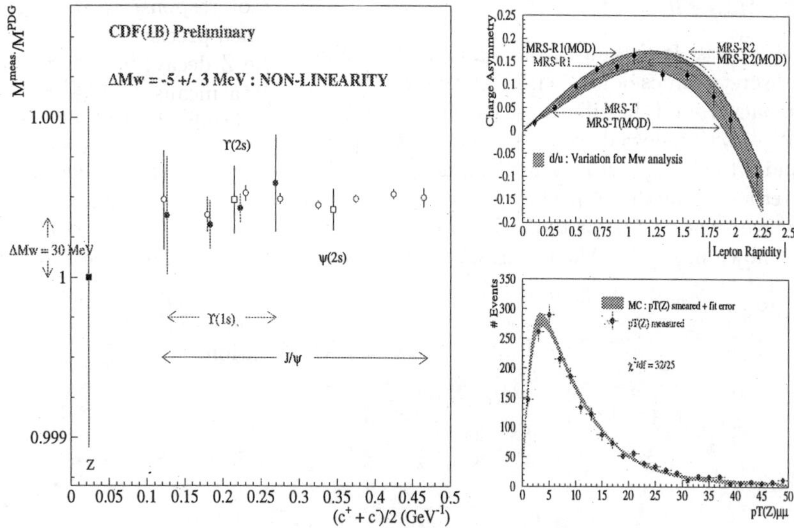

Figure 1. LEFT: The CDF determination of the momentum scale and non-linearity using dimuon resonances. RIGHT UPPER: The modified PDF sets used in the Mw analysis, which span CDF's W charge asymmetry measurement. RIGHT LOWER : The Z P_T distribution as measured by CDF in the $Z \to \mu^+\mu^-$ channel.

of the CDF $Z \to \mu^+\mu^-$ sample is shown in Fig. 1. The detector response and resolution functions to the W-recoil and underlying event products are determined by both experiments using Z and minimum bias events. Since the W-recoil products are typically produced along the direction of the vector boson P_T and the underlying event products are produced uniformly in azimuth, the response and resolution functions are determined separately in two projections – one in the plane of the vector boson and one perpendicular to it. Typically one finds the resolution in the plane of the vector boson is poorer owing to the presence of jets (initial state QCD radiation from the quark legs) which are absent in the perpendicular plane where the resolution function matches closely that expected from pure minimum bias events. The parton distribution functions (PDFs) determine the rapidity distribution of the W and hence of the charged lepton. Both experiments impose cuts on the rapidity of the charged lepton and so a reliable simulation of this cut is necessary if the W mass determination is not to be biassed. On average the u quark is found to carry more momentum than the d quark resulting in a charge asymmetry of the produced W i.e. $W^{+(-)}$ are produced preferentially along the $(p\bar{p})$ direction. Since the V-A structure of the W decay is well understood, a measurement of the charged lepton asymmetry therefore serves as a reliable means to constrain the PDFs. To determine the uncertainty in the W mass arising from PDFs, MRS PDFs were modified to span the CDF charged lepton asymmetry measurements [7]. This is illustrated in Fig. 1.

4.3 Mass Fits

The W mass is obtained from a maximum likelihood fit of M_T templates generated at discrete values of M_W with Γ_W fixed at the SM value. The templates also include the background distributions, which are small ($< 5\%$) and have three components : $W \to \tau\nu$, followed by $\tau \to \mu/e\nu\nu$, QCD processes where one mis-measured jet mimics the $\not{E}_T$ signature and the other jet satisfies the charged lepton identification criteria and finally Z events where one of the lepton legs is not detected. The transverse mass fits for the D∅ end-cap electrons and the two CDF measurements are shown in Fig. 2. The uncertainties associated with the measurements are listed

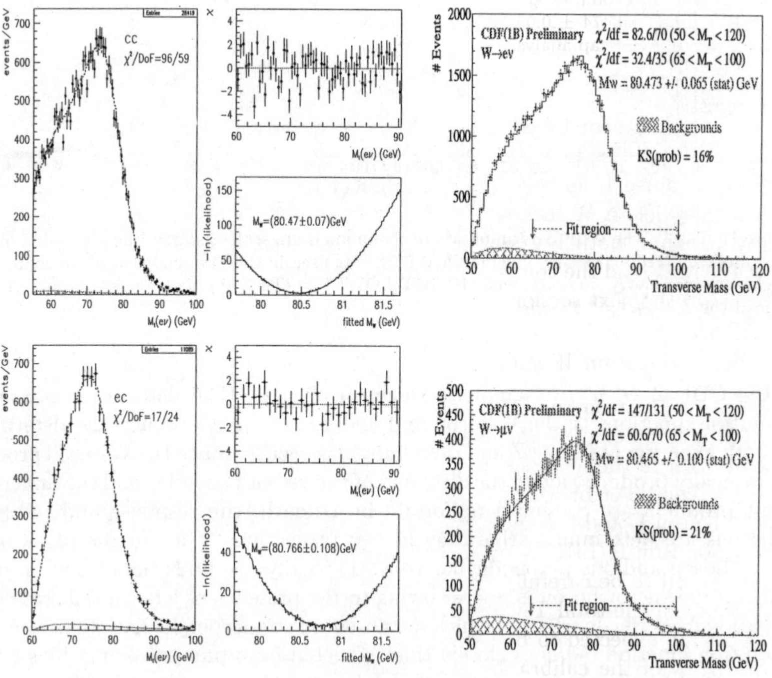

Figure 2. Transverse mass distributions compared to the best fit. LEFT : D∅'s published central-electron analysis and preliminary end-cap analysis. RIGHT : CDF's electron and muon channel analyses. The fit likelihood and residuals are also shown for the two D∅ distributions.

in Table 1. The uncertainties of the published D∅ central-electron analysis are also listed. For both experiments the largest errors are statistical in nature, both from the statistics of the W sample and also the statistics of the Z samples which are used to define many of the systematic uncertainties e.g. the uncertainties in the lepton energy/momentum scales and the W P_T model. The CDF and D∅

Error Source	DØ (EC)	DØ (C)	CDF (e)	CDF (μ)
Statistical	70	105	65	100
Lepton Scale+Resolution	70	185	80	90
$P_T^W + \not{E}_T$ Model	35	50	40	40
Other experimental	40	60	5	30
Theory (PDFs, QED)	30	40	25	20
Total Error	120	235	113	143
Mass Value	80.440	80.766	80.473	80.465
Combined Mass Values	80.497 ± 0.098 GeV		80.470 ± 0.089 GeV	

Table 1. The mass values and uncertainties of the CDF and DØ W mass analyses using the 1994–1995 data. The uncertainties are quoted in MeV. The mass values when the 1992–1993 data are included become : 80.474 ± 0.093 GeV for DØ and 80.430 ± 0.079 GeV for CDF. (EC) denotes the large rapidity end-cap analysis and (C) denotes the central rapidity analysis.

measurements are combined with a 25 MeV common uncertainty which accounts for the uncertainties in PDFs and QED radiative corrections which by virtue of being constrained from the same source are highly correlated. Together the two experiments yield a W mass value of 80.450 GeV with an uncertainty of 63 MeV. For the first time, both Tevatron experiments have measurements with uncertainties below 100 MeV and the combined uncertainty is comparable with the LEP2 results described in the next section.

4.4 Future Tevatron W mass measurements

In the next Tevatron run, the statistical size of the W mass event samples will increase by a factor of ~ 25. This means that the statistical part of the Tevatron W mass error in the next run will be ~ 10 MeV, where this also includes the part of the systematic error which is statistical in nature e.g. the determination of the charged lepton E and p scales from Z events. However to obtain reliable estimates for the total Run-II/LHC W mass error, the sources of error which do not scale with statistics need to be carefully evaluated. At present these error sources contribute 25 MeV out of the total Tevatron W mass error of 60 MeV. The four areas where the errors are expected to be dominated by systematic effects (not determined by the statistics of the calibration/control samples) and thus could be the limiting factors in achieving a W mass uncertainty of < 30 MeV are [8] :

- Knowledge of detector non-linearity i.e. the energy and momentum scales determined at the Z mass must be extrapolated to the momentum range relevant to the W mass analysis.

- The model for QCD radiation is derived from Z events and must be extrapolated to W events.

- QED radiative corrections.

- Parton distribution functions.

5 LEP2 Measurement

The first LEP2 W mass determinination [9] was made from a measurement of the WW cross section at the W pair threshold energy ($\sqrt{s} = 161$ GeV). This is shown in figure 3 and yields a W mass of : $80.40\pm^{0.22}_{0.21}$ GeV, where the error is almost entirely due to the statistical error on the cross section measurement.

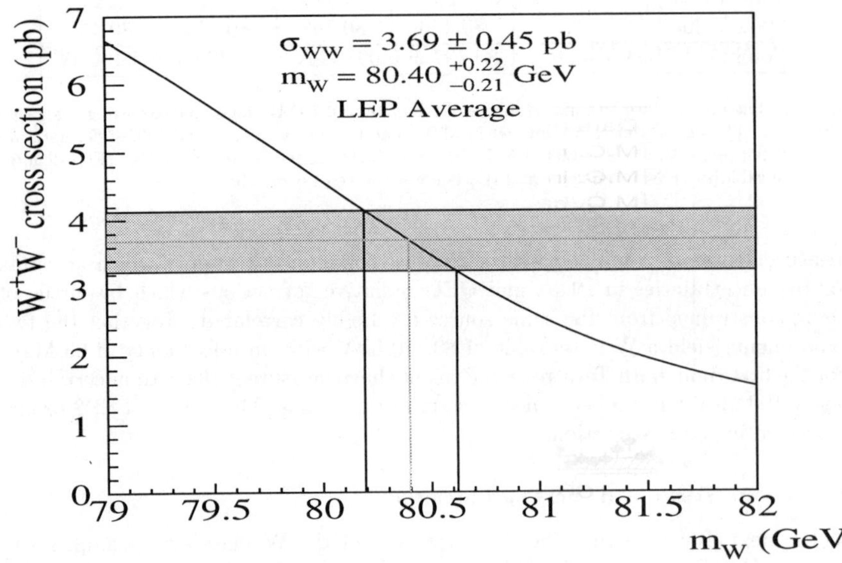

Figure 3. The SM prediction for the $e^+e^- \to$ WW cross section as a function of the W mass at $\sqrt{s} = 161$ GeV. The LEP2 measurement is shown by the shaded region.

The most precise LEP2 measurements [10] have come from the direct reconstruction of the W invariant mass in the all-hadronic and semi-leptonic decay modes. ALEPH has also recently made a measurement in the all leptonic mode. A comparison of the all-hadronic and semi-leptonic measurements allows valuable cross checks to be made on the integrity of the results because the two measurements have different systematic errors.

In the semi-leptonic analysis, two hadronic jets are selected by the Durham algorithm in addition to the large $\not{E}_T$ and isolated lepton. The lepton selection is generally augmented by imposing additional topological requirements e.g $cos\theta_{l\nu}$. The backgrounds are small and arise from single W production and also the so-called "radiative return" events. In these events, a hard photon is emitted from the final state such that the sub-process centre of mass energy is returned to the Z-pole where the cross section is large. The $qq'l\nu$ signature is faked by lepton mis-identifiucation or leptons arising from heavy quark decay. The all-hadronic channel event selection makes uses of neural networks to exploit the difference between the 4 quark final state arising from the decay of two Ws in comparsion to the 4 quark configuration arising in radiative-return Z events. In these events one of the quarks produced by the Z radiates a gluon which then undergoes : $g \to q\bar{q}$. Another complication in the all-hadronic analysis is that one must choose the correct pairings of the jets i.e. one must try and choose the 2 jets from the same W. This is done, for instance in the OPAL analysis, by forming a jet-pairing likelihood which is derived from such quantities as the difference in mass between the possible pairings and the sum of the opening angles in the two pairings. A failure to choose the correct pairing basically removes any sensitivity to the W mass. This is illustrated in figure 4.

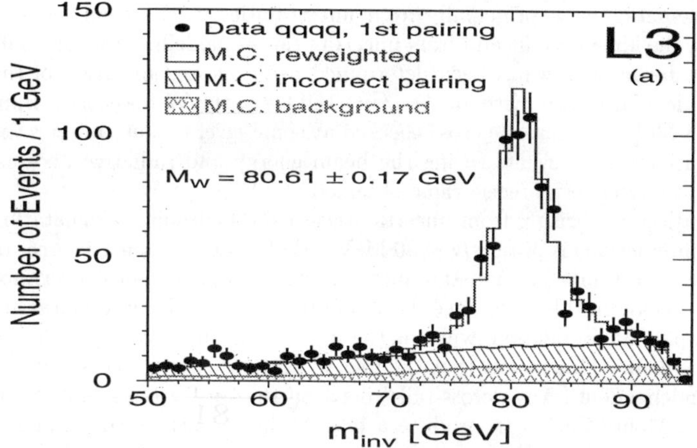

Figure 4. The jet-jet invariant mass distribution for WW $\to$ 4q events. Around 15% of the events have an incorrect jet pairing – the mass distribution of which is shown for Monte Carlo events.

After the event selection and preliminary invariant mass measurements, kinematic fits are performed which greatly increase the resolution of the final W mass and reduce the level of background in the events used to extract the W mass. In the semi-leptonic channel, all four experiments require that the W mass from the $l\nu$ and the $q\bar{q}'$ decays be the same and impose energy and momentum conservation. This results in a 2C fit and yields one mass per event. The details for the all-hadronic channel depend on the experiment e.g. DELPHI permits the presence of a fifth jet and ALEPH rescales the jet masses to the beam energy. Nevertheless, all constraints result in a 4C fit which produce 2 masses per event.

How the event by event masses from the kinematic fit are then used to determine a final W mass for the event sample again varies with each experiment. But broadly speaking two techniques are used :

- A Monte-Carlo re-weighting technique. Large samples of events are generated at a given M_w, Γ_w and are reweighted using the CC03 matrix elements to any M_w, Γ_w such that a likelihood minimisation of the data with respect to M_w and Γ_w can be performed.

- A Monte-Carlo convolution technique. This tends to exploit more information from the events. For each event a 2D probability curve is formed and convoluted with the signal and background probability. The offsets in the resulting mass e.g. due to initial state radiation are corrected in the final result using Monte-Carlo calibrations.

The experiments typically use one method for the published results and the second as a cross check.

5.1 LEP2 Systematics

The LEP2 systematics arise principally from uncertainties in the beam energy, which is used in the kinematic fit and from uncertainties in modelling the hadronic final state. The beam energy has been determined using an extrapolation of the LEP1 resonant depolarisation method. At $\sqrt{s} = 189$ GeV, the W mass error from this source is 18 MeV. This can be cross checked at some level by determining the Z mass in radiative return events (using the beam energy and radiative photon) and comparing it to the very precise value obtained at LEP1.

The systematic error arising from uncertainties in the hadronic fragmentation and final state interactions is presently ~ 30 MeV out of a total systematic error of ~ 45 MeV. The uncertainty in final state interactions are a particular concern for the all-hadronic decay. In this case, since the W lifetime is so small, the 4 quarks are only ~ 0.1 fm apart at production, whereas the fragmentation process takes place over a scale of ~ 1 fm. This means that the two W systems cannot necessarily be considered as independent. Any "cross-talk" between the two systems that is not simulated in the Monte-Carlo can produce a bias in the reconstructed jet angles and thus shift the W mass. Energy shifts are generally less significant because of the imposition of the beam energy constraint. Two effects occuring between the two W systems have been considered. Firstly, colour re-combination effects – non perturbative colour flux tubes extend between the two W systems. Secondly, Bose-Einstein (BE) correlations between the fragmentation products of the two W bosons, whereby there are local enhancements in particle multiplicity. Presently, no strong consensus on these two effects has emerged and thus the systematic errors have been set conservatively. BE correlations have been observed between the fragmentation particles from a single W (in semi-leptonic events), as was the case for Z events, but the experiments are yet to agree on whether the effect is apparent between fragmentation particles from different Ws. DELPHI reports "evidence for BE correlations between different Ws" [11], whilst ALEPH states that they are "disfavoured at the 2.7 σ level" [12] and OPAL concludes that it is "not established whether they exist or not".

In the case of colour re-combination effects, the experiments compare distributions from the hadronically decaying W in the semi-leptonic event sample with the same distribution for Ws in the all-hadronic event. However at the moment, there do not appear to be distributions which, with the present statistical uncertainty, are incisive enough to see the size of effect that the models of colour re-combination predict. For example, by comparing the multiplicity from a single W decay in the two event samples, one finds they are different by 0.1 ± 0.4 whereas the models tend to predict only a difference $\lesssim 0.1$. Regrettably, it appears that the most incisive discriminator of the colour-recombination effects is the W mass itself. At the moment the W mass from the all-hadronic and semi-leptonic event samples differ by 2.1 σ.

This area of final state interactions is potentially the area which could limit further significant improvements in the LEP2 W mass error. With larger statistics and the use of different, more incisive distributions, it is hoped that such effects can be better understood and hence their influence on the W mass minimised.

The LEP2 mass values are compared with the Tevatron values in figure 5. They are in excellent agreement despite being measured in very different ways with widely differing sources of systematic error. These direct measurements are also in very good agreement with the indirect measurement from NuTeV and the prediction based on fits to existing, non W, electroweak data.

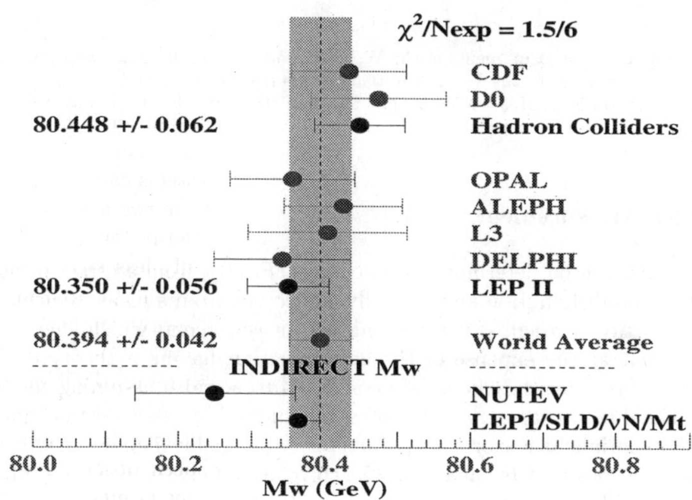

Figure 5. The direct determinations of the W mass from the Tevatron and LEP2 experiments are compared with the indirect measurement from NuTeV and the prediction based on fits to existing electroweak data.

The precision of these measurements has increased the sensitivity that one now has to the mass of the Higgs Boson. Indeed now it is the uncertainty on the top quark mass that is now becoming the limiting factor in the determiniation of the Higgs mass. As figure 6 shows, the available data tends to favour a Higgs boson with a mass < 250 GeV.

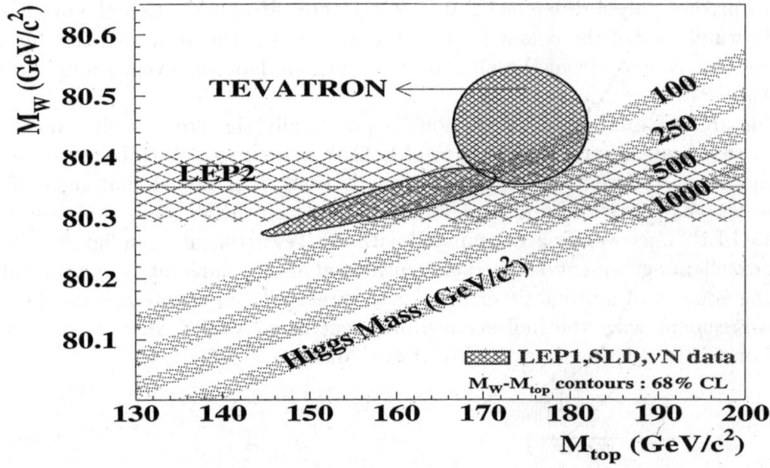

Figure 6. The direct measurements of the W boson and top quark mass from the Tevatron experiments are compared to the W mass measurements from LEP2 and the predictions based on electroweak fits to LEP1/SLD/νN data. The SM predictions for the Higgs mass as a function of the W and top masses are also shown.

6 Width Measurement

The W width can be determined in two ways. Firstly, from a direct measurement of the W mass distribution and secondly, indirectly, from a measurement of the W branching ratios. Presently, it is the indirect measurements which have the greater precision, but at the expense of the determination having a theory dependence, since the indirect measurements are converted to a width assuming the Standard Model. The direct measurement are presently limited by the statistical uncertainty. The LEP experiments perform a 2 parameter, M_W, Γ_W, likelihood fit to the invariant mass distributions to determine Γ_W. The correlation between the fitted M_W and Γ_W is small. The LEP2 results give a direct Γ_W of 2.12 ± 0.2 GeV. This is only based on $\sim$ 25 % of the total integrated luminosity, consequently a substantial improvement in the precision is anticipated.

The Tevatron experiments determine the width by a one parameter likelihood fit to the high transverse mass end of the transverse mass distribution. Detector resolution effects fall off in a Gaussian manner such that at high transverse masses ($M_T \gtrsim 120$ GeV), the distribution is dominated by the Breit-Wigner behaviour of the cross section (see figure 7). In the fit region, CDF has 750 events, in the electron and muon channels combined.

At LEP2, the W branching fractions are determined by an explicit cross section measurement whilst at the Tevatron they are determined from a measurement of

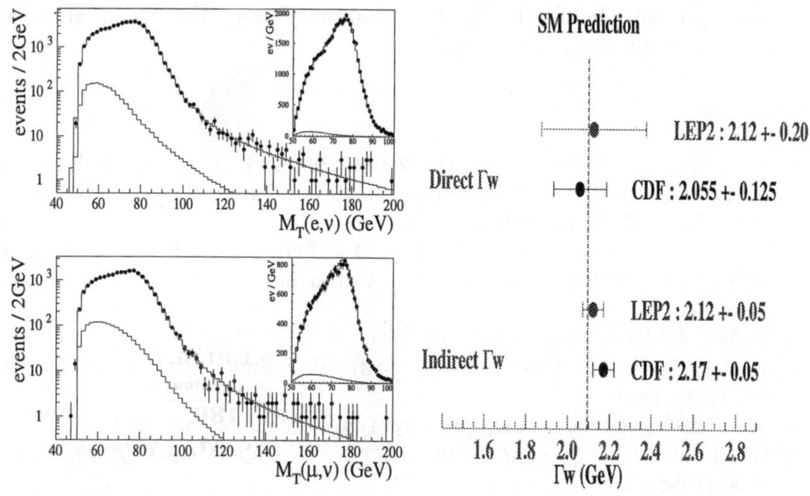

Figure 7. LEFT: The transverse mass distribution of the CDF W $\to$ $e\nu$ (upper) and W $\to$ $\mu\nu$ (lower) data showing the events at high transverse mass from which the W width is determined. (b) A comparison of the direct and indirect W width determinations from LEP and the Tevatron. The LEP indirect determination has been based on the W $\to$ $e\nu$ branching fraction only.

a cross section ratio. Specifically, the W branching fraction can be written as :
$\frac{\Gamma(W \to e\nu)}{\Gamma(W)} = \frac{\sigma_W}{\sigma_Z} \cdot \frac{\Gamma(Z \to ee)}{\Gamma(Z)} \cdot \frac{1}{R}$; where $R = \frac{\sigma_W \cdot Br(W \to e\nu)}{\sigma_Z \cdot Br(Z \to ee)}$ is the measurement made at the Tevatron. This determination thus relies on the LEP1 measurement of the Z branching fractions and the theoretical calculation of the ratio of the total Z and W cross sections. The indirect Tevatron measurement, $\Gamma_w = 2.171 \pm 0.051$ GeV, is now becoming systematics limited. In particular, the uncertainty due to QED radiative corrections in the acceptance calculation and in $\frac{\sigma_W}{\sigma_Z}$ contributes 0.040 GeV to the total systematic uncertainty of 0.047 GeV in the indirect W width determination. The corresponding measurement from LEP2, using only the W $\to$ $e\nu$ branching fraction, is $\Gamma_w = 2.120 \pm 0.050$ GeV.

7   Conclusions

The Tevatron and LEP2 experiments now for the first time have W mass measurements with errors < 100 MeV per experiment. The determinations from the two colliders are in excellent agreement and show no deviation from the Standard Model expectations. These measurements when combined with other electroweak measurements have a predeliction for a Higgs boson of mass < 250 GeV. Further significant reductions in the W mass error will require progress to be made on a number of issues e.g. final state interactions at LEP2. The W width measurements are presently limited by the statistical uncertainty, but, again, are in good agreement with the Standard Model.

References

1. G. Arnison *et al.*, Phys. Lett. **B122** 103 (1983), M. Banner *et al.*, Phys. Lett. **B122** 476 (1983), C. Albajar *et al.*, Z. Phys. **C44** 15 (1989), J. Aliti *et al.*, Phys. Lett. **B277** 354 (1992).
2. F. Abe *et al.*, *Phys. Rev. Lett.* **82**, 271 (1999); S. Abachi *et al.*, *Phys. Rev.* D**58** 052001 (1998). The current Tevatron uncertainty on the top mass is 5.1 GeV.
3. NuTeV Collaboration, hep-ex/9906024.
4. V. Barger *et al.*, *Z. Phys.* **C21** 99 (1983) ; B. J. Smith *et al.*, *Phys. Rev. Lett.* **50**, 1738 (1983).
5. See http://www-cdf.fnal.gov/physics/ewk/wmass_new.html.
6. B. Abbott *et al.*, hep-ex/9909030 (1999) ; B. Abbott *et al.*, Phys. Rev. 80 3000 (1998).
7. F. Abe *et al.*, *Phys. Rev. Lett.* **81**, 5754 (1998).
8. R. Jones, E. Thompson, M. Lancaster, D. Waters, Proceedings of the UK Phenomenology Workshop on Collider Physics, Durham, September 1999.
9. K. Ackerstaff *et al.*. Phys. Lett. **B389** 416 (1996); R. Barate *et al.*. Phys. Lett. **B401** 347 (1997); M. Acciarri *et al.*. Phys. Lett. **B398** 223 (1997); P. Abreu *et al.*. Phys. Lett. **B397** 158 (1997).
10. ALEPH Collaboration, CERN-EP/99-027; DELPHI Collaboration, DELPHI 99-41 CONF 240; L3 Collaboration, CERN-EP/99-17; OPAL Collaboration, CERN-EP/98-197; ALEPH Collaboration, ALPEH 99-015 CONF 99-010; ALEPH Collaboration, ALPEH 99-017 CONF 99-012; DELPHI Collaboration, DELPHI 99-51 CONF 244; L3 Collaboration, L3 Note 2377; OPAL Collaboration, Physics Note PN385.
11. Z. Metreveli *et al*, DELPHI Collaboration, DEL 99-22 MORIO CONF 221, March 1999.
12. ALEPH Collaboration, ALEPH 99-027, CONF 99-021, March 1999.
13. G. Abbiendi *et al*, OPAL Collaboration, Eur. Phys. J. C8 539 (1999).

W COUPLINGS AND PROPERTIES

R. TENCHINI

*INFN – Sezione di Pisa, via Livornese 1291, I-56010 S. Piero a Grado, Pisa
ITALY
E-mail: Roberto.Tenchini@cern.ch*

Electroweak decay and production properties of the W are reviewed, following the most recent measurements at LEP and Tevatron. New preliminary measurements from the four LEP collaborations determine the cross section in e^+e^- collisions up to 196 GeV. The W branching ratios, measured both in the individual leptonic channels and in the hadronic channel, yield a direct test of lepton universality and a measurement of the CKM matrix element $|V_{cs}|$. The WWZ and WWγ Triple Gauge Couplings are tested both in e^+e^- and hadron collisions. No sign of anomalous couplings is seen.

1 Introduction

In the past few years the data collected at the LEP and Tevatron colliders has significantly increased our knowledge of the electroweak properties of the W boson. One important electroweak parameter, the W mass, is discussed elsewhere [1]. In this paper the electroweak decay and production properties of the W, namely the W branching ratios and the Triple Gauge Couplings, are described.

Decays of W bosons to the three lepton species have been copiously detected at LEP and Tevatron. Comparing the electron, muon and τ branching ratios is a direct test of lepton universality. Hadronic decays of the W have been measured for the first time at LEP. The W hadronic branching ratio is related to CKM matrix elements, in particular to one of the less well-known elements, $|V_{cs}|$. By using the world average value for the other elements it is possible to considerably increase the precision of the $|V_{cs}|$ determination. Furthermore, the possibility to perform charm tagging permits a less precise, but more direct, measurement of the same matrix element.

The non-Abelian structure of the Standard Model of the electroweak interactions is tested by the measurement of the WWZ and WWγ Triple Gauge Couplings. The detailed study of the WW cross section, the WW production and decay angular distributions, single W and single γ events at LEP, together with the analysis of high p_T leptons at the Tevatron, probes this part of the Standard Model to the 10^{-1}, 10^{-2} level.

This paper is organized as follows. As the most recent data is coming from the ongoing four LEP experiments, the WW event selection and classification at LEP is briefly described. In this way the measurements of the total e+e- cross section (important for the TGC tests) and of the W branching ratios are naturally introduced. The lepton universality test and $|V_{cs}|$ measurements are then described. In the second part of the paper the TGC measurements are discussed and the most recent averages for the TGC parameters are given.

2 WW Event Selection and Total Cross Section at LEP

2.1 Selecting W's.

W production in e^+e^- collisions at centre-of-mass energies greater than 161 GeV is dominated by the so-called CC03 diagrams (Fig.1) [2,3,4,5]. Therefore W bosons are produced in pairs and events can be classified in three main topologies: fully leptonic, semileptonic and fully hadronic, depending on the number of W's decaying to leptons (2, 1 or 0 respectively). Event selections are designed to tag the three topologies with the highest possible efficiencies and purities (Table 1). The main selection criteria can be summarized as follows [6,7,8,9]:

- Fully leptonic selection. In this case low multiplicity events are selected, in particular two isolated acoplanar leptons with missing transverse momentum are normally required. Note well that in 5 cases out of 9 at least one of the leptons is a τ, so the lepton candidate can be a low multiplicity jet. This also accounts for final state radiation. Background is low and it is mainly due to low multiplicity two-photon events and ZZ production.
- Semileptonic selection. The presence of two hadronic jets, one isolated, high energy lepton and missing momentum from the undetected neutrino gives these events very distinct signatures. Again, background is low and due to $qq(\gamma)$, Wev and Zee production.
- Fully hadronic decays. Here the selection is more difficult due to the $qq(\gamma)$ background, which is about one order of magnitude larger than the signal. The criteria aim to select high multiplicity four jet events . Selections are based on event properties such as sphericity, jet angles, Fox-Wolfram momenta, etc., which are often combined with Neural Network techniques to enhance purity and efficiency. The idea is to lower as much as possible the 4-jet QCD background, where two jets (normally at higher energy) come from the two original quarks, while two more jets originate from gluon radiation and therefore tend to be at lower energy and smaller opening angles than jets in fully hadronic WW events.

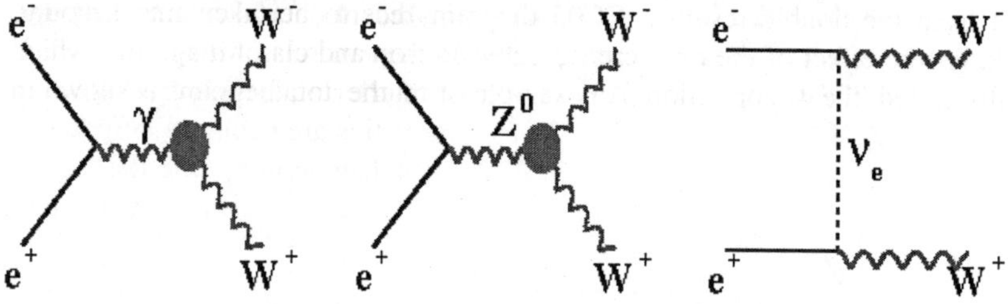

Figure 1. W pair production at LEP. The main production channels (CC03 diagrams).

Channel		efficiency	purity
$l\nu\, l\nu$		7 0 - 7 5 %	8 5 - 9 0 %
	e ν e ν	7 5 - 8 0 %	
	e ν μ ν	7 5 - 8 0 %	
	e ν τ ν	6 5 - 7 0 %	
	μ ν μ ν	7 5 - 8 0 %	
	μ ν τ ν	7 0 - 7 5 %	
	τ ν τ ν	5 0 - 6 0 %	
$q q\, l\nu$		7 5 - 8 5 %	9 0 - 9 5 %
	q q e ν	8 0 - 9 0 %	
	q q μ ν	8 5 - 9 0 %	
	q q τ ν	5 5 - 6 5 %	
$q q q q$		7 0 - 8 0 %	7 5 - 8 5 %

Table 1. W pair production at LEP. Typical performance of the event selections in the various channels.

2.2 The $e^+e^- \to W^+W^-$ Total Cross Section

The measurement of the $e^+e^- \to W^+W^-$ cross section requires a correct definition of the four fermion final state. The definition which is used by the four LEP collaborations is based on the CC03 diagrams (Fig.1). As the events selected by the experiments depend, through the experimental cuts, only on the final state itself, the effect of non-CC03 diagrams [2,5] yielding the same four fermion final state that from the double resonant CC03 diagrams has to be taken into account. This includes the effect of the interference between the two sets of diagrams, which is normally called the 4f correction. An example of interfering diagrams is shown in Fig. 2.

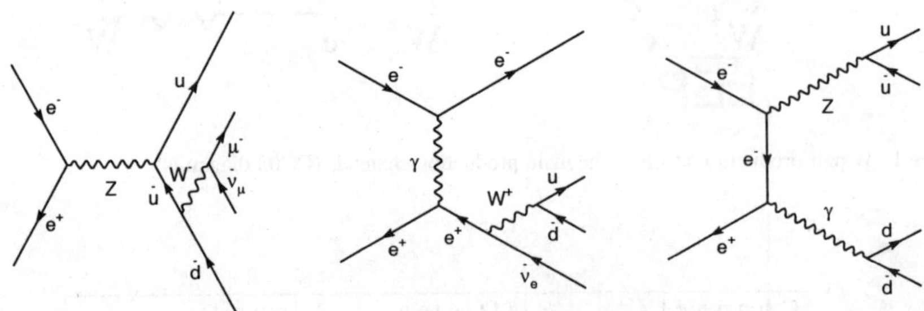

Figure 2. The WW Cross Section. Three examples of four fermion processes yielding the same final state as from CC03 WW production and therefore interfering with it.

The σ_i^{WW} cross sections (where i stands for the individual WW channel, i.e. fully leptonic, semileptonic, fully hadronic) can be computed from the number of events (N_j) tagged by selection j, the ε_{ij} efficiency matrix, the background cross section σ_j^{bkg} and the $\delta\sigma_j^{4f}$ interference correction described above. This can be done, for instance, using a simple Poissonian likelihood as

$$\prod_j P_j(\sigma_i) = \prod_j \frac{\mu_j^{N_j}}{N_j!} e^{-\mu_j}$$

where

$$\mu_j = L(\varepsilon_{ij}\sigma_i^{WW} + \sigma_j^{bkg} + \delta\sigma_j^{4f}) \quad .$$

The value of the interference correction depends on the experimental cuts and is non-negligible, though typically less than 10%, only for final states with electrons and positrons. The background correction is larger for the fully hadronic channel (see Table 1). The total cross section can be readily computed from the cross sections in the individual channels, usually by assuming the Standard Model branching ratios.

The WW cross section has been measured at 6 centre of mass energies between 161 and 196 GeV, with a total integrated luminosity of about 320 pb-1 for each LEP experiment [10,11,12,13]. The result is summarized in Fig.3 where the 6 points are compared to the Standard Model prediction. At higher energies, the points are a clear signature of the the presence of the ZWW triple gauge vertex. The theoretical prediction has an uncertainty of about 2% due to missing higher order electroweak corrections.

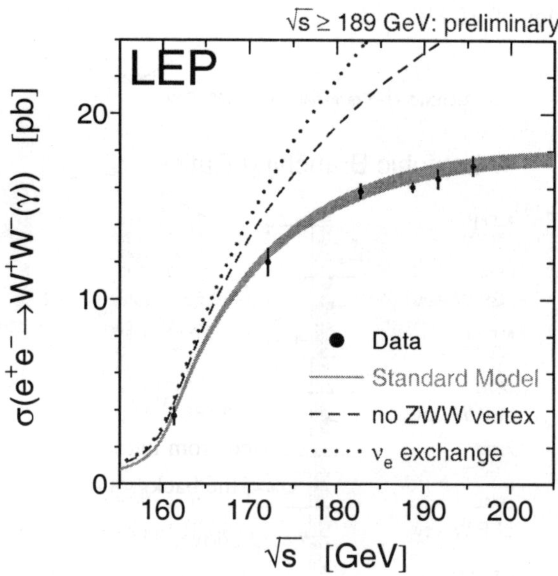

Figure 3. Total W pair production cross section at LEP. The measurements at the 6 centre of mass energies are compared to the Standard Model. The band on the theoretical prediction reflects the uncertainty due to missing higher order electroweak corrections. The dashed line shows the prediction without ZWW triple gauge vertex, while the dotted line indicates the expectation from neutrino exchange only.

3   Branching Ratios, Lepton Universality test and $|V_{cs}|$.

3.1   W decays to leptons and universality.

Leptonic decays of W's to e, μ and τ can be efficiently tagged both at LEP and Tevatron. This can be seen, for LEP, in table 1 where the typical performance of the selections in individual leptonic sub-channels are given. The method described in the previous section for the total cross section can be easily extended to the measurement of the W decay branching ratios to the individual lepton species. The hadronic branching ratio is set to $1 - B_e - B_\mu - B_\tau$. The measurements from the four LEP Collaborations, together with the resulting LEP averages [14], are shown in Fig.4. All data collected up to the centre of mass energy of 189 GeV are used [6,7,8,9]. No sign of lepton universality violation is seen and the branching ratios from the 3 species can be averaged to $(10.64 \pm 0.13)\%$ in agreement with the Standard Model expectation of 10.8%.

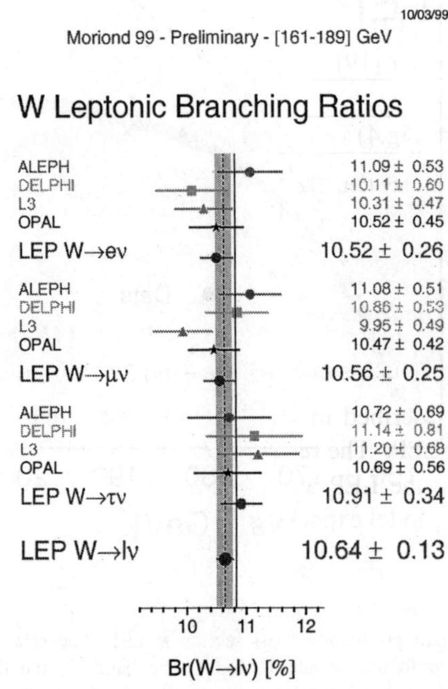

Figure 4. Measurements of the individual W leptonic branching ratios at LEP . The values from the individual experiments, as well as the LEP average, are shown.

At the Tevatron, W bosons are produced mainly through quark annihilation and are detected by the search for an high p_T isolated lepton plus missing transverse energy [15]. A test of lepton universality can be performed by taking the ratio of the cross sections to two lepton species, as in this way many systematic uncertainties cancels. The test is performed using the τ's and it is given in terms of $g(\tau)/g(e)$, similarly to what is done for the measurement of the τ lifetime [16] with the advantage of being free from external inputs as the τ branching ratios. One can use hadronic decays of the τ and tag isolated, narrow jets [17,18] or select electron final states and analyze the track impact parameter to separate direct $W \rightarrow e\nu_e$ decays from the $W \rightarrow \tau\nu_\tau \rightarrow e\nu_e\nu_\tau$ cascade, by taking advantage of the long τ lifetime and, therefore, larger impact parameter [18]. The results are shown in table 2. The LEP measurements of the W branching ratio to electrons and τ have been converted to a g(τ)/g(e) measurement for comparison.

CDF τ three prongs [17]	g(τ)/g(e) = 0.97 +/- 0.07
CDF impact parameter [19]	g(τ)/g(e) = 1.01 +/- 0.17 +/- 0.09
D0 τ jet [16]	g(τ)/g(e) = 1.004 +/- 0.019 +/- 0.029
LEP average (from Fig.4)	g(τ)/g(e) = 1.017 +/- 0.017

Table 2. Lepton Universality test. The measurements at the Tevatron and the LEP average. The results are given in terms of the ratio of the strength of the τ and electron charged currents.

3.2 *W decays to hadrons and $|V_{cs}|$*

If lepton universality is assumed, one can write $B_e = B_\mu = B_\tau = (1 - B_q)/3$ and use the methods described in section 2 to perform a measurement of the hadronic branching ratio at LEP. The results of the four LEP experiments [6,7,8,9] are shown in Fig. 5, yielding the LEP average value [14] of (68.07 $\pm$ 0.42)% in agreement with the Standard Model expectation of 67.5% .

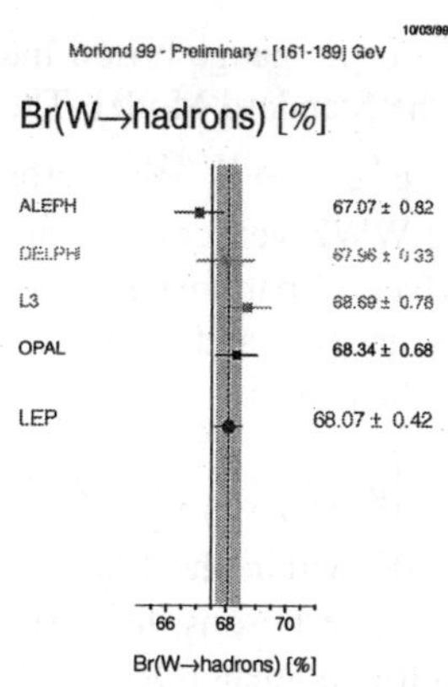

Figure 5. Measurements of the W hadronic branching ratios and the LEP average.

The hadronic branching ratio (B_h) is related to six $|V_{ij}|$ elements of the CKM matrix through the relation

$$\frac{B_h}{1-B_h} = \sum_{i=u,c\ j=d,s,b,} \left|V_{ij}\right|^2 (1+\frac{\alpha_s}{\pi})$$

where α_s is the strong coupling constant. One can use this relation to determine the less well-known of the six matrix elements, $|V_{cs}|$, by constraining the other five elements and α_s to their world average values. This procedure, together with the LEP average of B_h, yields [14]

$$|V_{cs}| = 1.002 \pm 0.020$$

which shows a sizeable improvement with respect to the previous determination based on D meson semileptonic decays ($|V_{cs}|=1.01 \pm 0.18$,[20]).

As the matrix elements involving the b quark have a negligible role in W decays and therefore charm is in practice the heaviest flavour, the $|V_{cs}|$ matrix element can also be measured in a more direct way by selecting W decays to charm hadrons. Typical signatures are the presence of a secondary vertex, of a high energy lepton within a jet, of D mesons and jet shape variables modified with respect to lighter quark decays. Since none of these features can be used on its own to separate charm from background, many observables are combined, often using neural network techniques, to enhance the discriminating power [21,22,23,24]. This direct method yield the less precise, but still very interesting, value of ($|V_{cs}|=0.99 \pm 0.11$,[25]).

4   Triple Gauge Couplings.

The presence of tree level triple gauge boson interactions is a distinct signature of the non-abelian nature of the Standard Model. The effect of such interactions can be already seen on the total $e^+e^- \rightarrow W^+W^-$ cross section (Fig. 1) whose shape is modified by the WWγ and WWZ vertices. A Lorentz invariant description of triple gauge boson vertices involves 14 parameters which can be reduced to 5 by requiring electromagnetic gauge invariance and conservation of C and P [26]. The five parameters are normally indicated as

$$\left\{g_1^z, \kappa_Z, \kappa_\gamma, \lambda_Z, \lambda_\gamma\right\} .$$

and are equal to $\{1,1,1,0,0\}$ within the Standard Model. They can be related to physical properties of the vector bosons, for instance the W magnetic moment (μ_W) and the W quadrupole electric moment (Q_W) can be written as

$$\mu_w = \frac{e}{2m_W}(1+\kappa_\gamma+\lambda_\gamma) \quad , \qquad Q_w = -\frac{e}{m_W^2}(\kappa_\gamma-\lambda_\gamma)$$

respectively. Anomalous couplings are constrained by the precision measurements performed at the Z pole [27], motivating the SU(2)xU(1) relationships

$$\Delta\kappa_Z = -\Delta\kappa_\gamma \tan^2\theta_W + \Delta g_Z^1$$

$$\lambda_Z = \lambda_\gamma$$

where Δ indicates the deviation with respect to the Standard Model value. Due to these further constraints, analyses of Triple Gauge Couplings (TGC) in e^+e^- collisions are done in terms of the three parameters

$$\left\{\Delta g_Z^1, \Delta\kappa_\gamma, \lambda_\gamma\right\}$$

which are equal to zero within the Standard Model. A deviation from zero indicates the presence of anomalous couplings and a signature of new physics.

In e^+e^- collisions the W^+W^- helicity states have different sensitivity to the anomalous couplings. As a consequence, anomalous couplings affect both the total cross section and the W^+W^- production and decay angular distribution. Neglecting the effects of initial state radiation and the finite W width, a $e^+e^- \to W^+W^-$ decay can be described by 5 angles: one W^+W^- production angle, two polar decay angles and two azimuthal decay angles. Therefore the Triple Gauge Couplings can be determined by measuring the total cross section and the angular distributions. As an example, in Fig.6 the measured W production angular distribution is shown and compared to expected distributions for various values of λ_γ.

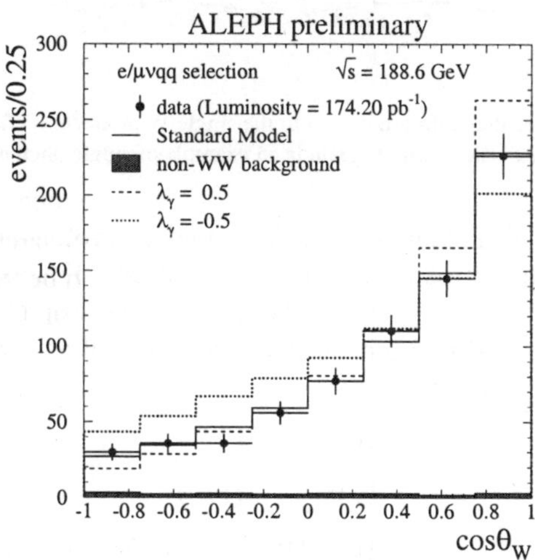

Figure 6. The W production angle in $e^+e^- \to W^+W^-$ as measured by the ALEPH experiment. The dots are the experimental data and the histograms the expectations for various values of λ_γ.

The maximal sensitivity on the TGC's is obtained when the direction and the charge of the W's and their decays products is known. This means that semileptonic decays are best suited to perform this measurement, as the lepton determines the W charge and the two jets the W direction. Analyses based on fully hadronic decays have to take into account the ambiguities in charge and direction. These can be partially overcome by using jet charge techniques and by pairing jets with invariant mass closest to the W mass. Fully leptonic events can also be used, but with reduced sensitivity due to the presence of at least two neutrinos. In all cases kinematic fits, which force the total momentum to be equal to zero and the total energy to be equal to the centre of mass energy, improve the measurements of the four momenta and therefore of the W directions. The most recent measurements of the four LEP Collaborations can be found in [30,31,32,34].

The production of W pairs is not the only process sensitive to TGC at LEP. Both the production of a single W, through $e^+e^- \to W^+ e\nu$, and of a single photon, through $e^+e^- \to \gamma\nu\bar{\nu}$, can occur through the WWγ vertex, as can be seen in Fig. 7. The measurement of these processes substantially improves the constraints on $\Delta\kappa_\gamma$ and λ_γ [28,29,33].

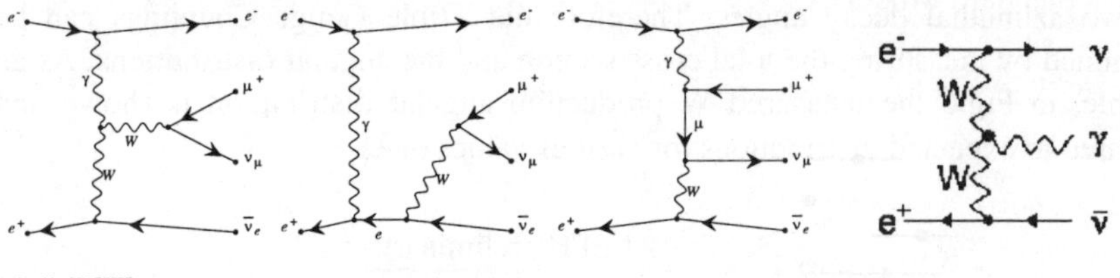

Figure 7. The first three diagrams are related to the single W production. The very first diagram occurs through the WWγ vertex. The fourth diagram is an example of single photon production sensitive to the WWγ vertex.

Triple Gauge Couplings can be measured in proton-antiproton collisions by detecting associated production of gauge bosons [35,36,37]. A signature of anomalous couplings is an excess of W/Z/γ pairs, especially at large invariant pair masses and high p_T. Typical channels for detection of associated production at the Tevatron are the following :

$$W\gamma \to \ell\upsilon\gamma$$
$$WW \to \ell\upsilon\ell\upsilon$$
$$WW/WZ \to \ell\upsilon q\bar{q}$$
$$WZ \to q\bar{q}\ell^+\ell^-$$
$$WZ \to \ell\upsilon\ell^+\ell^- \quad .$$

It is interesting to see that is possible to isolate the WWZ vertex, by detecting jet-jet-dilepton or tri-lepton events for instance, as can be seen by the last channels in the list. As the amplitudes for gauge boson production increase with energy, in order to avoid unitarity violation at the Tevatron a cutoff scale (Λ) of the order of a TeV is introduced. The anomalous coupling is related to the low energy limit by the form

$$\Delta\kappa(\hat{s}) = \frac{\Delta\kappa}{\left(1 + \hat{s}/\Lambda^2\right)^2}$$

where $\hat{s}$ is the invariant mass of the diboson pair. The CDF and D0 collaborations also use the so-called HISZ relations [38] in their TGC analyses, which introduces an additional constraint between Δg^Z_1 and $\Delta\kappa_\gamma$ reducing the number of TGC parameters to two, namely $\Delta\kappa_\gamma$ and λ_γ.

The four LEP experiments and D0 have combined their TGC results [39] to provide the most precise constraints on the anomalous couplings. The combination is done by combining the log-likelihood curves provided by each experiment and it is shown in Fig. 8 and Fig. 9. All three TGC parameters are consistent with zero, i.e. with the Standard Model expectation.

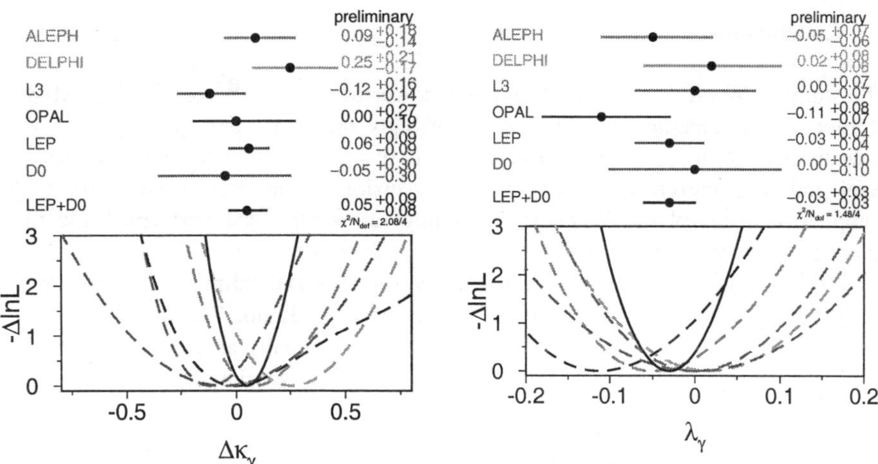

Figure 8. Single parameter fits to $\Delta\kappa_\gamma$ and λ_γ. The LEP and D0 log-likelihood curves are combined. The one-standard deviation limits, as obtained from the curves, are indicated.

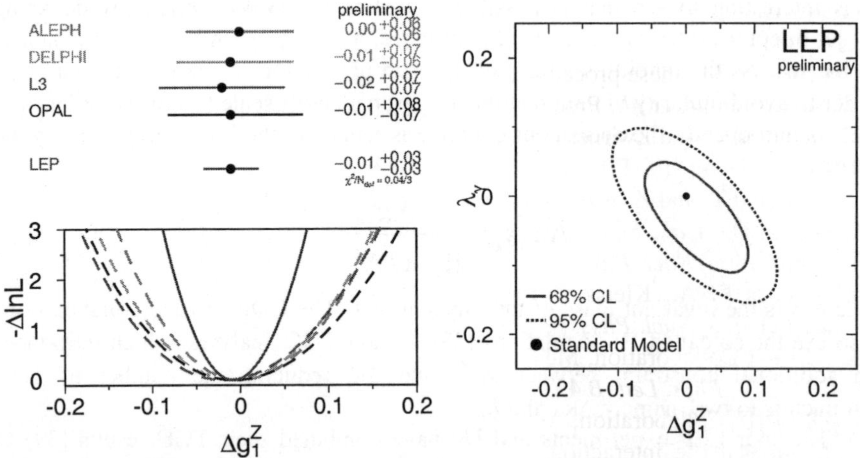

Figure 9. Single parameter fits to Δg^Z_1 (left plot). The LEP log-likelihood curves are combined. The one-standard deviation limits, as obtained from the curves, are indicated. The right plot show a two parameter fit to the LEP data obtained by varying simultaneously Δg^Z_1 and λ_γ. The 68% and 95% confidence levels are indicated.

5   Conclusions

The study of W production and decay is a powerful tool to test the Standard Model of Electroweak Interactions. Thanks to the high luminosity provided by LEP, the measurement of the total WW cross section in e^+e^- interactions has reached the 1% level and it is currently challenging the precision of theoretical calculations. A complete analysis of the W decay branching ratios has been performed and this sector is now known at the few percent level. The era of precision measurements of the Triple Gauge Couplings has started. They have been tested at the 10^{-1}, 10^{-2} level by the LEP and Tevatron experiments. No sign of new physics is seen.

6   Acknowledgements

I am grateful to John Ellison, Douglas Glenzinski, Eric Lancon and Darien Wood. I am greatly indebted to the LEP Electroweak Working Group and in particular to the W Mass Working Group and Triple Gauge Couplings Working Group. Special thanks to Evelyn Thomson for reading the manuscript.

References

1. Lancaster M., these proceedings.
2. Beenakker W. *et al.*, Proceedings of CERN LEP2 Workshop, eds Altarelli G., Sjostrand T. and Zwirner F. , CERN 96-01 (1996) , Vols 1 p. 79.
3. Bardin D. *et al.*, Proceedings of CERN LEP2 Workshop, eds Altarelli G., Sjostrand T. and Zwirner F. , CERN 96-01 (1996) , Vols 2 p. 3.
4. Bardin D. *et al.*, Semi-analytical approach to four fermion production in e^+e^- annihilation, *Nucl. Phys. (Proc. Suppl.) B* **37** (1994) 148.
5. Berends F. A., Kleiss R. and Pittau R. , Four fermion production in e^+e^- annihilation, *Nucl. Phys. (Proc. Suppl.) B* **37** (1994) 163.
6. ALEPH Collaboration, Measurement of W-pair production in e^+e^- collisions at 183 GeV, *Phys. Lett. B* **453** (1999) 107.
7. DELPHI Collaboration, W pair production cross-section and W branching fractions in e^+e^- interactions at 183 GeV, *Phys. Lett. B* **456** (1999) 310.
8. L3 Collaboration, Measurement of W-Pair Cross Sections in e^+e^- Interactions at root(s)= 183 GeV and W-Decay Branching Fractions, *Phys. Lett. B* **436** (1998) 437.
9. OPAL Collaboration, W+W- production and triple gauge boson couplings at LEP energies up to 183 GeV, *Eur. Phys. J.* **C8** (1999) 191.
10. ALEPH Collaboration, Measurement of W-pair production in e+e- collisions at 189 GeV, ALEPH 99-064, CONF 99-038, see also Preliminary results on Standard Model processes in e+e- collisions above 190 GeV, ALEPH 99-077, CONF 99-049, http://alephwww.cern.ch/ALPUB/conf/conf.html
11. DELPHI Collaboration, Measurement of the W-pair Production Cross-section and W Branching Ratios at sqrt(s) = 189 GeV, DELPHI 99-61 Conf 248, see also Measurement of WW production in e^+e^- collisions with data collected in 1999, DELPHI99-66 Conf253, http://home.cern.ch/~pubxx/www/delsec/papers/public/papers.html
12. L3 Collaboration, Preliminary Results on the Measurement of W-Pair Cross Sections in e^+e^- Interactions at sqrt{s} = 189 GeV and W-Decay Branching Fractions, L3 note 2376, see also Preliminary Standard Model Measurements at sqrt{s}=192 and 196 GeV by the L3 Experiment, L3 note 2440, http://l3www.cern.ch/conferences/EPS99/
13. OPAL Collaboration, W+W- Production in e+e- Collision at 189 GeV, OPAL PN378, see also Standard Model Measurements in e^+e^- Collisions at sqrt(s)>190 GeV, OPAL PN409. http://www.cern.ch/Opal/pubs/summer99_sub.html
14. The LEP Electroweak Working Group, A Combination of Preliminary Electroweak Measurements and Constraints on the Standard Model, CERN-EP/99-15 and LEPEWWG/WW/99-01, http://www.cern.ch/LEPEWWG.
15. CDF Collaboration, Measurement of the W boson mass, *Phys. Rev. D* **52** (1995) 4784 and references therein.

16. ALEPH Collaboration, Updated measurement of the τ lepton lifetime, *Phys. Lett. B* **414** (1997) 362.

17. CDF Collaboration, Measurement of the ratio $B(W \to \tau\nu)/B(W \to e\nu)$, in p anti-p collisions at sqrt(s)=1.8-TeV, *Phys.Rev.Lett.* **68** (1992) 3398.

18. D0 Collaboration, Measurement of $\sigma(p\bar{p} \to W) \cdot B(W \to \tau\nu)$ at DØ, Contributed paper to the XXIX International Conference on High Energy Physics, July 23-29, 1998, Vancouver (Canada), http://www-d0.fnal.gov/public/wz/proceedings/proceedings.html

19. Rimondi F., $B(W \to \tau\nu)$ and Γ_W studies at the Tevatron Collider, Proceedings of the XIII Topical Conference on Hadron Collider Physics, Mumbai (India) January 14-20, 1999 , CDF/PUB//CDFR/4922 .

20. Particle Data Group, Review of Particle Properties, *Phys. Rev. D* **54** (1996) 1.

21. ALEPH Collaboration, Measurement of $|V_{cs}|$ in hadronic W decays, ALEPH/99-062, CONF/99-037, submitted to Physics Letters B.

22. DELPHI Collaboration, Measurement of $|V_{cs}|$ using W-decays at LEP 2, DELPHI 98-107, CONF 174.

23. L3 Collaboration, Measurement of $W \to cs$ and direct determination of $|V_{cs}|$, L3 note 2232.

24. OPAL Collaboration, Measurement of $R_c(W)$ in $e^+e^- \to W^+W^-$ at sqrt(s) = 183 GeV, OPAL Physics Note PN355.

25. Obraztsov V., Measurements of $|V_{cs}|$ in W decays at LEP, Proceedings of the XXIX International Conference on High Energy Physics, July 23-29, 1998, Vancouver (Canada), World Scientific Publishing Company, see also http://ichep98.triumf.ca/info/talks/talklist.asp?sessionID=3

26. Hagiwara K., Hikasa K., Peccei R. D. and D. Zeppenfeld, Probing the weak boson sector in $e^+e^- \to W^+W^-$, *Nucl. Phys. B* **282** (1987) 253.

27. Gounaris G. *et al.*, Proceedings of CERN LEP2 Workshop, eds Altarelli G., Sjostrand T. and Zwirner F. , CERN 96-01 (1996) , Vols 1 p. 525.

28. ALEPH Collaboration, Measurement of triple gauge WWγ couplings at LEP2 using photonic events, *Phys. Lett. B* **445** (1998) 239.

29. ALEPH Collaboration , A study of single W production in e+e- collisions at sqrt(s) = 161 - 183 GeV, *CERN EP/99-086*, submitted to *Physics Letters B* .

30. ALEPH Collaboration, Measurement of Triple Gauge-Boson Couplings at 183-189 GeV, ALEPH 99-019 CONF 99-014 .

31. DELPHI Collaboration, Measurement of Trilinear Gauge Boson Couplings WWV in e+e- Collisions at 189 GeV, DELPHI 99-36 CONF 235.

32. L3 Collaboration, Preliminary Results on the Measurement of Triple-Gauge-Boson Couplings of the W Boson at LEP, L3 Note 2378.

33. L3 Collaboration, Preliminary Results on Single W Boson Production at sqrt{s} 189 GeV, L3 note 2367.

34. OPAL Collaboration, Measurement of triple gauge boson couplings from W+W- production at LEP energies up to 189 GeV, Opal Physica Note PN375.

35. CDF Collaboration, Limits on WWZ and WWγ couplings from WW and WZ production in p anti-p collisions at s1/2 = 1.8-TeV, *Phys.Rev.Lett.* **75** (1995) 1017.

36. D0 Collaboration, Search for Anomalous WW and WZ Production in p anti-p Collisions at sqrt s=1.8 TeV, *Phys. Rev. Lett.* **77**, (1996) 3303.

37. D0 Collaboration, Studies of WW and WZ Production and Limits on Anomalous WWgamma and WWZ Couplings, FERMILAB PUB-99/139-E, submitted to Phys. Rev. D .

38. Hagiwara K., Ishihara S., Szalapski R. and Zeppenfeld D., *Phys. Rev. D* **48** (1993) 2182.

39. http://www.cern.ch/LEPEWWG/tgc

TOP QUARK RESULTS FROM THE TEVATRON

H. GREENLEE
FOR THE CDF AND DØ COLLABORATIONS
Fermilab, P.O. Box 500, Batavia, IL 60510, USA

The CDF and DØ collaborations have collected top quark data samples approaching 100 events. Using these data samples, the two Tevatron experiments have measured the top quark mass and pair production cross section in a variety of decay channels. The Tevatron average of the top quark mass is $m_t = 174.3 \pm 5.1$ GeV/c^2. The Tevatron experiments have also begun to measure properties of the top quark besides its mass and production cross section.

The top quark was originally observed by the CDF and DØ collaborations at the Tevatron in 1995.[1,2] Since the announcement of that discovery, the experiments have doubled their data samples and improved their analysis techniques. Initially, the experiments focused on measuring the top quark pair production cross section and on extracting the top quark mass. As these analyses matured, the top quark data samples have been used for other studies, such as:

- Search for top quark decays to a charged Higgs boson and a b quark.

- Search for flavor changing neutral current decays of the top quark.[3]

- Measurement of $|V_{tb}|$.

- Measurement of the polarization of W bosons arising from top quark decay.

- Measurement of top quark pair spin correlation.

- Search for heavy particles decaying into top quark pairs and other anomalous $t\bar{t}$ production mechanisms.

- Search for single top production.

Results from some of these newer analyses are included below. Most have not achieved very high sensitivity because of the small sample of top quarks. However, they can be viewed as a preview of possibilities in the next Tevatron run, where the experiments expect to increase their top quark samples by at least a factor of 30.

Table 1. Present world samples of $t\bar{t}$ events and cross sections in various channels.

Channel	Events	Background	$\sigma_{t\bar{t}}$ (pb)
		DØ	
Dilepton (ee, $e\mu$, $\mu\mu$)	5	1.4 ± 0.4	5.0 ± 3.3
$\ell +$ jets (lepton b-tag)	11	2.4 ± 0.5	8.3 ± 3.6
$\ell +$ jets (topological)	19	8.7 ± 1.7	4.1 ± 2.1
All jets	41	24.8 ± 2.4	7.1 ± 3.2
$e\nu$	4	1.2 ± 0.4	9.6 ± 7.5
Combined			5.9 ± 1.7
		CDF	
Dilepton (ee, $e\mu$, $\mu\mu$)	9	2.4 ± 0.5	$8.2^{+4.4}_{-3.4}$
$\ell +$ jets (vertex b-tag)	34	9.2 ± 1.5	$6.2^{+2.1}_{-1.7}$
$\ell +$ jets (lepton b-tag)	40	22.6 ± 2.8	$9.2^{+4.3}_{-3.6}$
All jets	187	142 ± 12	$10.1^{+4.5}_{-3.6}$
$e\tau$, $\mu\tau$	4	2.0 ± 0.4	15.6^{+19}_{-13}
Combined			$7.6^{+1.8}_{-1.5}$

1 Introduction

At the Tevatron, top quark pairs are produced predominantly (about 90%) through QCD quark-antiquark annihilation. According to the Standard Model, top quarks decay 100% of the time into a W boson and a b quark. The W bosons decay into charged leptons and neutrinos, or into a quark-antiquark pairs. Top quark pair events are classified according to the decay of the two W bosons into the dilepton, lepton plus jets, and all jets channels.

2 Top Quark Pair Production Cross Section

The main interest in the $t\bar{t}$ production cross section stems from whether it supports the interpretation of any excess of events over background as being exclusively due to Standard Model top quark pair production. The present world samples of top quark pairs and top quark cross sections in various channels are summarized in Table 1.[5,6] A comparison with theoretical predictions is shown in Fig. 1.

Clearly, the top quark results support the interpretation that the excess is due to the Standard Model top quark. The only slightly anomalous result has been the CDF total cross section, which is about one and a half standard deviations above theory. (Since this was presented, CDF has revised their top

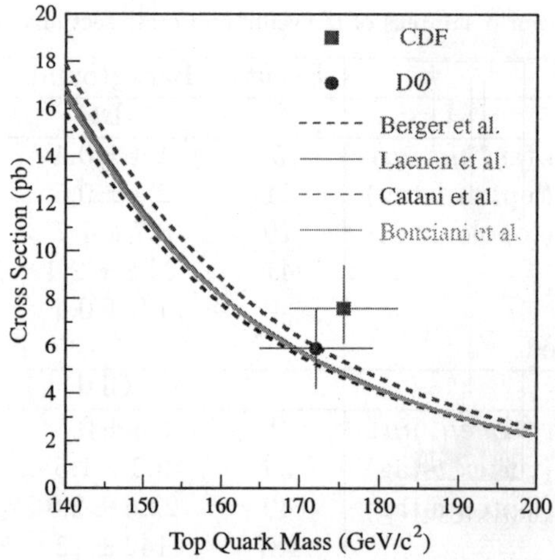

Figure 1. Measured vs. theoretical cross section for top quark pair production.

quark cross section downward from $7.6^{+1.8}_{-1.5}$ pb to $6.5^{+1.7}_{-1.4}$ pb.[7] The latter is in better agreement with theory.)

3 Top Quark Mass

The top quark mass has been measured in a variety of channels, using a variety of techniques. It has been measured by direct reconstruction of events in the lepton plus jets channel and the all jets channel, and by likelihood methods in the dilepton channel.

3.1 Top Quark Mass in the Lepton Plus Jets Channel

Both experiments have measured the top quark mass in the lepton plus jets channel using similar direct reconstruction techniques. These reconstruct a single estimated top mass for each candidate event using kinematic fitting with two constraints (2C). This provides a likelihood as a function of mass, based on a fit to the observed distribution of reconstructed masses consisting of a linear combination of expected distributions (templates for fixed m_t) for top quark signal and background.

CDF has optimized their result by dividing their top quark sample into four distinct subsamples having different intrinsic purities and resolu-

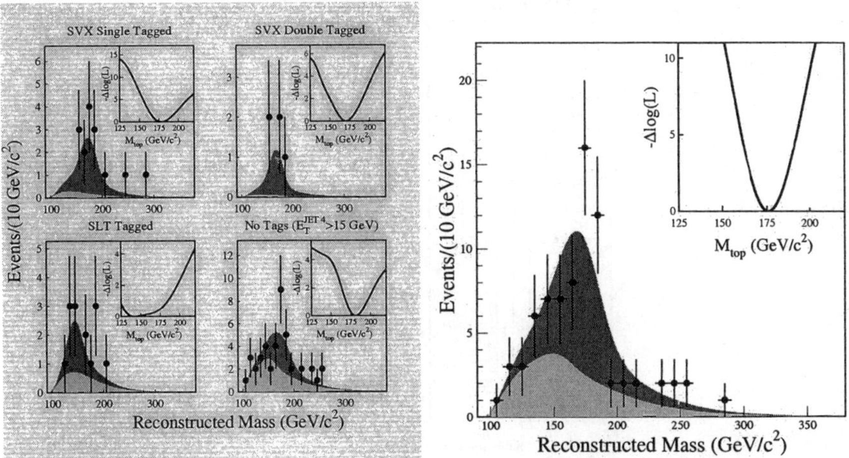

Figure 2. Reconstructed top quark mass for four lepton plus jets $t\bar{t}$ subsamples from CDF (left) and results (right).

tions: single displaced vertex b-tag (SVX tag), double SVX tag, soft lepton b-tag (SLT tag), no b-tags. A separate likelihood curve is generated for each subsample, and then the four likelihood curves are multiplied together to give the final likelihood curve (see Fig. 2). The CDF result is $m_t = 175.9 \pm 4.8$ (stat.) ± 5.3 (syst.) GeV/c^2.[8] DØ uses a likelihood fit with a floating background normalization, and does a simultaneous fit for signal rich and background rich regions (see Fig. 3). The DØ result is $m_t = 173.3 \pm 5.6$ (stat.) ± 5.5 (syst.) GeV/c^2.[9]

3.2 Top Quark Mass in the All Jets Channel

The technique used to measure the top quark mass in the all jets channel is similar to that for lepton plus jets. In the all jets case typically, a 3C fit is used that allows any arbitrary transverse momentum for the $t\bar{t}$ system. The all jets likelihood fit is identical to one used in the lepton plus jets channel. Of course, the background is much larger in the case of all jets. The reconstructed top quark mass and likelihood fit for CDF are shown in Fig. 4. The result is $m_t = 186 \pm 10$ (stat.) ± 5.7 (syst.) GeV/c^2.[8]

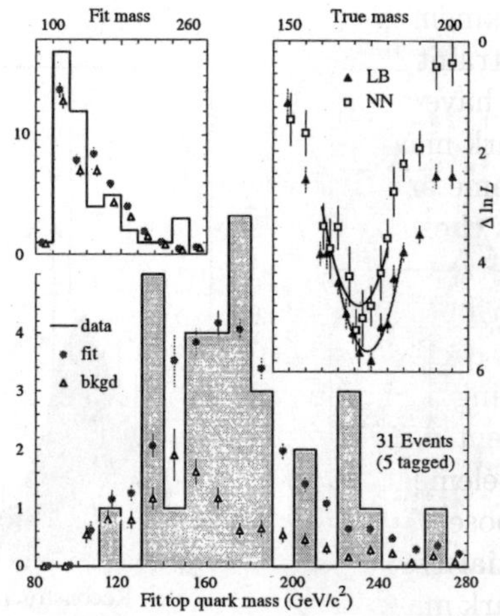

Figure 3. Reconstructed top quark mass in the lepton plus jets channel at DØ.

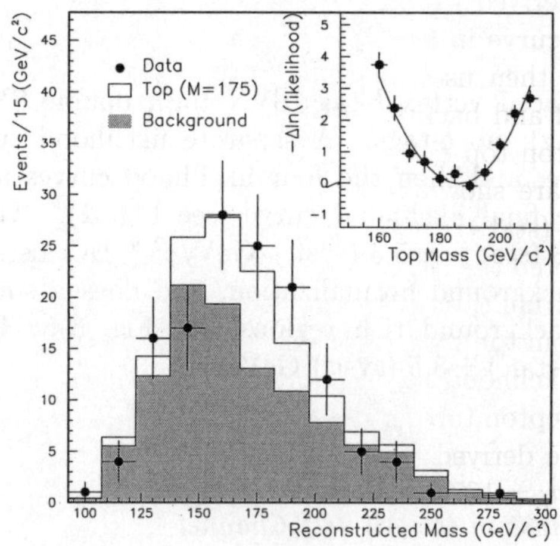

Figure 4. Reconstructed top quark mass in the all jets channel at CDF.

3.3 Top Quark Mass in the Dilepton Channel

The dilepton channel is fundamentally different from the other channels in
that it is not possible to recontruct the top quark mass on an event by event

basis. The amount of kinematic information available in the dilepton channels is one kinematic constraint short $(-1C)$ of full event reconstruction. Therefore, the experiments have resorted to using mass estimators other than the reconstructed top quark mass. The likelihood fit based on such mass estimators is basically the same as in the other channels.

Both experiments chose to base their estimators on the "dynamic likelihood" method of Dalitz, Goldstein, and Kondo,[12,13] in which one derives a dynamic likehood distribution, or "weight curve," for each event, as a function of top quark mass. At each point, the weight curve reflects the consistency of the observed event being a $t\bar{t}$ event of that top quark mass.

The DØ experiment uses two different weight curves. The first, which DØ calls the "matrix element weight" (MWT), is an improved version of the weight originally proposed by Dalitz and Goldstein.[13] This weight is proportional to the differential cross section for producing the observed event, for some assumed top quark mass. The second weight used by DØ is the so-called "neutrino weight" (νWT), which is a measure of the volume of neutrino phase space that is consistent with the $t\bar{t}$ hypothesis for some assumed top quark mass. For either choice of weight curve, DØ characterizes each event through four parameters derived from the weight curve, namely, the fractional area under the weight curve in five contiguous mass bins. This four-dimensional mass estimator is then used in a four-dimensional likelihood fit with four-dimensional signal and background templates. DØ's dilepton mass result is based in six dilepton top event candidates. The summed weight curves and likelihood curves are shown in Fig. 5. The resulting DØ dilepton mass is $m_t = 168.4 \pm 12.3$ (stat.) ± 3.6 (syst.) GeV/c^2.[14]

CDF has adopted the neutrino weight method, similar to DØ's. However, instead of using a multidimensional mass estimator, CDF extracts a single top quark mass from the weight curve for use as mass estimator in a standard one-dimensional likelihood fit. The CDF dilepton mass result is based on a sample of eight dilepton top quark event candidates. The distribution of mass estimators and the derived likelihood curve are shown in Fig. 6. The CDF dilepton mass is $m_t = 167.4 \pm 10.3$ (stat.) ± 4.8 (syst.) GeV/c^2.[8]

3.4 Systematic Errors on Top Quark Mass

Table 2 summarizes the results on m_t and their systematic errors. The Tevatron results are even currently limited by systematics. It is clear that any future progress in measuring the top quark mass with increased statistics will depend on the ability of experiments to reduce their systematic uncertainties. The largest systematic error is from the jet energy scale. The next largest

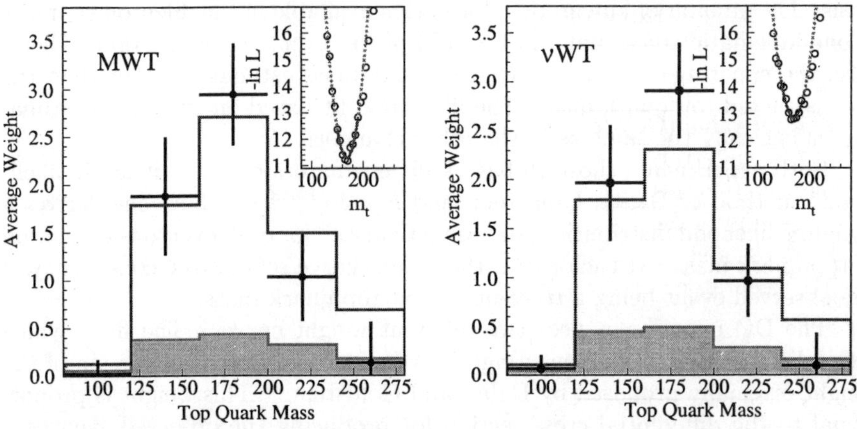

Figure 5. Summed weight curves for DØ dilepton events, and the likelihood as a function of m_t for MWT and νWT weights.

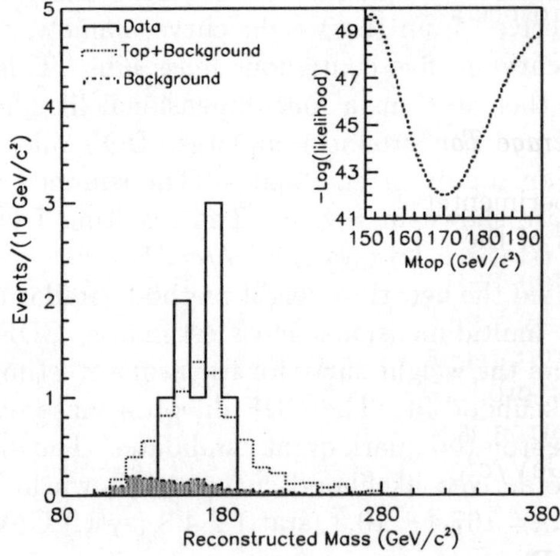

Figure 6. Distributions in the top quark mass estimator and likelihood as a function of m_t for CDF dilepton events.

is from the model for signal. The uncertainty on the jet energy scale is at present limited by the statistics of the data samples used for calibration. It

Table 2. Extracted masses and systematic uncertainties on m_t.

	CDF			DØ	
	$\ell + $ jets	$\ell\ell$	All jets	$\ell + $ jets	$\ell\ell$
Central value (GeV/c^2)	175.9	167.4	186	173.3	168.4
Statistical error	4.8	10.3	10	5.6	12.3
Systematic error	5.3	4.8	5.7	5.5	3.6
Jet energy scale	4.4	3.8	5.0	4.0	2.4
Signal model	2.6	2.8	1.8	1.9	1.7
MC generator	0.1	0.8	0.6	0	0
Noise	0	0	0.0	1.3	1.3
Background model	1.3	0.3	1.7	2.5	1.0
Method	0	0.6	0.7	1.5	1.1

may be more problematic to reduce the uncertainties on modeling signal, but with larger top quark data samples, it should be possible to limit and exclude variations in models of top quark production and decay. Therefore, there appear to be good prospects for reducing the systematic error on m_t in future experiments.

3.5 *Tevatron Average Top Quark Mass*

A subgroup of experimenters from CDF and DØ has analyzed the measurements of the mass of the top quark in both experiments for the purpose of producing a Tevatron average, taking into account the correlations in the systematic errors.[15] The "top averaging group" took as its input the five results and systematic errors reported in Table 2 (in fact, the top averaging group was responsible for producing Table 2 in its present form). The final average from the Tevatron is $m_t = 174.3 \pm 3.2$ (stat.) ± 4.0 (syst.) GeV/c^2 or $m_t = 174.3 \pm 5.1$ GeV/c^2.

4 Hadronic W Boson Decays in Top Quark Events

The UA2 collaboration was previously able to observe a mass peak (actually a shoulder) in the process $\bar{p}p \to W \to jj$ over an almost overwhelming QCD background.[10] So far, the same type of analysis has not succeeded at the Tevatron. However, at the Tevatron and higher energy machines, top quark events in the lepton plus jets channel can provide samples of hadronically decaying W bosons that are relatively free of QCD background. The CDF

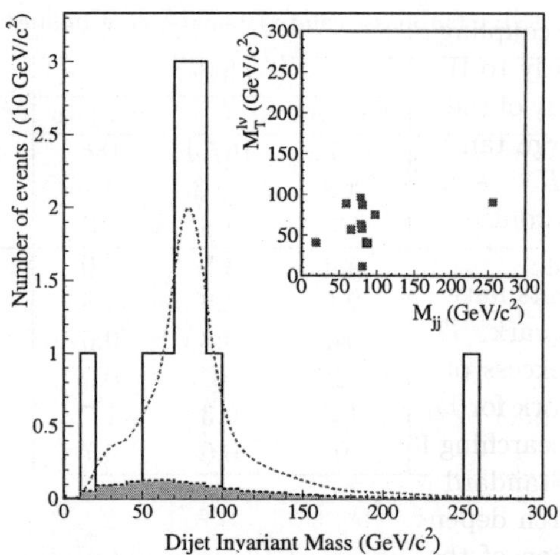

Figure 7. Reconstructed W quark mass in SVX double tagged $t\bar{t}$ events from CDF.

experiment has analyzed their lepton plus jets top quark events with the goal of reconstructing the hadronic W boson decays. A dijet mass peak at the W mass is observed in SVX double tagged events (see Fig. 7) with mass $M_W = 77.2 \pm 4.6$ GeV/c^2. Although the statistics are poor, this type of analysis confirms the interpretation of top quark candidate events as being due to the Standard Model top quark, and can provide an important check on the jet energy scale.[11]

5   Search for Top Quark Decaying to Charged Higgs

The Standard Model contains one complex Higgs doublet, which leaves one neutral Higgs boson after symmetry breaking. Extensions of the Standard Model to more than one Higgs doublet provide more Higgs bosons. Two Higgs doublet models (2HDM), which include the minimal supersymmetric model, have five physical Higgs bosons, denoted by H^0, h^0, A^0, and $H^{\pm}$. If the charged Higgs bosons are light enough, the top quark decay via the charged Higgs, $t \to H^+ b$, can compete with, or even dominate, the standard decay, $t \to Wb$.

The charged Higgs phenomenology of the 2HDM is characterized by two parameters: the mass of the charged Higgs boson, m_{H^+}, and the ratio of the vacuum expectation values of the two Higgs doublets, $\tan\beta$. At $\tan\beta =$

$\sqrt{m_t/m_b} \approx 6$, the coupling of H^+ to the top quark is smallest, and top quarks decay predominantly to W. At $\tan\beta$ values larger and smaller than this, the charged Higgs decay of the top quark becomes more important, and eventually dominates. At large $\tan\beta$ ($\gg 6$), the charged Higgs decays predominantly leptonically (i.e., $H^+ \to \tau\nu$). At small $\tan\beta$ ($\ll 6$), the charged Higgs boson decays mainly hadronically (i.e., $H^+ \to c\bar{s}$ or $H^+ \to t^*\bar{b} \to Wb\bar{b}$, depending on m_{H^+}).

The experiments have used two stategies to search for the charged Higgs decay of the top quark. The "appearance" or direct search strategy involves searching for an excess of $t\bar{t}$ events involving τ's. Obviously, such a direct search can only work for large $\tan\beta$. The "disappearance" or indirect search strategy involves searching for a deficit of top quark events relative to expectations from the Standard Model. The limits in parameter space derived for either type of search depend on the assumed theoretical $t\bar{t}$ production cross section. A variation of the indirect search is the cross section independent indirect method, which searches for a deficit of dilepton top quark events relative to lepton plus jets events. At the present level of statistics, the cross section independent method is not competitive with the cross section dependent methods, but it may become more important in the future when higher statistics become available. Figure 8 shows DØ's exclusion contour for the (cross section dependent) indirect search for three assumed top quark cross sections.[16] Figure 9 shows CDF's exclusion contours for all three methods for two assumed top quark cross sections.[17]

6 W Polarization in Top Quark Decay

The Standard Model predicts that W bosons in top quark decay are produced with an admixture of left handed and longitudinal polarization (the right handed W polarization always vanishes for $V - A$ interactions). The longitudinal polarization fraction is predicted to be $F_0 = m_t^2/(m_t^2 + 2M_W^2) = 0.70$. Information about the W polarization can be obtained by analyzing the p_T spectrum of charged leptons from W decay in top quark events (left handed W bosons tend to emit charged leptons in the backward direction). CDF has done a W polarization likelihood analysis using their lepton plus jets and dilepton event samples (see Fig. 10). They obtain $F_0 = 0.97\pm0.37$ (stat.)±0.12 (syst.), in agreement with the Standard Model prediction.

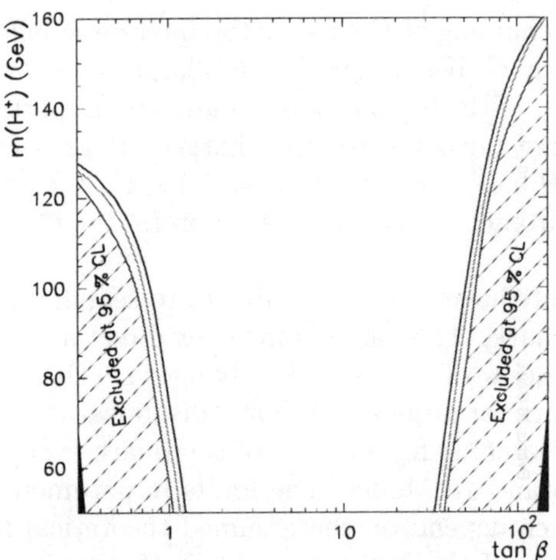

Figure 8. Exclusion contours from DØ for charged Higgs, for an assumed $t\bar{t}$ cross section of $\sigma_{t\bar{t}} = 4.5$, 5.0 and 5.5 pb. (Upper curve is for lowest $\sigma_{t\bar{t}}$.)

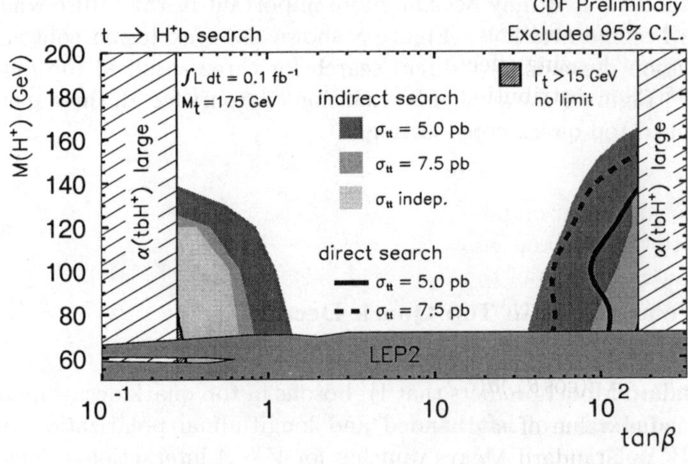

Figure 9. Exclusion contours for charged Higgs from CDF.

7 Top-Antitop Spin Correlation

At the Tevatron, pair produced top quarks are not polarized individually, but their spins are correlated. The top quark lifetime is sufficiently short that

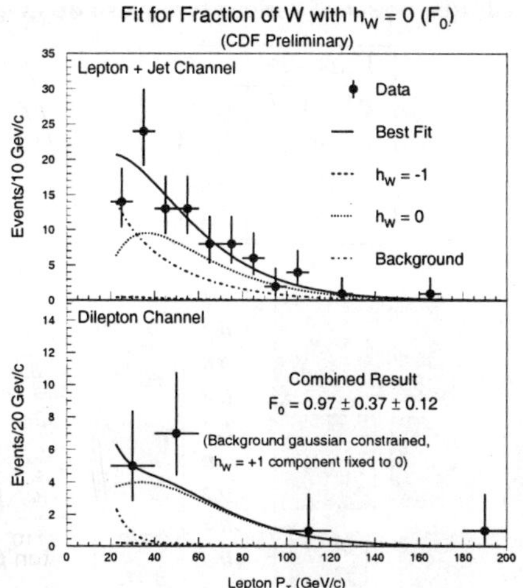

Figure 10. W polarization in top decay.

information about top quark spin is transmitted to top quark decay products before the depolarizing effects of hadronization can occur.

The angular distribution for decay products of a polarized top quark is

$$\frac{1}{\Gamma}\frac{d\Gamma}{d(\cos\theta)} = \frac{1 + \alpha\cos\theta}{2}.$$ (1)

The analyzing power of different decay products is shown in Table 3. Charged leptons are the best top quark polarimeters. Top quark spin correlation can be described in terms of the decay angles of the analyzing particles (θ_+ and θ_-), relative to some reference axis, through the following equation:

$$\frac{1}{\sigma}\frac{d\sigma}{d(\cos\theta_+)d(\cos\theta_-)} = \frac{1 + \kappa\alpha_+\alpha_-\cos\theta_+\cos\theta_-}{4}.$$ (2)

The predicted value of κ depends on the choice of reference axis. At the Tevatron, the Standard Model predicts $\kappa = 0.9$ for the optimal off diagonal reference axis.[18]

DØ has analyzed top quark spin correlation in the dilepton mode using its six dilepton events. The result of a two-dimensional likelihood fit is shown in Fig. 11. The fit weakly favors the prediction of the Standard Model over no correlation ($\kappa = 0$) or an anti-correlation ($\kappa = -1$, which would be the case for top quarks were produced through an intermediate scalar). Numerically, $\kappa > -0.2$ @ 68% CL.

Table 3. Analyzing power of various top quark decay products.

Particle	α
e^+ or d	1
ν or u	-0.31
W	0.41
b	-0.41

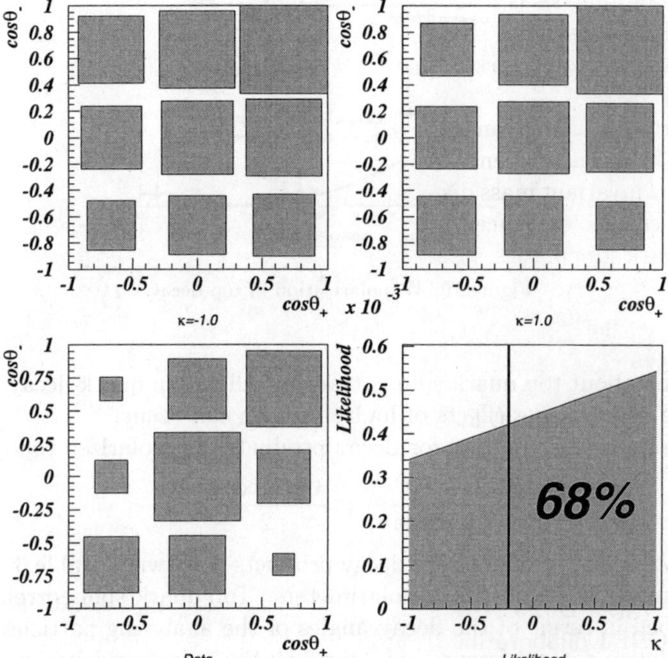

Figure 11. Results on top spin correlation from DØ.

8 Top Quark Production Kinematics

The Standard Model makes definite predictions for the differential cross section in $t\bar{t}$ production. A deviation from such predictions could correspond to evidence of physics beyond the Standard Model. An example of the type of non-Standard-Model physics that might be discovered through anomalous top quark pair production is production and subsequent decay of a heavy intermediate particle:

$$q\bar{q} \to X \to t\bar{t}. \tag{3}$$

Nonstandard production mechanisms such as Eq. 3 can give rise to a higher than expected $t\bar{t}$ cross section, or an excess in certain kinematic regions, such as high $m(t\bar{t})$ or high $p_T(t)$.

The study of such effects is best done with fully reconstructed events. This type of analysis is most readily carried out in the lepton plus jets mode. The techniques used are similar to those used in the top quark mass analysis, except that the top quark mass can be constrained, yielding a 3C fit instead of a 2C fit.

Figure 12 shows an analysis by CDF of the p_T spectrum of top quarks in $t\bar{t}$ events. There is no evidence of excess top quarks at high p_T. Figures 13 and 14 show the $t\bar{t}$ invariant mass distribution for data from CDF and DØrespectively. Again there is no evidence of an excess at high $m(t\bar{t})$. Additional comparisons of top quark kinematics to Standard Model predictons can be found in the references.[19,9]

References

1. CDF Collaboration, F. Abe *et al.*, Phys. Rev. Lett., **74**, 2626 (1995).
2. DØ Collaboration, S. Abachi *et al.*, Phys. Rev. Lett., **74**, 2632 (1995).
3. CDF Collaboration, F. Abe *et al.*, Phys. Rev. Lett. **80**, 2525 (1998).
4. E. Laenen, J. Smith, and W.L. van Neerven, Phys. Lett. **B321**,254 (1994); E.L. Berger and H. Contopanagos, Phys. Rev. **D54**, 3085 (1996); S. Catani *et al.*, Phys. Lett. **B378**, 329 (1996); R. Bonciani *et al.*, Nucl. Phys. **B529**, 424 (1998).
5. The CDF Collaboration (F. Abe *et al.*), Phys. Rev. Lett. **80**, 2773 (1998); The CDF Collaboration (F. Abe *et al.*), Phys. Rev. Lett. **79**, 1992 (1997); The CDF Collaboration (F. Abe *et al.*), Phys. Rev. Lett. **79**, 3585 (1997).
6. The DØ Collaboration (S. Abachi *et al.*), Phys. Rev. Lett. **79**, 1203 (1997); the DØ Collaboration (B. Abbott *et al.*), Phys. Rev. **D60**, 012001 (1999).
7. F. Ptohos, talk presented at the International Europhysics Conference High Energy Physics 99, Tampere, Finland, July 17, 1999.
8. The CDF Collaboration (F. Abe *et al.*), Phys. Rev. Lett. **82**, 271 (1999), Erratum, *ibid.* **82**, 2808 (1999).
9. The DØ Collaboration (B. Abbott *et al.*), Phys. Rev. **D58**, 052001 (1998).
10. The UA2 Collaboration (J. Alitti *et al.*), Z. Phys. **C49**, 17 (1991).
11. The CDF Collaboration (F. Abe *et al.*), Phys. Rev. Lett. **80**, 5720 (1998).

Top Quark Differential Cross Section

$$R_1 + R_2 = 0.72^{+0.13}_{-0.13}(\text{stat})^{+0.06}_{-0.06}(\text{syst})$$
$$R_4 < 0.114 \text{ at } 95\% \text{ C.L.}$$

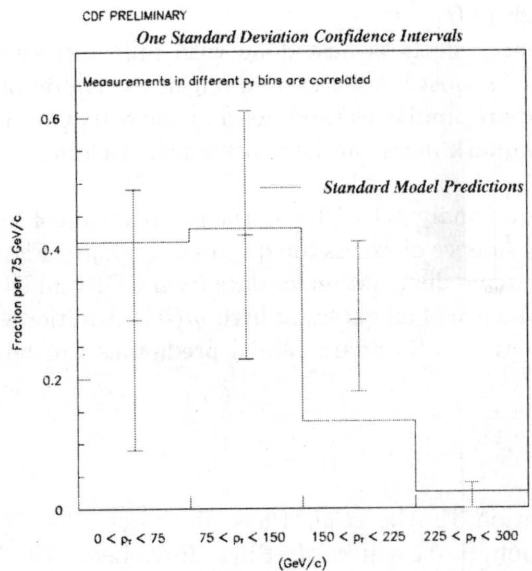

p_T Bin	Measured Fraction of Top Quarks
$0 < p_T < 75$ GeV/c	$R_1 = 0.29^{+0.18}_{-0.18}(\text{stat})^{+0.08}_{-0.08}(\text{syst})$
$75 < p_T < 150$ GeV/c	$R_2 = 0.42^{+0.18}_{-0.18}(\text{stat})^{+0.05}_{-0.07}(\text{syst})$
$150 < p_T < 225$ GeV/c	$R_3 = 0.29^{+0.12}_{-0.10}(\text{stat})^{+0.06}_{-0.05}(\text{syst})$
$225 < p_T < 300$ GeV/c	$R_4 = 0.000^{+0.035}_{-0.000}(\text{stat})^{+0.019}_{-0.000}(\text{syst})$

Figure 12. The p_T distribution of top quarks in $t\bar{t}$ events from CDF. R_n is the fraction of events in $p_T(t)$ bin n.

12. K. Kondo, J. Phys. Soc. Japan **57**, 4126 (1988); *ibid.* **60**, 836 (1991).

13. R.H. Dalitz and G.R. Goldstein, Phys. Rev. **D45**, 1531 (1992).

14. The DØ Collaboration *et al.*, Phys. Rev. **D60**, 052001 (1999).

15. The Top Averaging Group, CDF and D0 Collaborations (L. Demortier *et al.*), Fermilab-TM-2084, 1999 (unpublished).

16. The DØ Collaboration (B. Abbott *et al.*), Phys. Rev. Lett. **82**, 4975 (1999).

17. The CDF Collaboration (F. Abe *et al.*), Phys. Rev. Lett. **79**, 357 (1997.)

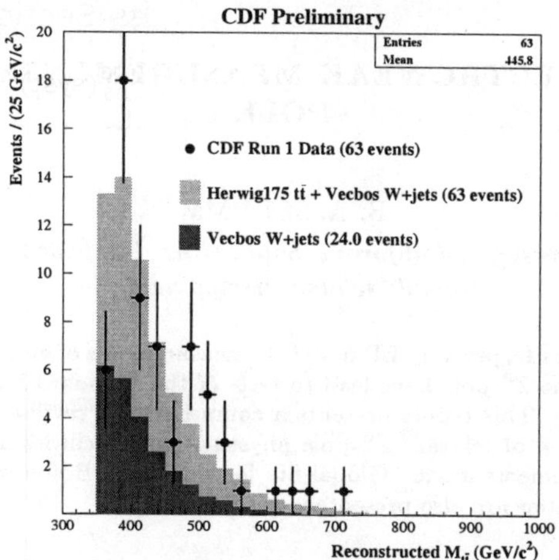

Figure 13. Distribution in $m(t\bar{t})$ from CDF.

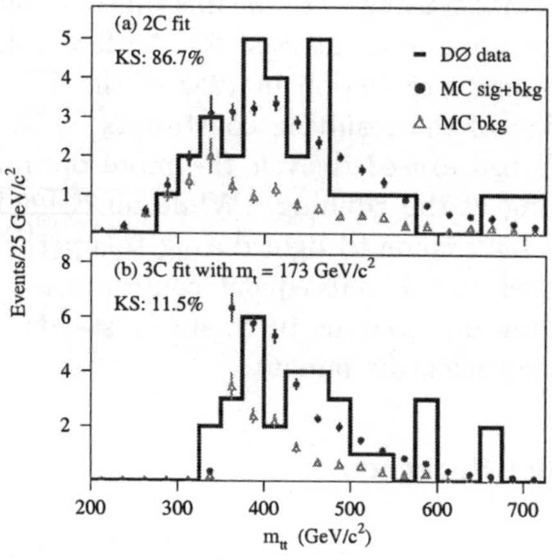

Figure 14. Distribution in $m(t\bar{t})$ from D∅.

18. G. Mahlon and S. Parke, Phys. Lett. **B411**, 173 (1997).
19. The CDF Collaboration (F. Abe *et al.*), Phys. Rev. **D59**, 092001 (1999).

PRECISE ELECTROWEAK MEASUREMENTS AT THE Z^0 POLE

B. A. SCHUMM

University of California, Santa Cruz, CA 95064, USA
E-mail: schumm@scipp.ucsc.edu

Over the last decade, precise LEP and SLC measurements of electroweak coupling parameters at the Z^0 pole have lead to tests of the Standard Model to unprecedented precision. This report presents a comprehensive review of these studies, including a review of relevant Z^0 pole physics issues, facilities, instrumentation, and the measurements made. Global fits for the Higgs Boson mass and $Z^0 - b$ coupling parameters are also presented.

1 Introduction

With the end of the SLC/SLD run at SLAC in June, 1998, the era of precision tests of the standard model using e^+e^- beams tuned to the energy of the Z^0-boson resonance came to a close, at least for the foreseeable future. Within the decade since the first production in 1988 of the Z^0 with lepton beams, however, the precision of the resulting constraints on the phenomonolgy of the Standard Model has exceeded even the more optimistic estimates put forth at the beginning of the running. While no clear disagreements with the Standard Model have come to light during this period, the development of the Standard Model and its subsequent confirmation at the 1-loop level, to accuracies exceeding one part in 10^{-3}, surely stands as one of the great achievements of human scientific pursuit.

2 Standard Model Background

Precise tests of the Standard Model at the Z^0 pole are performed via measurements of the couplings of the various fermions to the Z^0, and primarily through the quantitative extent of parity violation in those couplings. In the Standard Model, the neutral components of the parity-violating $SU(2)_L$ and partially parity-conserving $U(1)$ interactions are mixed according to an angle θ_W, the 'weak mixing angle'. The resulting physical states are the weak Z^0 boson and the electromagnetic γ, the latter of which mediates a purely parity conserving interaction, while the Z^0 thus mediates a partially parity-violating interaction. The quantitative extent of the Z^0 parity violation thus depends on the value of θ_W, as well as the electric (Q) and weak isospin (t_3) charges

Table 1. Standard Model Couplings and Parity Violation (assuming $\sin^2 \theta_W = 0.231$)

Fermion	t_3^f	Q^f	g_L^f	g_R^f	A_f	$dA_f/d\sin^2\theta_W$
e, μ, τ	-1/2	-1	-0.269	0.231	0.151	-7.8
ν	+1/2	0	0.500	0.000	1.000	-0.0
d, s, b	-1/2	-1/3	-0.423	0.077	0.935	-0.6
u, c, t	+1/2	+2/3	0.346	-0.154	0.669	-3.5

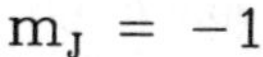

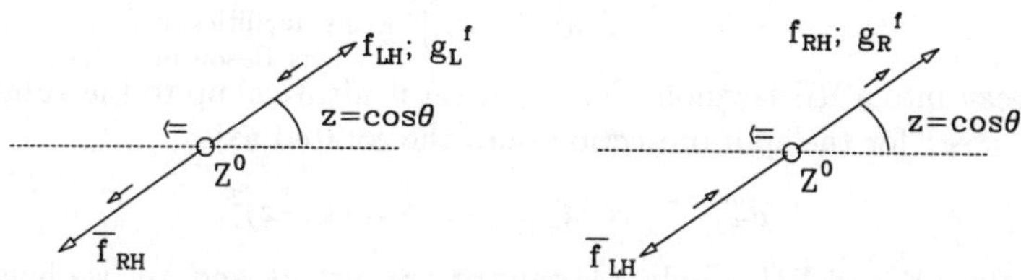

Figure 1. Decay of an $m_J = -1$ Z^0 boson into left- and right- handed fermions.

of the interacting fermion. Specifically, in terms of the left- and right-handed Z^0-fermion couplings g_L^f and g_R^f, the quantitative extent of parity violation is expressed via the parity-violation parameters A_f:

$$A_f = \frac{(g_L^f)^2 - (g_R^f)^2}{(g_L^f)^2 + (g_R^f)^2} \tag{1}$$

where $g_L^f = t_3^f - Q^f \sin^2\theta_W$ and $g_R^f = -Q^f \sin^2\theta_W$. Table 1 shows the electroweak charges, couplings and parity violation parameters A_f for the various fermion species. Also shown is the resulting sensitivity to the weak mixing angle. It is seen that the extent of parity violation in the coupling of the Z^0 to charged leptons is very sensitive to the value of $\sin^2\theta_W$, and thus to the vacuum polarization effects which renormalize the boson propagator. On the other hand, the extent of parity violation to down-type quarks is quite insensitive to $\sin^2\theta_W$, and thus measurements of parity violation in these couplings are interesting in that they constrain $Z^0 - f$ vertex corrections independently of the propagator effects which renormalize $\sin^2\theta_W$.

The angular differential cross-section for fermion production at the Z^0

pole is given by

$$\frac{d\sigma^f}{dz} \propto (1 - A_e P_e)(1 + z^2) + 2A_f(A_e - P_e)z, \tag{2}$$

where P_e is the longitudinal polarization of the electron beam, and $z = \cos\theta$. This relation can be understood entirely from angular momentum arguments. Figure 1 shows that the angular cross section for an $m_J = -1$ Z^0 boson decaying to a LH fermion and RH antifermion will be proportional to square of the spin-1 rotation matrix element corresponding to the transformation between an $m_J = -1$ state with spin projection on the z axis and an $m_J = -1$ state with spin projection along an axis rotated by $\cos\theta$ relative to the z axis: [1]

$$\sigma_{LH}^{m_J=-1} \propto |d_{-1,-1}^{(1)}(z)|^2 = (1 + z)^2. \tag{3}$$

For decay into a RH fermion, the argument is identical up to the substitution of $m_J = +1$ for the spin projection onto the rotated axis:

$$\sigma_{RH}^{m_J=-1} \propto |d_{+1,-1}^{(1)}(z)|^2 = (1 - z)^2. \tag{4}$$

Since the LH and RH coupling strengths are just g_L and g_R, we have for an $m_J = -1$ Z^0 sample:

$$\sigma_{tot}^{m_J=-1}(z) \propto g_L^2(1+z)^2 + g_R^2(1-z)^2 = (g_L^2 + g_R^2)(1+z^2) + 2(g_L^2 - g_R^2)z. \tag{5}$$

The presence of the term odd in z leads to a forward-backward asymmetry in the overall fermion production rate, provided that $|g_L| \neq |g_R|$, i.e., that parity is violated. With a purely left-handed electron beam ($P_e = -1$), the sample of produced Z^0 bosons would be purely $m_J = -1$, and the forward-backward asymmetry would be

$$A_{FB}^{m_J=-1}(z) = \frac{\sigma^f(z) - \sigma^f(-z)}{\sigma^f(z) + \sigma^f(-z)} = \frac{g_L^2 - g_R^2}{g_L^2 + g_R^2}\frac{2z}{1+z^2} = A_f\frac{2z}{1+z^2} \tag{6}$$

where A_f is the parity violation parameter for fermion species f introduced above. However, by reproducing the above arguments for the case of $m_J = +1$, it's easy to see that the sign of the asymmetry flips. Thus, a forward-backward asymmetry develops only if there are *different numbers* of $m_J = +1$ and $m_J = -1$ Z^0's in the overall sample. For an unpolarized electron beam (equal numbers of LH and RH electrons), this will only be true to the extent that the LH and RH couplings of the *electron* differ. Thus, the unpolarized forward-backward asymmetry also depends upon the extent A_e of parity violation in the Z^0-e coupling:

$$A_{FB}^f(z) = \frac{\sigma^f(z) - \sigma^f(-z)}{\sigma^f(z) + \sigma^f(-z)} = A_e A_f\frac{2z}{1+z^2}, \tag{7}$$

in agreement with Eqn. (2). By convention, this observable is usually expressed in terms of the integrated forward ($z > 0$) and backward ($z < 0$) cross sections:

$$A_{FB}^f = \frac{\sigma_F^f - \sigma_B^f}{\sigma_F^f + \sigma_B^f} = \frac{3}{4} A_e A_f. \tag{8}$$

where $\sigma_F^f = \int_0^1 \sigma^f(z)dz$ and $\sigma_B^f = \int_0^1 \sigma^f(-z)dz$.

2.1 Polarized Electron Beams

With a polarized electron beam, parity violation in the Z^0-e coupling can be measured directly by forming the asymmetry A_{LR} in the left- and right-handed Z^0 production cross sections:

$$A_{\mathrm{LR}} = \frac{\sigma_{e_L^- e^+} - \sigma_{e_R^- e^+}}{\sigma_{e_L^- e^+} + \sigma_{e_R^- e^+}} = \frac{(g_L^e)^2 - (g_R^e)^2}{(g_L^e)^2 + (g_R^e)^2} = A_e. \tag{9}$$

The sensitivity of this parameter to $\sin^2 \theta_W$ (see Table 1), combined with the indifference of the quantity to the particular final state fermion, render this the single most accurate approach to the measurement of $\sin^2 \theta_W$.

It is also possible to form a polarized *final-state* asymmetry

$$\tilde{A}_{FB}^f(z) = \frac{[\sigma_L^f(z) - \sigma_L^f(-z)] - [\sigma_R^f(z) - \sigma_R^f(-z)]}{\sigma_L^f(z) + \sigma_L^f(-z) + \sigma_R^f(z) + \sigma_R^f(-z)} = A_f \frac{2z}{1+z^2}. \tag{10}$$

Unlike its unpolarized counterpart, this observable is sensitive to *final state* parity violation alone. In the case $f = b$, this provides a direct constraint on Z^0-b vertex effects, independent of the propagator effects which alter the value of $\sin^2 \theta_W$ (see Table 1).

Both of these observables, however, are mitigated by incomplete beam polarization ($|P_e| \neq 1$) according to

$$A_{meas} = |P_e| A_{true}.$$

Thus, in order to measure these observables accurately, it is necessary to have an electron beam with a large and well-measured polarization.

2.2 Z^0 Lineshape Parameters

At Born level, the total cross section for the production of fermion type f is given by

$$\sigma_{e^+ e^- \to f\bar{f}}(s) = \frac{12\pi}{M_Z^2} \frac{s\Gamma_{e^+ e^-}\Gamma_{f\bar{f}}}{(s - M_Z^2)^2 + s^2\Gamma_Z^2/M_Z^2} \tag{11}$$

where s is the square of the cms energy, and the partial width $\Gamma_{f\bar{f}}$ is given by

$$\Gamma_{f\bar{f}} = 2c_f[(g_L^f)^2 + (g_R^f)^2]\frac{G_F M_Z^3}{12\sqrt{2}\pi}. \tag{12}$$

Thus, measurements of the Z^0 resonance parameters M_Z, Γ_Z, and the peak cross section σ^0, are important for constraining the Standard Model. In particular, the relation

$$\sin 2\theta_W = [\frac{4\pi\alpha_{\rm em}}{\sqrt{2}G_F M_Z^2}]^{1/2} \tag{13}$$

exhibits the value of a precise measurement of M_Z, given our precise knowledge of $\alpha_{\rm em}$ and G_F. With the value of θ_W thus constrained, measurements of $\sin^2\theta_W$ via parity violation become exacting consistency checks of the Standard Model.

Finally, it should be pointed out that the discussion presented so far is strictly true only at Born level. At higher orders, radiative corrections act to change these relations somewhat. This is precisely what makes precision measurements interesting, in that new physics in radiative loops will potentially yield values for the observables in disagreement with Standard Model expectations. On the other hand, 'uninteresting' radiative corrections, to the extent that they are not empirically constrained, can enter in to obscure the above relations. As will be seen below, a particularly important example of this is the scale dependence of $\alpha_{\rm em}$, which receives contributions from intermediate-energy non-perturbative hadronic loops and thus is not known to arbitrary precision.

3 Z^0 Peak Facilities

Over the past decade, Z^0 peak running has been done at two facilities: the LEP electron-positron storage ring collider at CERN, and the SLC linear collider at SLAC. Table 2 shows the relative sample sizes (summed over the four LEP detectors) at the two facilities.

The SLC electron beam polarization, as discussed above, greatly increases the information content per hadronic event for many precision measurements, making the two facilities roughly comparable and nicely complementary. In addition, the small luminous region of the SLC (roughly 2 μm $\times$ 0.7 μm for the SLC, *vs.* 250 μm $\times$ 5 μm for LEP), as well as the smaller SLC interaction region beampipe radius (2.5 cm *vs.* 5.5 cm) give the SLC an intrinsic advantage for measurements making use of precision tracking – particularly those involving the production and decay of heavy flavor.

Table 2. Sample Sizes (No. of Hadronic Z^0's) and Beam Polarization for the LEP and SLC Colliders

Year	LEP Sample $\times 10^{-3}$	SLC Sample $\times 10^{-3}$	$< P_e >$ for the SLC
1988	0	0.5	0.0%
1989	0	0.3	0.0%
1990-91	1700	0.3	0.0%
1992	2800	10	22%
1993	2600	50	63%
1994	5800	70	78%
1995	2700	30	78%
1996	0	50	78%
1997-98	0	350	73%
Total	**15500**	**550**	

4 Z^0 Pole Detectors

The LEP facility is instrumented with four cylindrical geometry detectors (ALEPH, DELPHI, L3, and OPAL), while the single interaction region of the SLC was instrumented with the upgraded MARK-II detector through 1989, and the SLD detector thereafter. A schematic of the DELPHI detector is shown in Figure 2. The primary components of this detector, fairly typical of the six detectors that have run at the Z^0, are as follows. Precise charged particle reconstruction is done by a gaseous central tracker, with several precise space points near the IP provided by a silicon μ-strip vertex detector. Calorimetry is mounted inside the magnet coil, outside of which is placed coarse calorimetry and tracking to catch the hadronic tail and identify muons. For the DELPHI and SLD detectors, a Cerenkov ring-imaging detector resides between the tracking and calorimetry to provide dedicated particle identification. Forward instrumentation typically extends the $|\cos\theta|$ coverage to 0.9 or better for tracking and 0.95 to 0.99 for calorimetry. Far-forward position sensitive electromagnetic calorimetry, typically covering the angular region between 20 and 100 mrad, is used to measure the interaction region luminosity via electrons from t-channel Bhabha scattering.

Four advances in instrumentation technique which have been primarily instigated by Z^0 pole physics are discussed in more detail below.

52

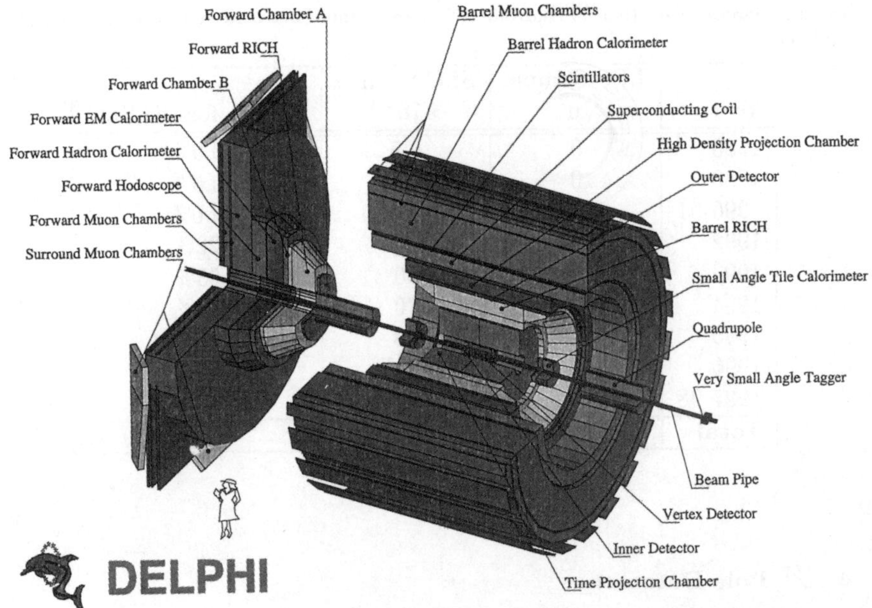

Figure 2. The DELPHI Detector at LEP

4.1 Cylindrical Geometry Silicon Tracking

The first cylindrical geometry silicon μ-strip detector was installed in the MARK-II detector at the SLC in 1989. By 1994 all of the Z^0 pole detectors featured precision silicon tracking, achieving impact parameter resolution at high momentum of better than 20 μm. The current state-of-the-art silicon tracking device is the SLD pixel CCD detector, shown in two different incarnations in Figure 3. The latest version (VXD3), taking advantage of 8cm silicon wafer technology, comprises three layers between 2.8 and 4.8 cm in r, providing 3 precise 3-d space points out to $|\cos\theta| = 0.85$. The CCD pixel dimension is $22 \times 22 \times 300$ μm, providing a single-hit resolution of approximately 4 μm. Including the uncertainty in the beam spot location, VXD3 achieves an $r - \phi$ impact parameter resolution of 8 μm at high momentum, and 36 μm at $p_\perp \sqrt{\sin\theta} = 1$ GeV/c.

4.2 Cerenkov Ring Imaging Detectors

DELPHI and the SLD contain detectors which detect the photons emitted by charged particles in the well-defined Cerenkov cone as they pass through radiating materials. Photons from a low-threshold, compact liquid radiator are converted in a position-sensitive drift detector filled with a UV-sensitive

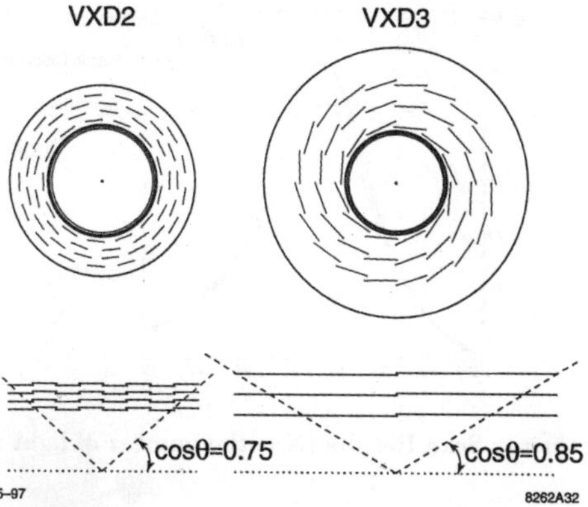

Figure 3. Schematic of the SLD CCD Pixel Tracking Detectors

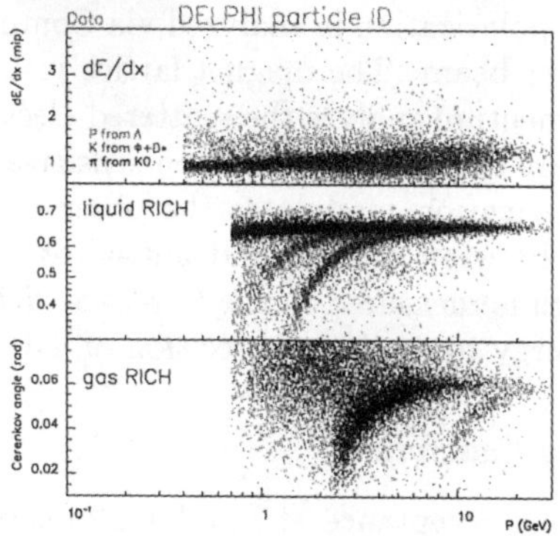

Figure 4. Performance of the DELPHI Particle ID System

gas. Photons from a high-threshold gas radiator are imaged back into the drift detector planes by curved mirrors mounted to the outer radius of the Cerenkov detector. The radius of the reconstructed Cerenkov photon ring is a measure of the angle of the Cerenkov cone, and thus the velocity of the radiating particle, which, when combined with momentum measured in the tracking system, provides a measure of the particle's mass. Figure 4 shows the performance of the liquid and gas Cerenkov systems, as well as the dE/dX information from the gaseous tracker, for the DELPHI detector. Taken together, these systems provide reliable particle species separation over much of the available kinematic range.

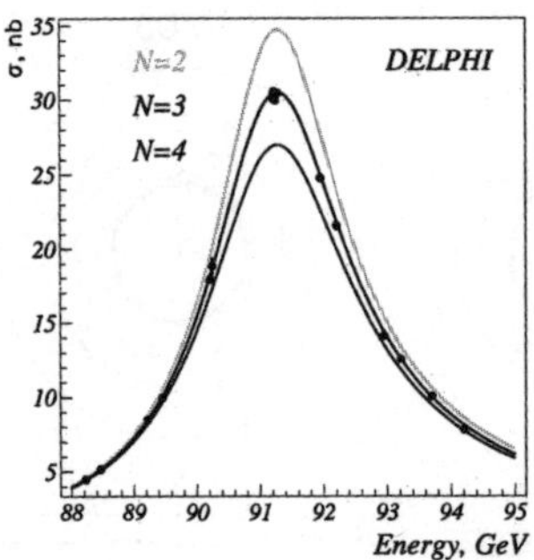

Figure 5. DELPHI Energy Scan Results (N is the number of light ν species assumed).

4.3 Electron Beam Polarimetry

The SLD electron beam polarization is analyzed via Compton scattering from a polarized Nd:YAG laser beam. The magnet lattice just downstream of the final focus serves to momentum analyze the scattered electrons, beyond which the scattered electrons are detected in a position-sensitive Cerenkov detector. The electron beam polarization modulates the observed asymmetry in the Compton cross section between aligned and anti-aligned laser and electron beam polarization, which is measured in the Cerenkov detector as a function of scattered electron energy to a relative precision of $\pm 0.7\%$.

4.4 Precise Luminosity Calibration

The leading-order Bhabha acceptance at small angles is given by

$$\sigma_{e+e-\to e+e-} = \frac{1040nb - GeV^2}{s}(\frac{1}{\theta^2_{min}} - \frac{1}{\theta^2_{max}}). \tag{14}$$

Typical acceptance regions of between 20 and 100 mrad lead to cross sections 3-10 times that of the Z^0 peak hadronic cross section. The combination of precise, well-calibrated Si/W calorimeters and higher order calculations of the Bhabha scattering cross-section has lead to an overall relative error on the luminosity scale approaching $\pm 0.1\%$.

5 Z^0 Lineshape Measurements

The precise knowledge of the luminosity scale permits a correspondingly precise measurement of the energy dependence of the Z^0 production cross-section,

Table 3. Measured Z^0 Lineshape Parameters

M_Z	91.1867 ± 0.0021 GeV/c^2
Γ_Z	2.4939 ± 0.0024 GeV/c^2
σ^0_{had}	41.491 ± 0.058 nb

from which accurate values of the lineshape parameters can be extracted. Figure 5 shows the results of such an energy scan, making use of all available off-peak running, from the DELPHI collaboration. Table 3 shows the resulting lineshape parameters, averaged for the four LEP experiments [2]. The relative accuracy on the Z^0 mass (2×10^{-5}) rivals that of the Fermi coupling constant G_F (1×10^{-5}) and thus places a very tight constraint on the value of the weak mixing angle via equation (13).

6   Forward-Backward Asymmetry Measurements

In measuring the forward-backward asymmetries (polarized or unpolarized) there are two basic challenges to be met, both of which are relatively straight-forward for final state fermions $f = lepton$. First, the $Z^0 \rightarrow f\bar{f}$ final state of interest must be identified. Secondly, the fermion direction must be determined – usually by finding the thrust axis, and then signing it according to the best estimate of which hemisphere contains the fermion.

Calculating the mass of identified secondary vertices can tag events with heavy final state quarks (b, c) as well as discriminate between the b and c final states themselves. $Z^0 \rightarrow b\bar{b}$ events are tagged in this way with an efficiency as high as 60%, and with a purity as high as 98%. The fermion direction of such 'inclusively tagged' events can be determined by the net momentum-weighted hemisphere track charge, the sign of identified kaons from the $b \rightarrow c \rightarrow s$ cascade, or simply the net charge of the tracks forming the vertex. Heavy quark events can be tagged by leptons from semileptonic decay, or by the exclusive reconstruction of a heavy meson state, such as $D^+ \rightarrow K^- \pi^+ \pi^+$. For these latter approaches, the charge of the lepton or reconstructed state indicates the charge of the underlying heavy quark.

Figure 6 shows an example of the angular distribution of the estimated b quark direction, in this case for the polarized electron beam at SLAC. Clear forward-backward asymmetries are observed for both left- and right-handed beam, from which the coupling parameter A_b can be measured according to (2) and (10). Table 4 shows the resulting electroweak parameter results from the various forward-backward asymmetry measurements [2].

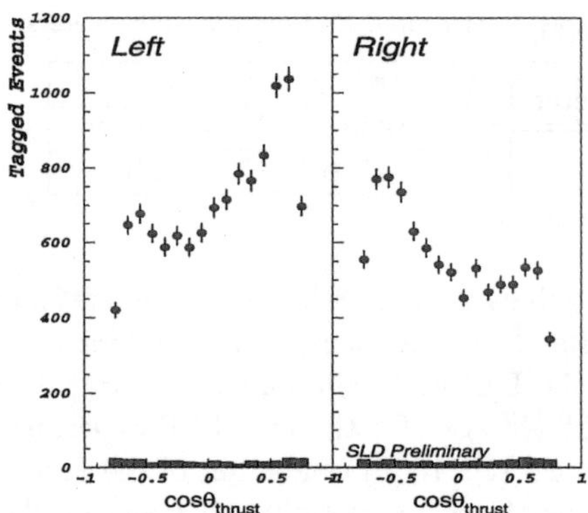

Figure 6. Distribution of the b decay angle in SLD $Z \to b\bar{b}$ events, separately for left and right handed electron beam.

Table 4. Forward-backward asymmetry measurements and associated extracted electroweak parameters. For the unpolarized asymmetries, the final state coupling parameters A_f are derived under the assumption that $A_e = 0.1489 \pm 0.0017$, which is the LEP/SLC combined value for A_{lepton}.

Observable	Measured Value	Extracted EW Parameter
A_{FB}^{l}	$.01683 \pm .00096$	$\sin^2 \theta_W = .23117 \pm .00055$
A_{FB}^{b}	$.0991 \pm .0020$	$\sin^2 \theta_W = .23225 \pm .00037$
A_{FB}^{c}	$.0713 \pm .0043$	$\sin^2 \theta_W = .2321 \pm .0010$
A_{FB}^{had} (Q_{FB})		$\sin^2 \theta_W = .2321 \pm .0010$
$\tilde{A}_{FB}^{l}$	$.1456 \pm .0063$	$\sin^2 \theta_W = .2317 \pm .0008$
$\tilde{A}_{FB}^{b}$	$.898 \pm .029$	$A_b = .898 \pm .029$
$\tilde{A}_{FB}^{c}$	$.634 \pm .027$	$A_c = .634 \pm .027$
A_{FB}^{b}	See caption	$A_b = .887 \pm .021$
A_{FB}^{c}	See caption	$A_c = .637 \pm .039$

7 Final State Polarization of τ Leptons

An independent, direct measurement of the parity violation in the Z^0-τ coupling can be made by measuring the asymmetry in left- and right-polarized τ leptons in $Z^0 \to \tau\bar{\tau}$ events, which is experimentally accessible via the kinematic distributions of the τ decay products. For example, a diagram of the decay $\tau \to \nu_\tau \pi(K)$ similar to that of Figure 1 shows that the decay distri-

Table 5. τ Polarization Asymmetry Results

Parameter	Measured Value	Extracted $\sin^2 \theta_W$
A_τ	0.1431 ± 0.0045	0.23202 ± 0.00057
A_e	0.1479 ± 0.0051	0.23141 ± 0.00065

bution in the angle θ^* of the π (K) relative to the τ spin direction in the τ rest frame is proportional to $|d^{1/2}_{-1/2,1/2}|^2 = \cos^2(\theta^*/2)$. Boosting against the direction of the τ spin (to produce a right-handed τ) results in a distribution

$$\frac{d\Gamma^\tau_{RH}}{dx} \propto x \tag{15}$$

where $x = E_{\pi,K}/E_{beam}$ is the fractional energy of the pseudoscalar decay product. Similarly, for a boost in the direction of the τ spin (to produce a LH τ), the decay angular distribution is proportional to $1 - x$, and so the relative fraction of RH and LH polarized τ's can be extracted from the x distribution of 1-prong τ decays.

The relation between this τ polarization asymmetry and the electroweak coupling parameters is given by

$$P_\tau(z) = \frac{\sigma^\tau_L(z) - \sigma^\tau_R(z)}{\sigma^\tau_L(z) + \sigma^\tau_R(z)} = \frac{A_\tau(1 + z^2) + 2A_e z}{(1 + z^2) + 2A_\tau A_e z} \tag{16}$$

and so a measurement of $P_\tau(z)$ yields independent values for A_τ and A_e. Table 5 shows the τ polarization results, averaging over all analyzable τ decay modes for the four LEP experiments [2].

8   The Left-Right Asymmetry A_{LR}

The left-right asymmetry A_{LR} is measured by simply comparing the number of hadronic Z^0 decays produced with left- and right-handed electron beam:

$$A_{LR} = \frac{1}{|P_e|} \frac{N^L_{Z^0} - N^R_{Z^0}}{N^L_{Z^0} + N^R_{Z^0}}. \tag{17}$$

Corrections due to differences in left- and right-handed beam luminosity, energy, polarization, etc., are small and well understood. With a sample of 550,000 hadronic Z^0 decays produced with $\langle |P_e| \rangle \simeq 72\%$, the SLD experiment finds [2]

$$A_e = A_{LR} = 0.1511 \pm 0.0024$$
$$\sin^2 \theta_W = 0.23100 \pm 0.00031. \tag{18}$$

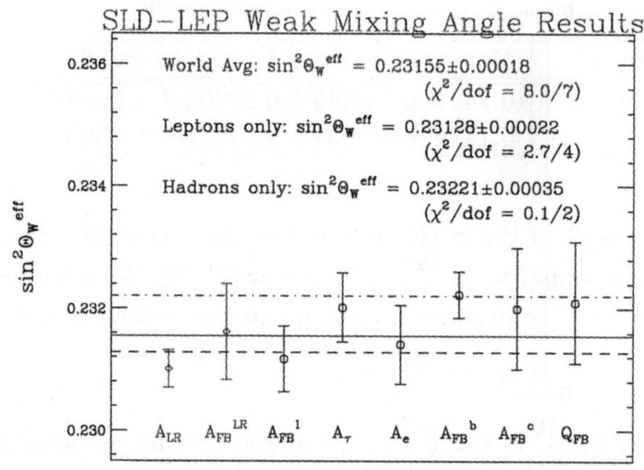

Figure 7. Results for $\sin^2\theta_W$ By Method

Table 6. Partial Width Measurements

Parameter	Measured Value	SM Expectation
R_b	0.2168 ± 0.0007	$0.2155 \pm 0.0003(m_t)$
R_c	0.1694 ± 0.0038	0.1723

In addition to being the single most precise determination of the weak mixing angle, the systematics of this measurement are almost entirely independent of that of other techniques discussed above.

9   Combined Weak Mixing Angle Results and Fits

Figure 7 shows the $\sin^2\theta_W$ results separately for each of the above approaches [2]. While the hypothesis that all of these measurements are consistent with a single mean value is reasonably well supported ($\chi^2 = 8.0$ for 7 d.o.f), there is a mild (2.2 σ) discrepancy between the values measured with leptonic final states and those measured with hadronic final states, the latter of which are dominated by the unpolarized A_{FB}^{b} measurement. Ignoring this issue for the moment, Figure 8 shows a fit for the Higgs Boson mass, assuming the Higgs contributes to radiative loops as prescribed by the Standard Model. At 95% confidence, the Higgs mass is restricted to be below about 270 GeV/c^2.

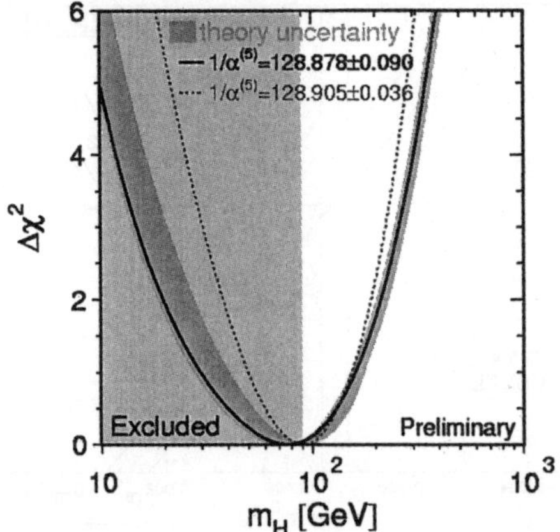

Figure 8. Fit of the SM Higgs Boson mass to electroweak precision data. The solid curve uses as input the value of $\alpha_{em}(Q^2 = M_Z^2)$ of Ref. [3], while the dashed curve uses that of Ref. [4].

10   Partial Width Measurements

Measurements of Z^0 decay partial widths (12) are particularly sensitive to vertex corrections. Due to the cancelation of both experimental and theoretical uncertainties (luminosity, fiducial volume, effects of gluon radiation, etc.), the partial widths are most cleanly constrained via hadronic branching fractions, e.g. $R_b = \Gamma_{Z^0 \to b\bar{b}}/\Gamma_{Z^0 \to \text{had}}$. In addition, with the pure and efficient tags provided by precise silicon tracking, it is possible to constrain the final-state tagging efficiency by comparing the single- *vs.* double-hemisphere tagging rate. The resulting precise constraints on the b and c quark partial widths are shown in Table 6.

11   Fit for $Z^0 - b$ Vertex Parameters

As discussed in Ref. [5], it is interesting to decouple potential anomalous $Z^0 - b$ vertex effects from the vacuum polarization corrections to the weak mixing angle. The explicit $\sin^2 \theta_W$ dependence of the $Z^0 - b$ coupling is removed by writing

$$\delta s = [\sin^2 \theta_W]_{meas} - [\sin^2 \theta_W]_{pred}$$

$$\tfrac{1}{3}\delta s + \delta g_L^b = [g_L^b]_{meas} - [g_L^b]_{pred} \tag{19}$$

$$\tfrac{1}{3}\delta s + \delta g_R^b = [g_R^b]_{meas} - [g_R^b]_{pred}$$

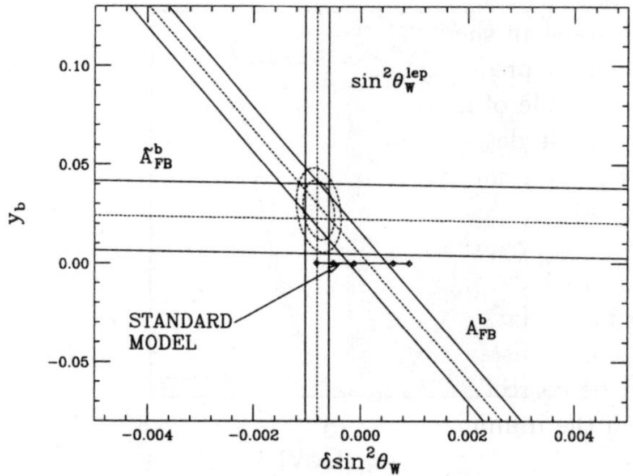

Figure 9. Fit for $Z^0 - b$ vertex parameters, projected into the $\delta\sin^2\theta_W - y_b$ (parity-violation) plane. The Standard Model range includes $169 < m_t < 179$ GeV/c^2, and $100 < m_H < 1000$ GeV/c^2.

Then, the linear combinations

$$x_b = \cos\phi\,\delta g_L^b - \sin\phi\,\delta g_R^b \simeq -\tfrac{1}{4}\delta R_b$$
$$y_b = \sin\phi\,\delta g_L^b + \cos\phi\,\delta g_R^b \simeq -\tfrac{3}{5}\delta A_b,$$

(20)

with $\phi = \tan^{-1}|g_R^b/g_L^b|$, are the deviations from the SM in the $Z^0 - b$ vertex partial width (x_b) and parity violation (y_b) strengths. These parameters are constrained by the precision measurements of $\sin^2\theta_W^{\rm lepton}$, R_b, R_c, $\tilde{A}_{FB}^b$, A_{FB}^b, $R_Z = \Gamma_{\rm had}/\Gamma_{\mu^+\mu^-}$, and $\sigma_{\rm had}^0$. Of particular interest, given the high value of $\sin^2\theta_W$ from A_{FB}^b, is the projection of this fit into the $(\delta s, y_b)$ plane, shown in Figure 9. It is seen that the high value of $\sin^2\theta_W$ from A_{FB} (see Table 4 or Figure 7) may possibly be due to an anomalous $Z^0 - b$ vertex coupling parameter, a hypothesis which is supported by the direct measurement provided by $\tilde{A}_{FB}^b$. The overall discrepancy in the strength of $Z^0 - b$ coupling parity violation with respect to the SM is just over 2.5σ.

12 Summary and Outlook

With the exception of relatively minor updates to existing results, the picture provided by Z^0 resonance precision measurements is essentially complete. The totality of accumulated data has allowed, with a number of systematically independent approaches, the test of the internal consistency of the Standard

Model to the unprecedented precision of $\delta \sin^2 \theta_W < \pm 0.0002$ – roughly 300 times more accurate than the $\pm \sim 0.06$ available from ν DIS in 1988. While this confirmation of the predictions of the Standard Model is remarkable, there are nonetheless a couple of issues to be noted. A mild discrepancy between the weak mixing angle determined via purely leptonic couplings relative to that made with hadronic couplings could possibly be due to anomalous $Z^0 - b$ vertex effects. This hypothesis is supported, albeit with somewhat inadequate experimental precision, by the SLD direct measurements of the $Z^0 - b$ vertex via the polarized forward-backward asymmetry $\tilde{A}^b_{FB}$. Secondly, the combination of all electroweak data, assuming only Standard Model contributions to radiative loops, is inconsistent with a heavy Higgs Boson. Should the minimal Standard Model be correct, the Higgs should easily be seen at the LHC, or perhaps earlier in the imminent RUN II of the Fermilab Tevatron.

Acknowledgment

This work was supported in part by the Department of Energy, Grant #DE-FG03-92ER40689.

References

1. A particularly clear presentation of angular momentum and rotation matrix elements is given in Chapter 8 and Appendix H of W. E. Burcham and M. Jobes, *Nuclear and Particle Physics*, Longman Scientific and Technical, 1995.
2. The Z^0 pole electroweak results presented throughout this report have been accumulated in the following two documents, which are products of the LEP Electroweak Working Group (http://www.cern.ch/LEPEWWG/Welcome.html): CERN-EP/99-15 and LEPHF 99-01.
3. S. Eidelmann and F. Jegerlehner, Z. Phys. **C67**, 585 (1995).
4. M. Davier and A. Höcker, Phys. Lett. **B419**, 419 (1998).
5. T. Takeuchi, A. K. Grant, and J. L. Rosner, hep-ph/9409211.

SEARCHING FOR SUSY AND HIGGS

W.MURRAY

Rutherford Appleton Laboratory, Chilton, Didcot, Oxon., OX11 0QX, UK
E-mail w.murray@rl.ac.uk

Searches for supersymmetric particles and Higgs bosons are reviewed. There are no convincing discoveries, although there is one intriguing hint. The search channels are outlined and the sensitivities of different experiments are contrasted.

1 Introduction

As has been detailed elsewhere (e.g.[1]), supersymmetry must be a broken symmetry as the superpartners are unobserved. For definiteness, and to reduce the parameter space, specific hypotheses for the symmetry breaking are frequently used, and these normally communicate the breaking from a 'hidden sector' in a flavour-blind way, through gravity or the gauge forces. These hypotheses include SUGRA, which is discussed in section 3 and Gauge Mediated Symmetry Breaking, in section 4. Searches for models where R parity is not conserved are in section 5, and sections 6 and 7 discuss the Higgs sector.

2 Experiments and data samples

Table 1. Experiments and data samples discussed here.

Accelerator	Experiment	Collision	$\sqrt{s}$	Luminosity
Tevatron	D0, CDF	$p\bar{p}$	1800 GeV	110pb^{-1}
LEP	ALEPH, DELPHI, L3, OPAL	e^+e^-	189 GeV	160pb^{-1}
Hera	ZEUS, H1	e^+p	320 GeV	50pb^{-1}
CESR	CLEO	e^+e^-	10.6 GeV	3100pb^{-1}

The data samples reported on here are summarized in table 1. The first three accelerators make direct searches, but the importance of CLEO lies in measurements such as[2,3,4]:

$$Br(b \to s\gamma) = 3.15 \pm 0.35 \pm 0.32 \pm 0.26 \cdot 10^{-4}$$

$$\frac{Br(D^0 \to K^+\pi^-)}{Br(D^0 \to K^-\pi^+)} = 0.50^{+0.11}_{-0.12} \pm 0.08\%$$

$$Br(b \to Kll) < 0.7 \cdot 10^{-5}$$

These are influenced by loops diagrams, which can have several different virtual particles entering with different signs. Because of this, they provide important constraints on a model, but not on an individual particle.

3 The Search for Minimal SUGRA

The minimal SUGRA model has 5 parameters, $(M_0, M_{\frac{1}{2}}, A, \tan\beta, \text{sign}\mu)$, from which the masses and mixings of all the superpartners can be computed. The neutralino provides the Lightest Supersymmetric Particle (LSP) in many cases, but is difficult to observe as it is a stable, weakly interacting particle. The next to lightest supersymmetric particle (NLSP) has a good chance of being the first to be discovered, and is typically a chargino, slepton or Higgs. The decay signatures depend upon δM, the mass difference between the particle in question and the LSP. There are many SUGRA-inspired models with similar properties.

3.1 Searches for squarks and gluinos:

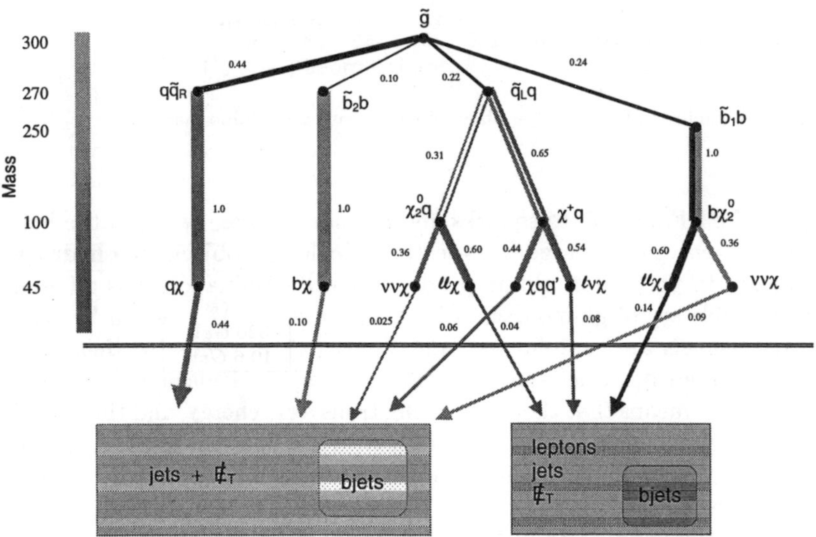

Figure 1. One possibly set of gluino decay chains, for a particular SUGRA point, $(M_0 = 100, M_{\frac{1}{2}} = 100, A = 0, \tan\beta = 2, \mu < 0)$,

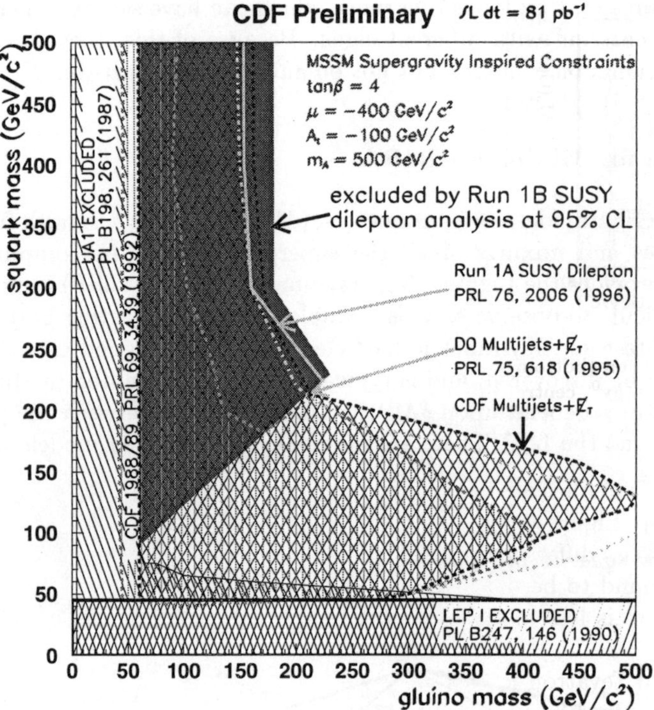

Figure 2. CDF exclusion limits in the plane of the squark and gluino masses.

The reach of CDF and D0 to squarks and gluinos is large, owing to the high centre of mass energy and large cross-section of coloured objects in a hadronic environment. In particular, gluino production will dominate if it is allowed, and one possible set of gluino decays is shown in figure 1[5]. As can be seen, the decay channels are complicated, and the search proceeds by identifying topologies of interest, rather than concentrating on particular particles. The unobserved LSP means that there is missing transverse energy, and the quarks arising in the gluino decay chain produce jets. There may or may not also be leptons. Thus there are two classes of search, looking for jets and $E\!\!\!/_T$ or jets, $E\!\!\!/_T$ and leptons.

Owing to the complexity of the decay chain, it is hard to make model-independent limits. In particular, the mass of the unobserved LSP has a major influence on the phenomenology. The typical region excluded can be seen in figure 2[5].

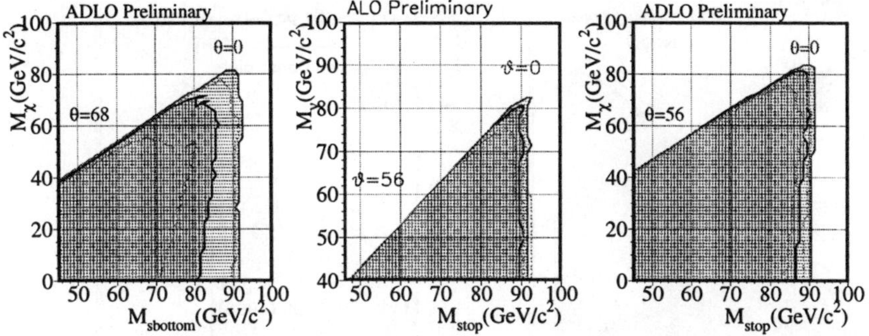

Figure 3. LEP limits on sbottom and stop quarks, as a function of the mass of the lightest neutralino. Left: $\tilde{b} \to b\chi^0$, centre: $\tilde{t} \to bl\tilde{\nu}$, right: $\tilde{t} \to c\chi^0$.

At LEP, while gluinos are not accessible, squarks might be pair produced, especially $\tilde{b}_1$ or $\tilde{t}_1$. Figure 3 shows the bounds on the sbottom and stop squark masses[6] For all mass differences over 10 GeV/c^2 and mixings, the mass of the stop squark is found to be over 87 GeV/c^2, and the sbottom squark is over 80 GeV/c^2. There is less model dependence in these limits, as the gluino is not involved.

The LEP and Tevatron squark limits probe different regions of the parameter space, and make different assumptions. They are clearly complementary, but it is reasonable to say that the Tevatron has a higher 'probability' (in some sense) of making a discovery in this region than LEP. Conversely the LEP limits are a little less model dependent.

3.2 Slepton searches

The largest sensitivity here is at LEP, due to the production mechanism. The $\tilde{\tau}$ might be expected to be the lightest slepton, but this is of course not guaranteed. There is a marginal excess of candidates in this channel, but no sign of a signal. The limits found[7] for the $\tilde{e}$ and $\tilde{\tau}$ are shown in figure 4, and if we impose $\delta M > 15$ GeV/c^2, then we find $m_{\tilde{e}} > 89 GeV/c^2$, $m_{\tilde{\mu}} > 84 GeV/c^2$ and $m_{\tilde{\tau}} > 71 GeV/c^2$.

3.3 HERA squarks and selectrons

HERA is not as well suited as the Tevatron or LEP. This is partially because the normal production process is of a squark and a selectron together, and a

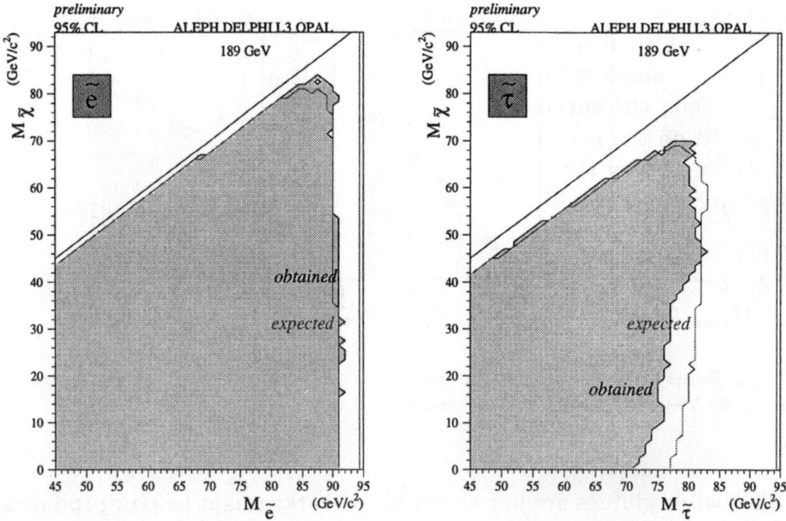

Figure 4. LEP limits on $\tilde{e}$'s (left) and $\tilde{\tau}$'s (right). The $\tilde{e}$ sensitivity is greater because there is a t-channel χ^0 exchange diagram, in addition to the s-channel production.

discovery would therefore require that both (s)particles have a low mass[8]. As usual, the results also depend critically upon the LSP mass.

3.4 Charginos and Neutralinos

These can be pair produced at LEP, in the s-channel or via $\tilde{\nu}/\tilde{e}$ exchange which have a negative interference, and therefore the production cross-section depends upon several parameters. The decay signatures depend upon δM, the mass difference to the LSP. The mass limits derived are of course only valid with a model.

For χ^+, the combined LEP limits are in table 2, based upon data up to 183 GeV[9]. For the low δm region the other model parameters are unimportant. Limits on the χ_2^0 are similar.

Table 2. Limits on chargino mass from LEP.

	Mass difference, $M_{\chi^+} - M_{\chi_1^0}$		
	$\geq 3\ GeV/c^2$	$\geq 4\ GeV/c^2$	$\geq 5\ GeV/c^2$
Limit	$81.7\ GeV/c^2$	$89.0\ GeV/c^2$	$90.1\ GeV/c^2$

The production of the 'invisible' χ_1^0 may also be seen through associated ISR, or the Z width if not a pure photino. Limits on it can be made by scanning the 5 parameters to see which points are excluded by any of a number of searches. Then the surviving point with the lightest neutralino gives an absolute bound on the neutralino mass in SUGRA. Examples from a scan of this sort, performed by the ALEPH collaboration, are in figure 5. (See [10] for an updated version of this result.)

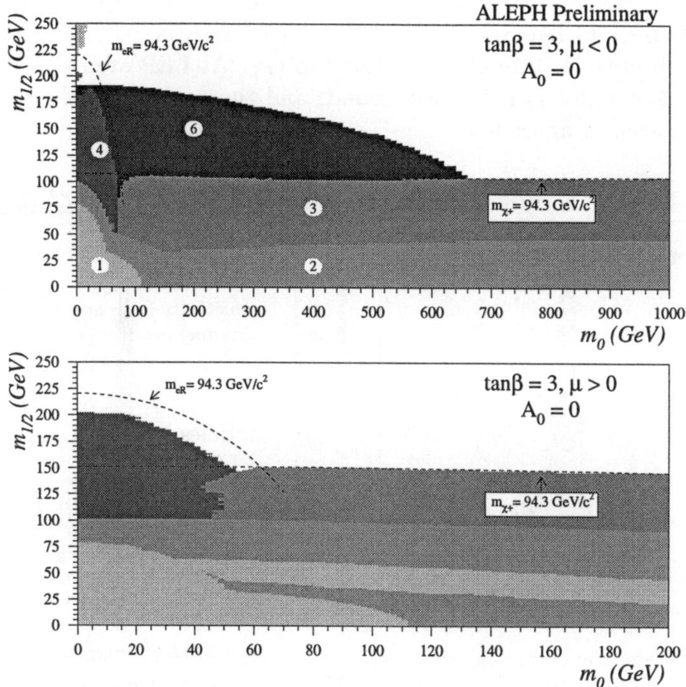

Figure 5. Chargino production processes at LEP. The shaded regions are excluded by (1) theoretical considerations, (2) The Z width at LEP 1, (3) Chargino searches (4) Slepton searches (5) Stable particle searches and (6) Higgs searches.

From this scan all neutralino masses below 36 GeV/c^2 are excluded. (Except that there are some regions of mixing to be checked for $\tan\beta$ greater than 10.) In the MSSM, with less constraints, $M_{\chi_1^0} > 32\ GeV/c^2$. These bounds constrain the region where the neutralino may be cosmologically significant.

4   Gauge Mediated Symmetry Breaking

There are several variations, but they all have the distinctive feature that the LSP is a Gravitino, $\tilde{G}$. This has very weak couplings, so the NLSP is very important, and most SUSY particles decay to it. The lifetime of this depends upon the parameters, but in a collider it ranges from prompt decay at production, through decay in flight in detector, to appearing stable. The χ_1^0 and right handed sleptons are good candidates for the NLSP.

In the case where the NLSP is a χ_1^0 with a long lifetime, appearing stable, the searches are very like SUGRA. On the other hand the pair production of χ_1^0 may be observable through their decay to $\tilde{G}\gamma$. At LEP events containing only one or two photons have been found, and the masses recoiling against them can be seen in figure 6.

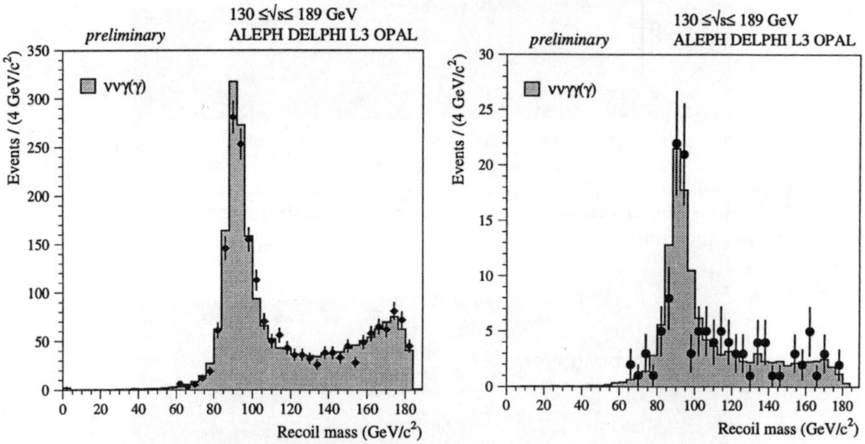

Figure 6. Mass spectrum recoiling against single and double photons. The (shaded) Standard Model background is dominated by $Z^0 \to \nu\bar{\nu}$ and there is no evidence for any additional contribution.

This search was originally motivated by the observation of a candidate for a pair of $\tilde{e} \to e\chi^0 \to e\tilde{G}\gamma$ by CDF, but the LEP limits almost exclude the parameter space where this was a possibility[11]. A search by $D0$, for two high photons with E_T greater than 12 and 20 GeV/c^2 and missing energy [12], in fact completely rules out this mode; the limits are shown in figure 7.

The other main direction suggested is that the NLSP is a charged slepton, which may appear stable. A LEP combined limit, using data up to 183 GeV, finds limits of 86.5 and 87.0 GeV/c^2 for right and left handed stable charged

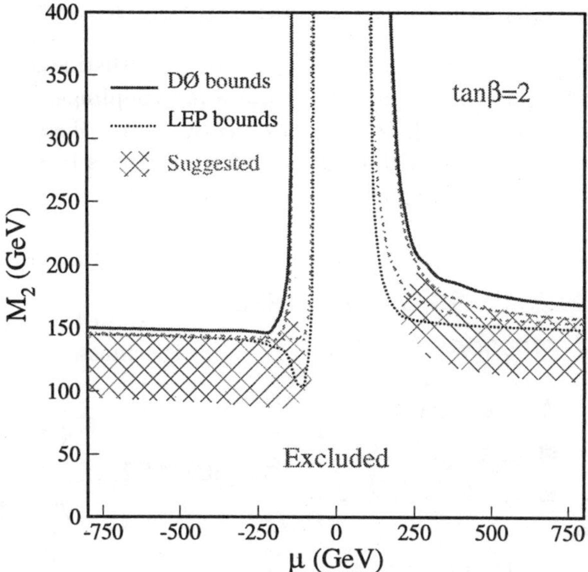

Figure 7. The parameter space region excluded by $D0$'s search for two high E_T photons SUGRA mass spectrum.

sleptons respectively, both close to the kinematic limit[13]. Studies of detached vertices and kinks in tracks[14] have excluded $\tilde{\tau}$'s of any lifetime with mass below 76 GeV/c^2, and in fact the sensitivity is greatest in the intermediate region.

5 R parity violating SuperSymmetry

R parity is defined as $-1^{3(B-L)+2s}$. This is $+1$ for the observed particles and -1 for their superpartners. Normally, it is assumed to be conserved; this leads to a stable LSP to which others will decay. If this is relaxed, we can introduce 3 extra couplings, λ, λ', λ'', which violate L, L and B respectively. Each couples one SUSY partner with two normal particles and has 3 generational indices to describe this. It cannot be the case that all the λ's are large, as this would allow proton decay, but sizeable single terms are certainly possible.

5.1 Karmen Anomaly

Karmen search for neutrino oscillations using pion decay at rest through $\pi^+ \to \mu^+ \nu_\mu$, followed by $\mu^+ \to e^+ \nu_e \overline{\nu_\mu}$. They use a pulsed pion source, and so the observed time spectrum of their neutrinos has two components: prompt, from the pion decay and delayed with the muon lifetime, from the muon decay. The observed ν_e spectrum is in figure 8.

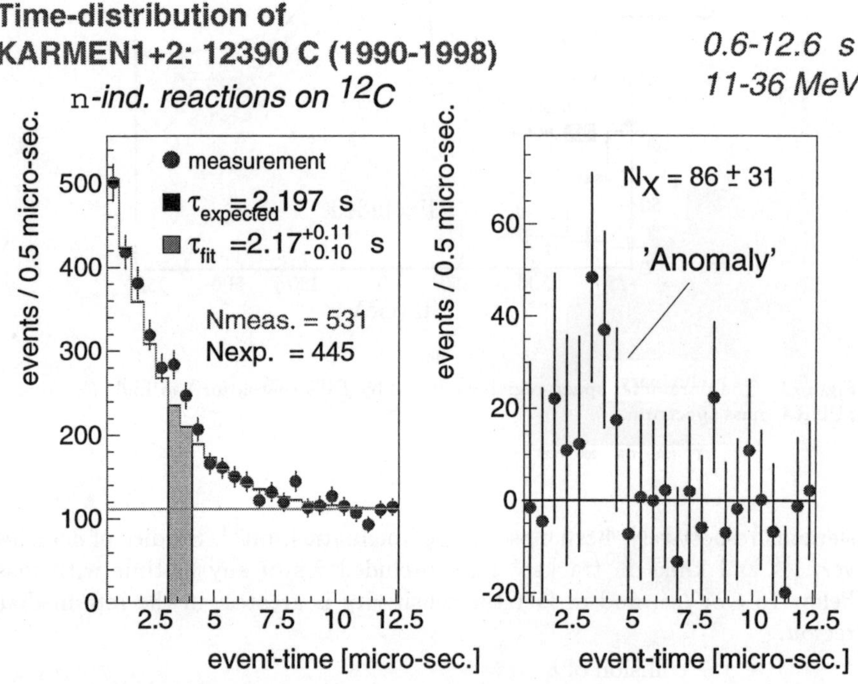

Figure 8. Left: The arrival time of neutrinos recorded by the Karmen 2 detector. Right: The residuals to the fit, showing a small excess at around $3.5\mu s$.

It is certainly possible that the observation represents a statistical fluctuation. There are two more scheduled years of data taking which may be informative. One possible explanation is that a slow moving (c/50) particle could have been produced in the pion decay. It would need a mass of $33.9 MeV/c^2$ to have the correct velocity. A fit testing this hypothesis has log-likelihood ratio of 9 at this mass.

One R parity violating SUSY explanation[15] is that the pion decays with a branching ratio of 10^{-8} into $\mu\chi_1^0$, and the χ_1^0 then decays with a lifetime of around 100 seconds into $e^+e^-\nu_e$. Note that avoiding the χ^0 mass constraint requires that it is essentially a pure photino, and that the SUGRA relations between netralino masses are removed.

5.2 HERA RPV searches

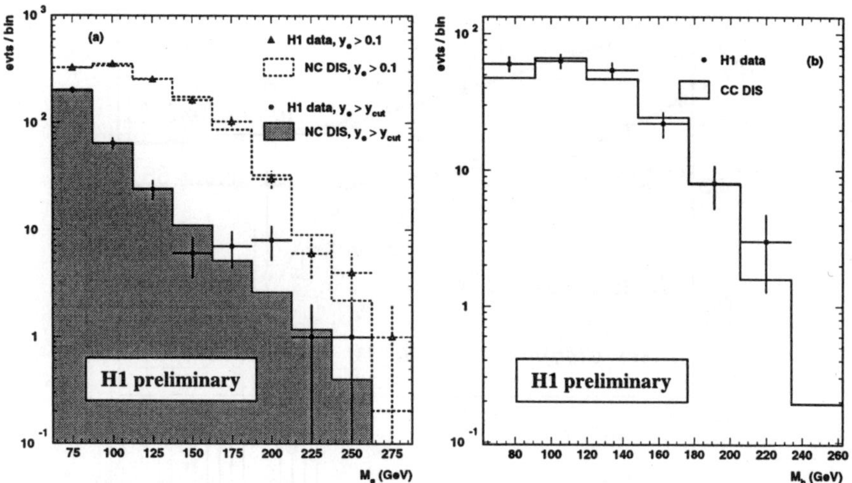

Figure 9. Left: H1's observed mass spectrum for neutral current $e - q$ scattering. The small excess at $M_{eq} = 200\ GeV/c^2$ has decreased in significance. Right: The spectrum for $eq \to \nu_e q'$ interactions.

At HERA, the collision of quarks and electrons means that there is potential to make a single squark (i.e. $e^+q \to \tilde{q}'$) if there is an appropriate λ' coupling. Such events could look like a lepto-quark, if the decay back to the intial particles dominates, or could have more complex signatures if other MSSM particles are involved in the decay chain. Excesses seen in H1 and ZEUS in 1996 have not been repeated – but are still present in the data. H1's observations can be seen in figure 9. Limits can be set on squark masses and couplings, and as the strength of the process depends upon the strength of the RPV coupling, these are limits on λ' as a function of the squark mass.

By contrast, at the Tevatron the squarks may be pair produced, and will then decay to eq or νq if there is *any* significant RPV coupling. Thus the CDF limits[16] in figure 10 depend upon the branching fraction into the eq mode,

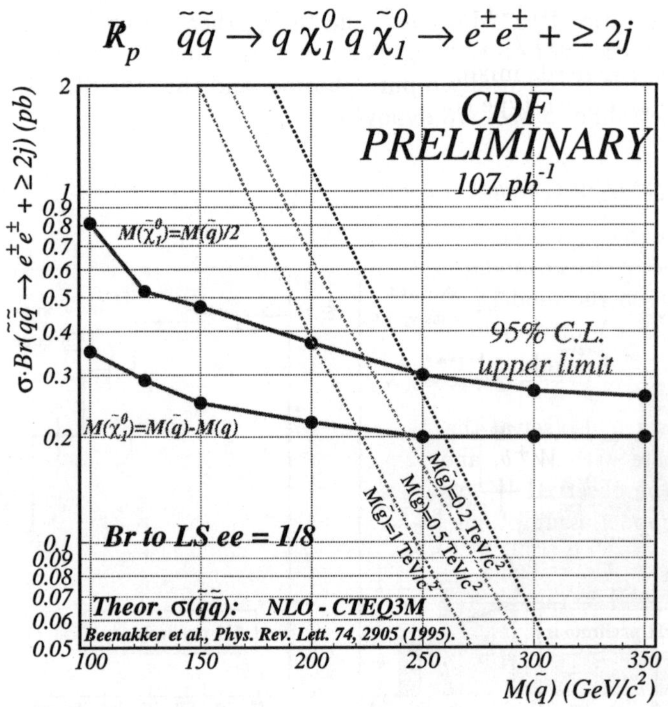

Figure 10. Limits on RPV squarks decaying to eq from CDF.

but not the precise value of λ'. Making the normal assumption that only one RPV coupling is large, these exclude the possible signal from H1.

The LEP experiments have also performed many searches for RPV SUSY particles such as charginos, neutralinos and sfermions. Decay modes such as multi-jets, multi-jets and leptons, or multi-leptons have all been searched searched for, with no evidence of an signals. In most cases the sensitivity extends to close to the beam energy, and as at the Tevatron is not sensitive to the exact value of the RPV coupling.

6   The MSSM Higgs sector

In the MSSM Higgs sector there are two Higgs doublets, which give rise to 5 physical particles; two are charged, the H^+, H^- and three neutral, the h^0, H^0

and A^0. The lightest, in general, is the h^0, which is constrained to weigh less than around 130 GeV/c^2, but the others may be much heavier. Description of the couplings needs mixing angles, especially $\tan\beta$ which controls the ratio of up type and down type couplings.

6.1 Charged Higgs

A direct search has been undertaken at LEP for pair produced H^+H^-. The two dominant decay modes are to $\tau^+\nu_\tau$ and $c\bar{s}$; the relative fraction of each depends upon the model parameters. By making searches for $q\bar{q}'q\bar{q}'$, $q\bar{q}'\tau\nu_\tau$ and $\tau\nu_\tau\tau\nu_\tau$ a limit can be made for any branching ratio with little model dependence. A combined LEP limit[17] of of $m_{H+} > 69.0 GeV/c^2$ has been found.

Limits are also set at the Tevatron, where the decay of the top into H^+b can compete with W^+b, and for very large or very small $\tan\beta$ this would suppress the observed W^+b signature. This is thus an indirect 'search'. but with few model assumptions[18].

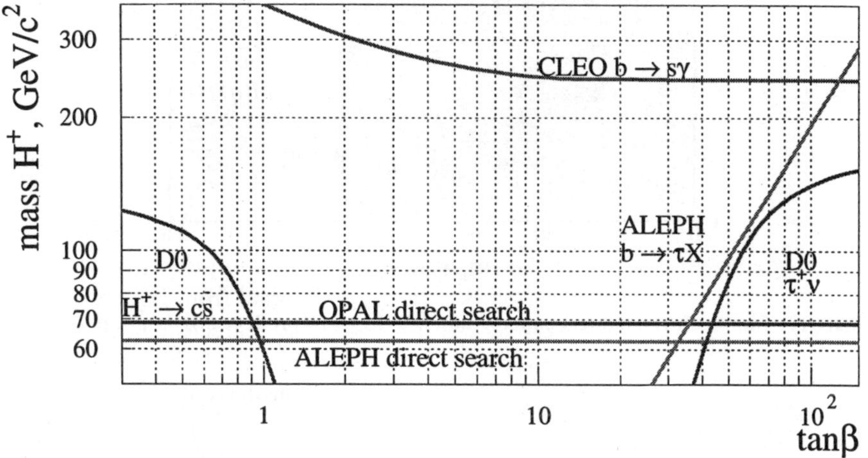

Figure 11. A comparison of the lower limits on H^+H^- from several different sources.

By contrast, limits placed from $b \to s\gamma$, which depend upon all virtual particles which might enter the loop, are extremely indirect, but very powerful. CLEO produce limits which assume only 2 Higgs doublets, and no SUSY particles.[19] They therefore cannot be used as absolute limits on the H^+ mass.

Figure 11 compares all these results, and also a limit from $b \to \tau X$ from ALEPH.[20]

6.2 Neutral MSSM Higgs

This search uses the Standard Model results which is are in section 7 for the hZ, and combined it with $hA \to bbbb$, and $hA \to \tau\tau bb$ results.

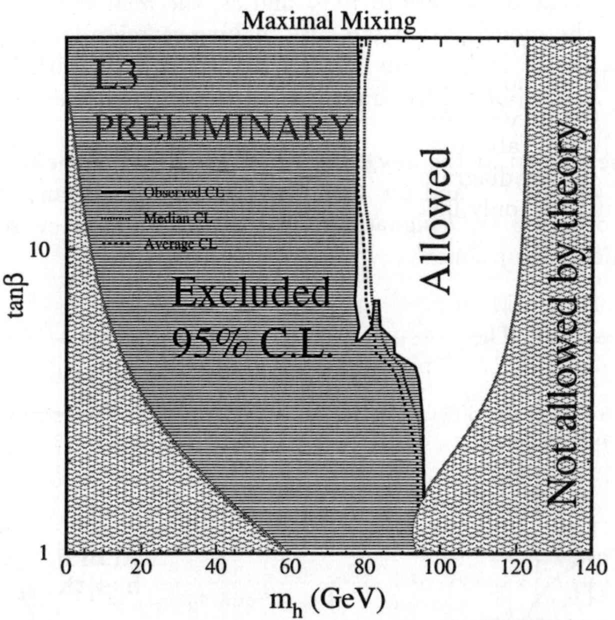

Figure 12. Limits on the lightest supersymmetric Higgs, h, produced by L3.

Figure 12 shows a 'benchmark' scan[17], by the L3 collaboration, where most of the MSSM parameters are fixed, and only $\tan\beta$ and m_h vary, in the fairly pessimistic scenario of maximal mixing in the stop sector. The highest limits reported are those from DELPHI[21], which are $m_A > 84.5 \ GeV/c^2$, $m_h > 83.5 \ GeV/c^2$, but it should be emphasized that in more complete scans, there are some unexcluded regions with lower masses.

These bounds are in the same region as the upper limit on m_h in minimal SUSY, but as the LEP beam energy will raised by only 15 GeV/c^2 or so, it will not be possible to search the whole of this region LEP. However, we should remain alert for the possibility of a discovery.

The Tevatron can also place limits on the h mass, through h, A or H radiation from b quarks, followed by decay into two b quarks. The signature is events with four b jets, and from this search CDF place limits[22] on the mass of the lightest Higgs at 84 GeV/c^2 for $\tan\beta = 40$ and 111 GeV/c^2 for $\tan\beta = 60$.

7 Standard Model Higgs search

The best results on the Standard Model Higgs come from LEP, where Higgs could be produced mostly by Higgs-strallung from a Z. There is a which has a kinematic limit at $\sqrt{s} - M_Z$; 10 GeV below it the cross-section is of order 0.5 pb, while 10 GeV above it is around 0.01 pb and falling fast. There are also WW and ZZ fusion diagrams which do not have a similar kinematic limit, but these contribute only 0.003 pb at the same point. As the combined data set of the four collaborations is less than $1000pb^{-1}$, the search range must be restricted to that accessible from the Higgs-strahlung process. The dominant decay mode of the S.M. Higgs is into $b\bar{b}$, 83%, but a 7.8% fraction of $\tau^+\tau^-$ have also been used. The channels searched for are shown in table 3.

Table 3. Comparison of the different Higgs search channels, using the search by DELPHI at 189GeV as an example.

Channel	Branching ratio	Background	Events	Signal	s/b
$HZ \to bbll$	5%	3.81	2	0.91	0.2
$HZ \to bb\nu\nu$	17%	4.11	2	1.27	0.3
$HZ \to bbqq$	58%	20.31	21	7.56	0.4
$HZ \to bb\tau\tau$	2.5%	0.55	0	0.10	0.2
$HZ \to \tau\tau qq$	5.5%	2.50	1	0.23	0.1
Total	88%	31.28	26	10.1	–

The $bbll$ channels have a good mass resolution, which compensates for the poor signal to background. The $bb\nu\nu$ channel potentially has a large contribution at the kinematic limit due to the WW fusion process, but the experimental signature is two back to back b jets with missing energy, and this can be faked by $e^+e^- \to bb$ events if the jet energies are mis-measured. In reducing this background, the efficiency suffers. The $bbqq$ channel suffers from a very large WW background, and it therefore relies heavily on b tagging, which must be well understood. The τ channels are rather difficult, but are always followed because of their relevance to the MSSM searches for hA.

The latest results are from data at $\sqrt{s} = 189\ GeV$, where there are several

Table 4. Results from the LEP collaborations on the S.M. Higgs

Experiment	ALEPH	DELPHI	L3	OPAL
Limit expected	$95.7 GeV/c^2$	$94.8 GeV/c^2$	$94.4 GeV/c^2$	$94.9 GeV/c^2$
Limit obtained	$90.2 GeV/c^2$	$95.2 GeV/c^2$	$95.3 GeV/c^2$	$91.0 GeV/c^2$
Probability	1%	64%	63%	4%

candidates. This is to be expected, as the irreducible ZZ background is large. The results[23,21,24,25] of the four collaborations can be seen in table 4. Two experiments have unusually low limits, which could indicate that a discovery is close, but a combined result was not ready for the conference[a]. LEP has set a lower limit on the mass of the Standard Model Higgs at $95.3 GeV/c^2$, removing the possibility that it is degenerate in mass with the Z.

8 Summary

There have been searches in very many 'SUSY' channels, testing a variety of models. These have tended to search for categories of signatures, in a fairly general way, which is a healthy sign that experimenters are open to new phenomena. Unfortunately, there are no convincing sign of any signals.

A limit of $95 GeV/c^2$ on the Standard Model Higgs has been set. Some of the limits are unusually low, but it seems that this does not come from a Higgs just beyond the limit.

References

1. S. Martin, *'A Supersymmetry Primer'* hep-ph/9707356
2. CLEO collaboration, *'b $\to$ sγ Branching Fraction and Asymmetry'* CLEO-CONF 99-10
3. CLEO collaboration, *'Search for $D^0 - \overline{D}^0$ mixing'* hep-ex/9908040
4. CLEO collaboration, *'Search for Electroweak Penguin decays $B \to K l^+ l^-$ and $B \to K^* l^+ l^-$ at CLEO'* CLEO-CONF 98-22
5. J. Nachtman, *'Searches for Squarks and Gluinos'* 'Higgs and Super-Symmetry: Search and Discovery' Univ. of Florida, March 1999. http://www.phys.ufl.edu/~rfield.hs_agn_frm.html
6. LEP SUSY W.G., *'Combined LEP 2 stop and sbottom limits at 189 GeV'* LEPSUSYWG/99-02.1, http://www.cern.ch/lepsusy/

---

[a]It was recently reported[26] that the 1999 data showed a limit on M_H of 102.6 GeV/c^2. Thus the hints in the 1998 dataset were not a sign of a signal.

7. LEP SUSY W.G., *'Combined LEP slepton results, 189 GeV'* LEPSUSYWG/99-01.1, http://www.cern.ch/lepsusy/

8. Zeus Collaboration, *'Search for Selectron and Squark Production in e^+p Collisions at HERA'* DESY-98-069

9. LEP SUSY W.G., *'Combined LEP chargino results at 183 GeV'* LEPSUSYWG/98-04.1, http://www.cern.ch/lepsusy/

10. The ALEPH collaboration *'Search for supsymmetric particles in e^+e^- collisions at $\sqrt{s} = 192 - 196 GeV$'* ALEPH 99-078 CONF 99-050.

11. S.Ambrosiano,G.Kane,G.Kribs,S.Martin & S.Mrenna Phys. Rev. D55(1997) 1372-1398

12. D0 collaboration, B. Abbott *et al.*, Phys. Rev. Lett. **80**, 1591 (1998).

13. LEP SUSY W.G., *'Stable Heavy Charged Particles'* LEPSUSYWG/98-07.1, http://www.cern.ch/lepsusy/

14. DELPHI collaboration, *'Update of the search for the lightest neutralino and stau pair production in the light gravitino secenarios with stau NLSP.'* DELPHI 99-13 CONF 213

15. D. Choudhury and S. Sarkar, Phys. Lett. **B 374**, 87-92 (1996).

16. M. Chertok, *'Lepton based SUSY searches at CDF'* Presented at SUSY 99, FNAL, June 1999.

17. LEP Higgs W.G., *'Limits on Higgs boson masses from combining the data of the four LEP experiments at $\sqrt{s}$ up to 183 GeV.'* CERN EP 99-060

18. D0 collaboration, *'Search for Charged Higgs Bosons in Decays of Top Quark Pairs'*, hep-ex/9902028

19. CLEO Collaboration, M.S.Alam *et al.*, Phys. Rev. Lett. **74**, 2885 (1995).

20. ALEPH Collaboration, D. Buskulic *et al.*, Phys. Lett. **B 343**, 444 (1995).

21. DELPHI collaboration, *'Search for neutral Higgs bosons in the Standard Model and the MSSM at 189GeV'* DELPHI 99-8 CONF 208, March 1999

22. J. Valls, *'Run 1 Tevatron Neutral Higgs Searches'* Presented at SUSY 99, FNAL, June 1999.

23. ALEPH collaboration, *'Search for the neutral Higgs bosons of the Standard Model and the MSSM in e^+e^- collisions at $\sqrt{s} = 189GeV$'* ALEPH 99-007

24. L3 collaboration, *'Search for the Standard Model Higgs bosons in e^+e^- collisions at $\sqrt{s} = 189GeV$'* CERN-EP/99-080, submitted to Phys. Lett. B

25. OPAL collaboration, *'Search for Neutral Higgs Bosons in e^+e^- Collisions at $\sqrt{s} = 189GeV$'* hep-ex/9908002, submitted to Euro. Phys. J. C

26. P. McNamara, *'Higgs particles'*, LEP working group report, CERN LEPC, Sept 7, 1999.

TWO PHOTON PHYSICS AT LEP

MANEESH WADHWA

University of Basel, Klingelbergstrasse 82,
CH-4056 Basel, Switzerland
E-mail:Maneesh.Wadhwa@cern.ch

LEP offers an excellent opportunity to measure two photon processes over a large kinematical range and thus study the complex nature of the photon. This article reviews the experimental status of "Two Photon Physics" at LEP. The recent results on resonances, multi-hadron production and photon structure functions are discussed.

1 Introduction

Over the past decade two photon physics has proven to be a very productive source of information about QED, QCD and hadron spectroscopy. The Feynman diagram responsible for a two photon collision process at LEP is shown in Figure 1, where the high energy incident electrons and positrons split off virtual photons and the scattered electrons take most of the beam energy.

These two photons then can interact to form a state X with mass $W_{\gamma\gamma}$. The four-momentum transfer q_i to the photons depends on the angle and energy of the scattered electrons[a]. When neither of the scattered electrons is detected (untagged events), the virtual photons are referred to as nearly real i.e. $q_1^2 \approx q_2^2 \approx 0$. This class of events allows several tests of QCD by studying hadronic resonances, the inclusive hadron cross section and jet production rates. If there is detection of one of the scattered electrons

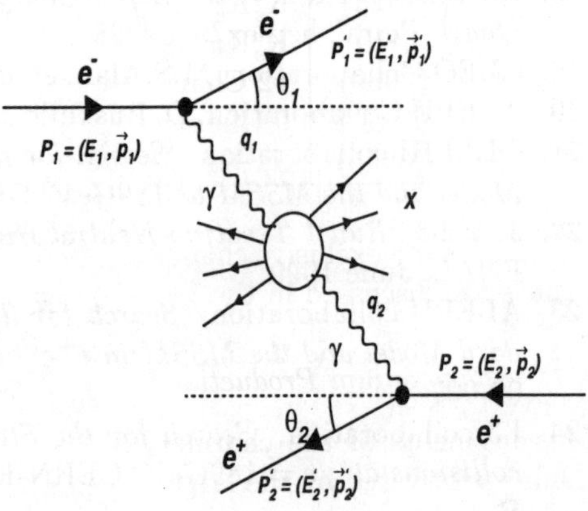

Figure 1. $\gamma\gamma$ *collision in* e⁺e⁻ *scattering*

$Q^2 = -q_1^2$ (single tagged events), it is possible to probe the other photon $q_2^2 \approx 0$ regarded as a "target" and study its structure. Finally, if both the

---

[a]Electron stands for electron and positron throughout this article

scattered electrons are detected $Q_i^2 = -q_i^2, (i = 1, 2)$ (double tagged events), the structure of the reaction of highly virtual photons is probed. In the following sections, a review is given of the $\gamma\gamma$ results obtained at LEP, with special attention to recent results.

2 Resonance production

Two photon formation of C-even meson resonances provides valuable information on the internal structure of mesons. In particular it is interesting to look for resonances whose $\gamma\gamma$ couplings are much smaller than quark-model predictions; e.g. glueball or hybrid quark-gluon states. One can also produce resonances in two-photon events in which one photon is far off mass shell. The interest in this case is twofold. First, the meson transition form factor can be measured and secondly spin-1 states can be produced.

Table 1. *List of resonances studied at LEP*

Resonance	Final state	J^{PC}	$\Gamma_{\gamma\gamma}(\text{keV})$
η' (958)	$\pi^+\pi^-\gamma$	0^{-+}	$4.17 \pm 0.10 \pm 0.27$ [1]
a_2 (1320)	$\pi^+\pi^-\pi^0$	2^{++}	$0.98 \pm 0.05 \pm 0.09$ [2]
a_2' (1750)	$\pi^+\pi^-\pi^0$	2^{++}	$0.29 \pm 0.04 \pm 0.02/(\text{BR})$ [2]
f_2' (1525)	$K_s^0 K_s^0$	2^{++}	$0.09 \pm 0.02 \pm 0.02/(\text{BR})$ [3]
η (1440)	$K_s^0 K\pi$	0^{-+}	$0.17 \pm 0.05/(\text{BR})$ [4]
η_c (2980)	12 Channels	0^{-+}	$8.0 \pm 2.3 \pm 2.4$ [5]
η_c (2980)	9 Channels	0^{-+}	$6.9 \pm 1.9 \pm 2.0$ [6]
$\chi_{c2}(3555)$	$J/\psi\,\gamma$	2^{++}	$0.97 \pm 0.40 \pm 0.36$ [7]

At LEP, many exclusive channels are studied as shown in Table 1. Two recent results are discussed in the following sections.

2.1 Charmonium Production

Measurements of the charmonium system in the two photon collisions are mainly motivated by the large quark mass, where the predictions are reliable, which provides a test of perturbative QCD. Using LEP I and LEP II data, with a total luminosity of 193 pb^{-1} , the charmonium resonance η_c is observed[6] and reconstructed in nine different decay modes. The two photon partial width of the η_c is extracted to be $\Gamma_{\gamma\gamma} = 6.9 \pm 1.9 \pm 2.0$ keV. Figure 2 (a) shows the invariant mass distribution of selected events with one of the scattered electron tagged in the forward calorimeter. The spectrum is fitted with a

80

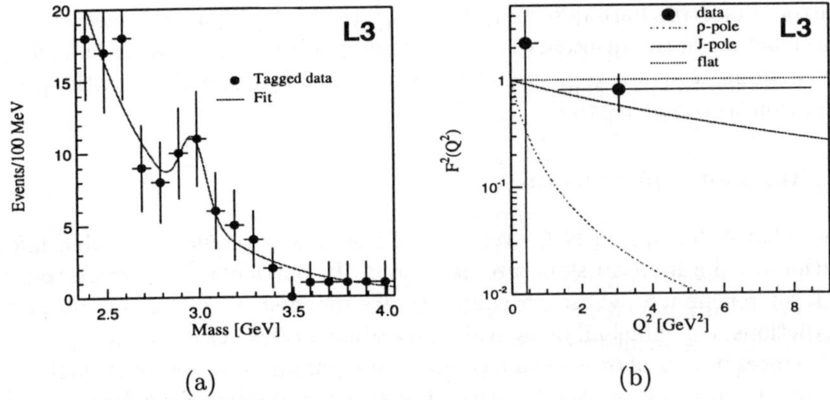

Figure 2. *(a) The η_c invariant mass spectrum , (b) the η_c form factor, fitted with a VDM pole form, with pole mass equal to* $M_{J/\psi}$.

Gaussian for the signal and a exponential for the background. These events allow to measure the η_c transition form factor in different Q^2 bins, (0.2 GeV^2 $< Q^2 < 9$ GeV^2). Figure 2(b) shows the η_c form-factor measurement by L3, which favors the form-factor with a J/ψ mass pole in the VDM model and are in agreement with theoretical calculations[10].

2.2 $K_s^0 K_s^0$ Resonances and GlueBall Search

The resonance formation process $\gamma\gamma \to R \to K_s^0 K_s^0 \to \pi^+\pi^-\pi^+\pi^-$ has been studied[3] with the L3 detec-tor. The $K_s^0 K_s^0$ mass spec-trum Figure. 3, shows clear ev-idence for the formation of the $f_2'(1525)$ tensor meson. Around 1300 MeV, $f_2(1270) - a_2(1320)$ destructive interference is ob-served consistent with theoret-ical predictions[11]. In addi-tion, there is an enhancement of ≈ 6 standard deviations around 1750 MeV which is possibly due to the formation of a radially ex-cited state of the f_2', according

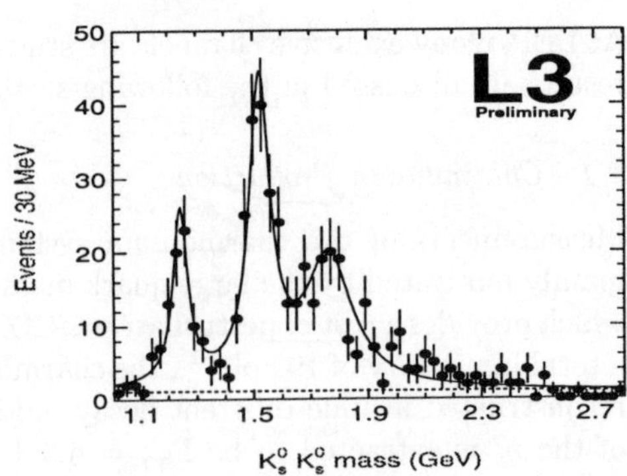

Figure 3. *The $K^0_s K^0_s$ mass spectrum*

to theoretical predictions[9]. The measured two photon partial width of the $f_2^{'}(1525)$ is shown in Table 1. A study of the angular distribution of the $f_2^{'}$ in the two-photon centre-of-mass system favours helicity-2 formation over helicity-0, consistent with theoretical predictions[14].

A search for the glueball candidate ξ (2230) has been performed at LEP in the $K_s^0 K_s^0$ decay channel. The search is motivated due to the previous observation of ξ (2230) by the Mark III Collaboration[13] which has been confirmed by BES Collaboration[12]. At LEP, non observation of signal gives an upper limit for $\Gamma_{\gamma\gamma}(\xi(2230)) \times \mathrm{Br}(\xi(2230) \to K_s^0 K_s^0) < 1.5\,\mathrm{eV}$ at 95% CL under the hypothesis it is a pure spin 2, helicity two state. This low value is most likely inconsistent with a $q\bar{q}$ assignment to the $\xi(2230)$.

3   The Two Photon Total Cross-section

At LEP II energies, the two photon process $e^+e^- \to e^+e^-\gamma^*\gamma^* \to e^+e^- hadrons$ is a copious source of hadron production. In this reaction the photons either interact as a point-like particle or undergo quantum fluctuation (resolved photon) into a resonant(VMD) or non-resonant virtual states opening up all the possibilities of hadronic interactions as shown in Figure 4. These interactions can be described in terms of Regge poles[15,16], (Pomeron or Reggeon exchange).

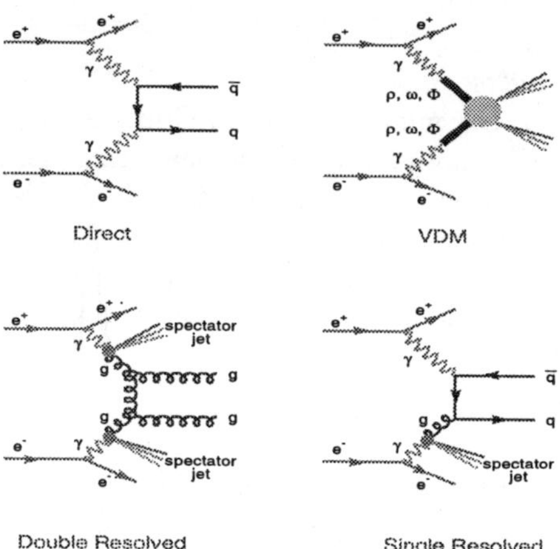

Figure 4. *Some diagram contributing to hadron production in $\gamma\gamma$ collisions at LEP.*

A measurement of the total hadronic cross section as a function of $\sqrt{s}$, improves our understanding of the hadronic nature of the photon. At LEP, using the high energy runs above the Z peak, L3 and OPAL have measured the cross section[17,18] $\sigma(\gamma\gamma \to hadrons)$ in the range $5 \leq W_{\gamma\gamma} \leq 145$ GeV as shown in the Figure 5.

The cross-section measurement of the two experiments show a clear rise at high energies, described by a "Soft Pomeron" and the data of the L3 experiment show a fast decrease at low energies due to "Reggeon exchange". The rise of $\sigma_{\gamma\gamma}$ is faster than the one observed in hadron-hadron or γp collisions; a simple factorization ansatz[19] $\sigma_{\gamma\gamma} = \sigma_{\gamma p}^2/\sigma_{pp}$ is excluded as can be seen in Figure 5 from the predictions of Schuler and Sjostrand[22]. The data are rather well described by the dual parton model of Engel and Ranft[23] or by analytical calculations which take into account the importance of QCD effects at high transverse momentum. In Figure 5, the minijet model of Godbole and Pancheri[24] is also represented. One has to notice that all models has some dependence which can change the cross section predictions by 10-30%. The Monte Carlo models PYTHIA and PHOJET which are used to correct the data, differ by $\approx 20\%$ in the absolute normalization. In future, improvements in the theoretical predictions especially the description of diffractive processes are desirable.

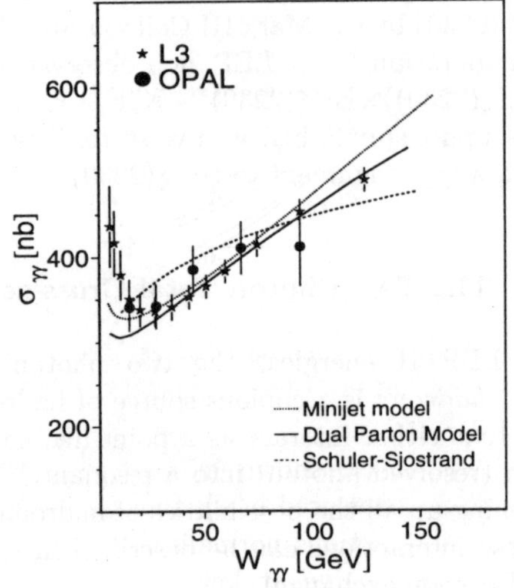

Figure 5. *The measured cross-section $\sigma(\gamma\gamma \to hadrons)$ as a function of $W_{\gamma\gamma}$.*

4 Single Particle and Jet Production

Inclusive production of charged hadrons, K_s^0 mesons, and jet studies has been performed at LEP by the OPAL experiment. Figure 6(a) shows a measurement of differential cross-section for charged hadrons produced in collision of the two quasi-real photons in the range 10 GeV$<$ $W_{\gamma\gamma}$ $<$125 GeV as a function of transverse momentum[25] p_T . The results are compared to NLO perturbative QCD calculations[26]. For lower values of $W_{\gamma\gamma}$, more charged hadrons than predicted are found at

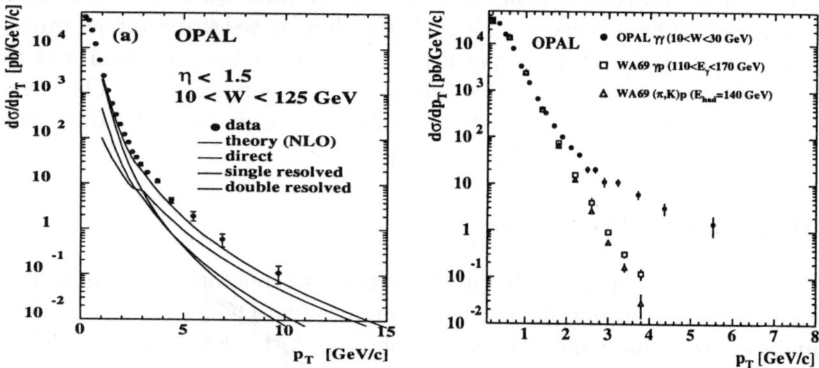

Figure 6. *a) Differential inclusive charged hadron cross-section and b) the* p_T *distribution measured in* $\gamma\gamma$ *interactions compared to the* γp *and* $(\pi,K)p$ *interactions.*

large p_T. Also shown in figure 6(b) is the comparison of the $\gamma\gamma$ data data to p_T measured in γp and (π, K)p interactions normalised at the same value at low p_T, one observes there is a significant increase of rates in the $\gamma\gamma$ process above a p_T of 2 GeV. The clear deviation from the hadronic interactions shows the effect of the direct component in the $\gamma\gamma$ interactions. Similar studies of p_T distributions of the K_s^0 mesons are in reasonable agreement with NLO calculations[25].

The OPAL experiment has performed a very nice measurement of dijet production in two-photon collisions at $\sqrt{s} = 161$ and 172 GeV. Their results[28] demonstrate that it is possible to distinguish between direct and resolved processes in the dijet events. With the help of the variable x_γ, which is the estimator of the fraction of the target pho-

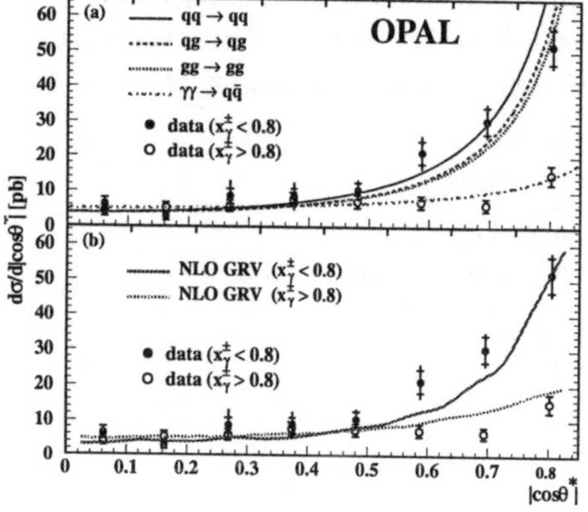

Figure 7. *The angular distribution in the dijet center-of-mass system for "Direct" and "Resolved" events*

84

ton's momentum carried by the parton which produces jets. Figure 7 shows the measured distribution of the parton scattering angle θ^* for direct and double-resolved processes, compared to the relevant QCD matrix element calculations[29]. One observes a clear distinction between the direct process $\gamma\gamma \to q\bar{q}$ ($x_\gamma > 0.8$), where a quark is exchanged in the t channel and the doubly resolved one ($x_\gamma < 0.8$), dominated by the gluon exchange. The strong rise in $\cos\theta^*$ distribution near $\cos\theta^*=1$ is due to a large double-resolved contribution, as expected from QCD.

5 Heavy Quark Production

The study of heavy quark (c,b) production in two photon collisions at LEP provides not only an excellent test of perturbative QCD but also gives an estimate of the gluon density in the photon. At LEP energies, the direct and resolved photon processes are predicted to give comparable contributions to the charm and beauty quark production cross-sections[30]. The resolved process is dominantly quark-gluon fusion: $\gamma g \to q\bar{q}$. The cross-section of the processes $e^+e^- \to e^+e^- c\bar{c}, b\bar{b}$ X has been measured by the L3[31] and OPAL[32] experiments. At L3, the charm and beauty quark are identified by tagging leptons (e, μ) from semileptonic charm and beauty decays. Charm quark were also identified by the reconstruction of $D^{\pm*}$ meson decays, where $D^* \to D^0\pi^\pm$, and OPAL tags charm quark with $D^* \to D^0\pi^\pm$ and $D^0 \to K^-\pi^+, K^-\pi^+\pi^o, K^-\pi^+\pi^-$.

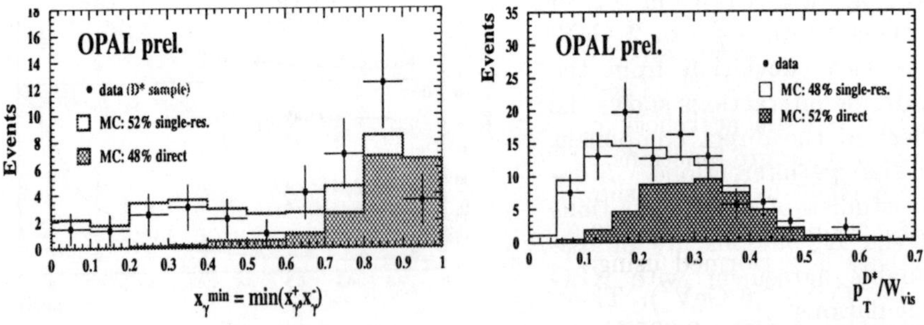

Figure 8. *The x_γ distribution of dijet events containing a D^* and the p_T distribution of the D^* normalised to the visible mass of the event.*

A good separation of direct and resolved processes is obtained by associating the D^* to a dijet analysis or by inspection of the p_T distribution of the D^* (See figure 8). As predicted the direct and resolved processes contribute roughly equally to the observed distribution. The differential D^* cross section agrees well with the NLO predictions and is independent of the Monte Carlo models used to correct the data over the range of detector acceptance. The total inclusive cross-sections are plotted in Figure 9 together with previous measurements. The data are compared to NLO QCD calculations[30]. The direct process $\gamma\gamma \to c\bar{c}, b\bar{b}$, shown with dotted line, is insufficient to describe the data, even if real and virtual gluon corrections are included. The cross sections requires contributions from the resolved processes which are dominantly $\gamma g \to c\bar{c}, b\bar{b}$. The data therefore requires a significant gluon content in the photon.

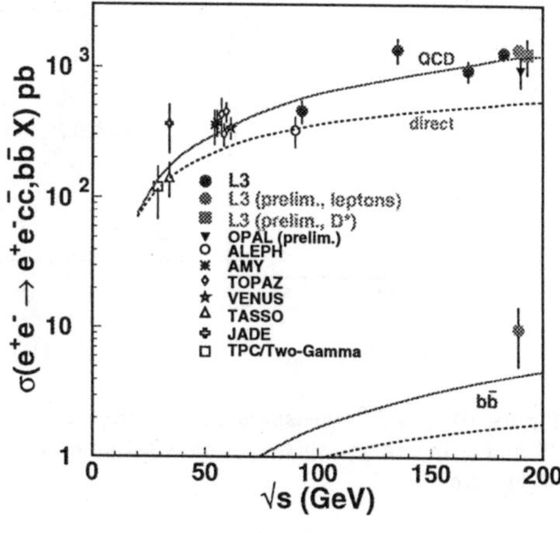

Figure 9. *The cross-section for heavy quarks production as measured at LEP and at previous e^+e^- colliders*

The $b\bar{b}$ cross section is measured for the first time in two photon collisions by the L3 experiment. The preliminary value of b cross section lie somewhat above QCD predictions.

6 Leptonic Structure Function, $F_2^{\gamma,QED}$

The leptonic structure function has been measured by all LEP experiments[33,34,35,36]. The measurement provides not only a QED test but also an experimental check for the procedures used in the study of the hadronic photon structure functions.

A result from L3, is shown as an example in figure 10 (a). It shows that it is possible to measure the effect of non-zero target photon virtuality. The analysis is performed using the $e^+e^- \to \mu^+\mu^-$ sample, for a range of Q^2 ($1.4 < Q^2 < 7.6$ GeV^2). The fit to $F_2^{\gamma,QED}$ corresponds to a target photon virtuality of 0.33 ± 0.005 GeV^2, in good agreement with QED predictions, if initial state radiative corrections are included.

Also shown in Figure 10 (c), is the measurement of the F_A and F_B structure functions , obtained by studying the azimuthal angle distribution of the μ^- in the $\gamma\gamma$ centre-of-mass system[38,39,40,41,42]. Assuming that the target photon direction is parallel to the beam axis, the polar angle θ^* of the μ^- and the azimuthal angle χ are defined as shown in Figure 10(b). Here χ is

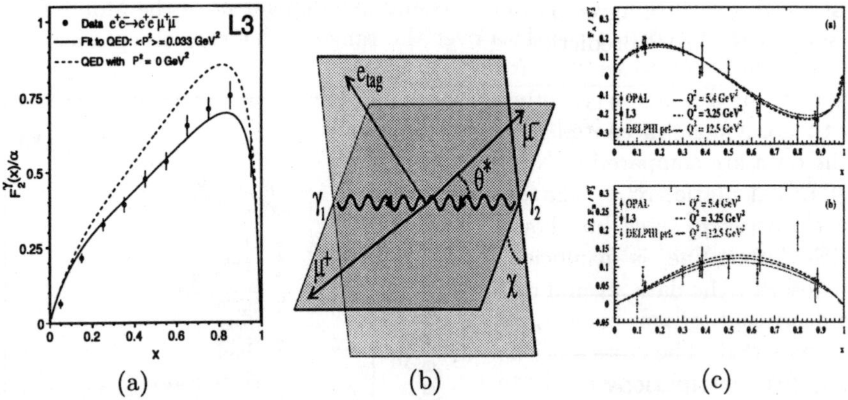

Figure 10. *(a)* F_2^γ *measured in the range* $1.4 < Q^2 < 7.6$ GeV^2, *(b) definition of the angles* θ^* *and* χ *in the* $\gamma\gamma$ *centre of mass frame and (c) measurement of the* F_A *and* F_B *structure of function.*

the angle between the plane defined by the μ^- direction and the $\gamma\gamma$ axis, and the scattering plane of the tagged electron. Both structure functions F_A and F_B, originate from the interference terms of the scattering amplitudes. The characteristic x dependence of the interference terms, as predicted by QED, is observed in the data as shown in figure 10 (c). In particular F_A is due to the interference between longitudinal-transverse and transverse-transverse photon amplitudes, thus providing information on the longitudinal component of the probe photon. With this measurement, LEP proves that the longitudinal leptonic photon helicity amplitude can be accessed by the study of azimuthal correlations and is significantly non-zero.

7 Hadronic structure function $F_2^{\gamma,QCD}$

The measurement of the hadronic structure function, $F_2^{\gamma,QCD}$ has been performed at LEP in the range $0.0025 < x < 1$ and 1.2 GeV$^2 < Q^2 < 279$ GeV^2 [46,47,48,49]. The physical interest in the analysis of the hadronic photon structure function is twofold. Firstly, to measure the shape of F_2^γ, especially at small values of x, at fixed Q^2, where HERA experiments observe a strong rise of the proton structure function. Secondly the Q^2 evolution of F_2^γ is investigated. The F_2^γ measurements from L3 and OPAL are shown in Figure 11 (a) in the Q^2 interval from 1.2 to 9.0 GeV^2. The x range is $0.002 < x < 0.1$ at $\langle Q^2 \rangle = 1.9$ GeV^2 and $0.005 < x < 0.2$ at $\langle Q^2 \rangle = 5.0$ GeV^2. For the low values of x, the data agree better with the parton density prediction of GRV[43], whereas SaS-1d[45] prediction is lower.

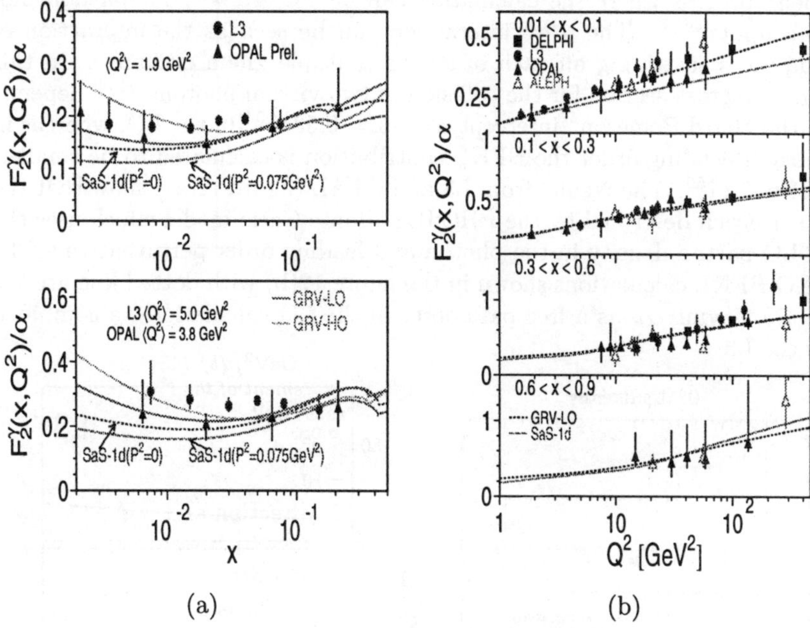

Figure 11. *(a) The measured F_2^γ at $\langle Q^2 \rangle = 1.9$ GeV^2 and 5.0 GeV^2 and (b) evolution of F_2^γ as a function of Q^2 for different range of x values.*

A compilation of the results for different experiments on the Q^2 evolution of F_2^γ in various ranges of x are shown in Figure 11 (b). The measured values of F_2^γ show clearly the linear growth with $\ln Q^2$ expected by QCD. The predictions of the GRV-LO[43] and SaS-1d[45] models are also shown. With all the statistics available at the end of LEP data taking, one hopes to extract the effective scale parameter Λ_{QCD} at large x.

8 $\gamma^*\gamma^*$ Collisions

The cross-section of $\gamma^*\gamma^*$ collisions has been measured at LEP with L3[50] and OPAL[51] experiments in the range of 3 GeV^2 < $Q_{1,2}^2$ < 37 GeV^2. Since the two photons are highly virtual and unlike the proton, they do not contain constituent quarks with an unknown density distribution, so one may hope to have a complete perturbative QCD calculation under particular kinematical conditions. An alternative QCD approach is based on the BFKL equation[52]. Here the highly virtual two-photon process, with $Q_1^2 \simeq Q_2^2$, is considered as the "golden" process where the calculation can be verified without phenomenological inputs[53,54]. The $\gamma^*\gamma^*$ interaction can be seen as the interaction of two $q\bar{q}$ pairs scattering off each other via multiple gluon exchange. In this scheme the cross-section for the collision of two virtual photons [53,54] depends upon the "hard Pomeron" intercept $\alpha_P - 1 = 0.53$[53,54] in the LO, whereas in the next-to-leading order the BFKL contribution is calculated to be smaller, $\alpha_P - 1 \simeq 0.17$[55]. The results from L3 and OPAL (figure 12(a)) show that the events are well described by the PHOJET Monte Carlo model which uses the GRV-LO parton density in the photon and leading order perturbative QCD. The LO BFKL calculations shown in the figure 12(b) with dotted line are too high. By leaving α_P as a free parameter in the LO calculations, a combined fit to the L3

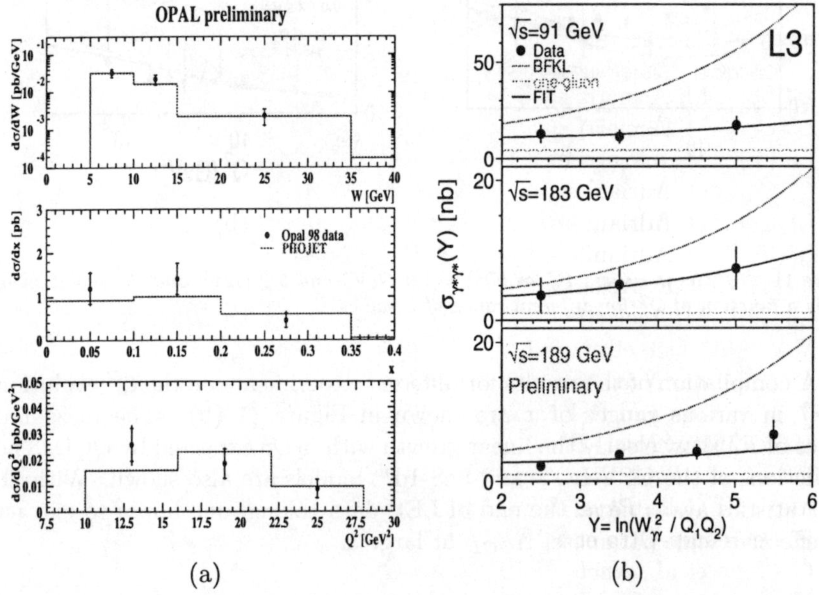

Figure 12. *(a) The differential cross-section of double tag events compared to PHOJET Monte Carlo predictions and (b) the two photon cross-sections at LEP1 and LEP2 compared to LO-BFKL calculations after subtraction of the direct contribution*

data obtained at $\sqrt{s} \approx 91$, 183 and 189 GeV gives a value of $\alpha_P - 1 = 0.29 \pm 0.025$ with χ^2/d.o.f =7/9.

Outlook

Progress in the field of the two photon physics at LEP is significant, most notable are multi-hadron production and photon structure functions. With the statistics of 500 pb^{-1} luminosity available at the end of LEP II data taking, we expect not only large improvements in the understanding of the photon structure function at small x values but also have possibility to actually observe glueball states with very low two photon widths.

Acknowledgements

I would like to thank all the LEP colleagues for their contribution. I am grateful to J.H. Field and M.N. Kienzle-Focacci for encouraging discussions. This work is supported by the Swiss National Science Foundation.

References

1. L3 Coll., M. Acciarri et al., *Phys. Lett.* B **418**, 399 (1998)
2. L3 Coll., M. Acciarri et al., *Phys. Lett.* B **413**, 147 (1997)
3. L3 Coll., M. Acciarri et al., *Phys. Lett.* B **363**, 118 (1995)
4. L3 Coll., O. Adriani et al., Submitted to EPS, Tampere, 1999
5. L3 Coll., O. Adriani et al., *Phys. Lett.* B **286**, 403 (1992)
6. L3 Coll., O. Adriani et al., CERN-EP/99-072
7. OPAL Coll.,K. Ackerstaff et al., *Phys. Lett.* B **439**, 197 (1998);
   L3 Coll., M. Acciarri et al., *Phys. Lett.* B **453**, 73 (1999)
8. V.V. Anisovich et al., hep-ph/9702383
9. C. R. Münz, *Nucl. Phys.* A **609**, 364 (1996)
10. T. Feldmann and P. Kroll, *Phys. Lett.* B **413**, 410 (1997)
11. H. J. Lipkin, *Nucl. Phys.* B **7**, 321 (1968)
12. BES Coll., *Phys. Rev. Lett.* **76**, 3502 (1996)
13. Mark III Coll., *Phys. Rev. Lett.* **56**, 107 (1986)
14. B. Schrempp-Otto et al., *Phys. Lett.* B **36**, 463 (1971);
    G. Köpp et al., *Nucl. Phys.* B **70**, 461 (1974);
    P. Grassberger and R. Kögerler, *Nucl. Phys.* B **106**, 451 (1976)
15. P. D. B. Collins, Introduction to Regge theory (Cambridge U.P., Cambridge, 1977)
16. Review of Particle Physics, *Phys. Rev.* D **54**, 191 (1996)
17. L3 Coll., M. Acciarri et al., *Phys. Lett.* B **408**, 450 (1997);
    L3 Coll., M. Acciarri et al., Submitted to EPS, Tampere, 1999
18. OPAL Coll., G. Abbiendi et al., CERN-EP/99-076.

19. V. N. Gribov and L. Ya. Pomeranchuk, *Phys. Lett.* B **8**, 343 (1962).

20. A. Donnachie and P. V. Landshoff, *Phys. Lett.* B **296**, 227 (1992).

21. Review of Particle Physics, *Eur. Phys. J.* **C3** 205 (1998)

22. G. A. Schuler and T. Sjöstrand, *Nucl. Phys.* B **407**, 539 (1993);
    G. A. Schuler and T. Sjöstrand, *Z. Phys.* C **73**, 677 (1997)

23. R. Engel, *Z. Phys.* C **66**, 203 (1995);
    R. Engel and J. Ranft, *Phys. Rev.* D **54**, 4246 (1996)

24. R.M.Godbole and G.Pancheri, hep-ph/9903331 (1999)

25. OPAL Coll., K. Ackerstaff et al., *Eur. Phys. J* **C6**, 253 (1999)

26. J. Binnewies, B. A. Kniehl and G. Kramer, *Phys. Rev.* D **53**, 6110 (1996)

27. OMEGA Coll., R. J. Apsimon et al., *Z. Phys.* C **43**, 63 (1989)

28. OPAL Coll., K. Ackerstaff et al. , *Z. Phys.* C **73**, 433 (1997);
    OPAL Coll., G. Abbiendi et al. , *Eur. Phys. J* C **6**, 253 (1999)

29. H. Kolanoski, Two-Photon Physics at e^+e^- Storage Rings, Springer-Verlag(1984);
    B. L. Combridge et al., *Phys. Lett.* B **70**, 234 (1977);
    D. W. Duke and J. F. Owens, *Phys. Rev.* D **26**, 1600 (1982)

30. M. Drees et al., *Phys. Lett.* B **306**, 371 (1993)

31. L3 Coll., M. Acciarri et al., Submitted to EPS, Tampere, 1999

32. OPAL Coll., G. Abbiendi et al. , Submitted to EPS, Tampere, 1999

33. ALEPH Coll., Contributed paper to Photon97 Egmond ann Zee, 1999.

34. DELPHI Coll., P. Abreu et al.,*Z. Phys.* C **96**, 199 (1994);
    DELPHI Coll., A. Zintchenko, PHOTON99, Freiburg, Germany

35. L3 Coll., M. Acciarri et al., *Phys. Lett.* B **436**, 403 (1998)

36. OPAL Coll., K. Ackerstaff et al., *Z. Phys.* C **60**, 593 (1993);
    OPAL Coll., K. Ackerstaff et al., *Z. Phys.* C **74**, 49 (1997);
    OPAL Coll., G. Abbiendi et al. , CERN-EP/99-010

37. V.M. Budnev et al., *Phys. Rep.* C **15**, 181 (1975)

38. CELLO Coll., H.J.Behrend et al., *Phys. Lett.* B **126**, 384 (1983)

39. P.Kessler, in Xth Int. Workshop on Gamma-Gamma Collisions, 1995, 281

40. Physics at LEP2, CERN96-01

41. C.Peterson,P.Zerwas and T.F.Walsh, *Nucl. Phys.* B **229**, 301 (1983)

42. R. Nisius and M. H. Seymour, *Phys. Lett.* B **452**, 409 (1999)

43. M. Glück, E. Reya and A. Vogt,*Phys. Rev.* D **45**, 3986 (1992);
    M. Glück, E. Reya and A. Vogt,*Phys. Rev.* D **46**, 1973 (1992)

44. H. Abramowicz, K. Charchula and A. Levy, *Phys. Lett.* B **269**, 458 (1991)

45. G. A. Schuler and T. Sjöstrand, *Z. Phys.* C **68**, 607 (1995);
    G. A. Schuler and T. Sjöstrand, *Phys. Lett.* B **376**, 193 (1996)

46. ALEPH Coll., R.Barate et al., *Phys. Lett.* B **458**, 152 (1999)
47. DELPHI Coll., Igor Tyapkin in " Workshop on Photon Interactions and the Photon structure", LUND 1998
48. L3 Coll., M. Acciarri et al., *Phys. Lett.* B **436**, 403 (1998);
L3 Coll., M. Acciarri et al., *Phys. Lett.* B **447**, 147 (1999)
49. OPAL Coll., K. Ackerstaff et al., *Z. Phys.* C **74**, 33 (1997);
OPAL Coll., K. Ackerstaff et al., *Phys. Lett.* B **411**, 387 (1997);
OPAL Coll., K. Ackerstaff et al., *Phys. Lett.* B **412**, 225 (1997). OPAL Coll., G. Abbiendi et al., Submitted to EPS, Tampere, 1999
50. L3 Coll., M. Acciarri et al., *Phys. Lett.* B **453**, 333 (1999);
L3 Coll., M. Acciarri et al., Submitted to EPS, Tampere, 1999
51. OPAL Coll., G. Abbiendi et al. , Submitted to EPS, Tampere, 1999
52. E.A. Kuraev, L.N. Lipatov and V.S. Fadin, Sov. Phys. JETP **45** (1977) 199;
Ya.Ya. Balitski and L.N. Lipatov, Sov. J. Nucl. Phys. **28** (1978)822
53. S.J. Brodsky, F. Hautmann and D.E. Soper, *Phys. Rev.* D **56**, 6957 (1997)
54. J. Bartels, A. De Roeck and H. Lotter, *Phys. Lett.* B **389**, 742 (1996);
J. Bartels, A. De Roeck, C. Ewerz and H. Lotter, hep-ph/9710500
55. S.J. Brodsky, V.S. Fadin, V.T. Kim, L.N. Lipatov and G.B. Pivovarov, hep-ph/99101229

PHOTOPRODUCTION AT LOW Q^2

P. J. BUSSEY

Department of Physics and Astronomy,
University of Glasgow,
Glasgow G12 8QQ, U.K.
E-mail: p.bussey@physics.gla.ac.uk

The past year has seen a number of important developments in hard photoproduction physics at the HERA collider. These are surveyed.

1 Introduction

The record luminosities obtained in recent years at the HERA collider have given many new results in deep inelastic scattering. However most of the virtual photons that mediate the *ep* collisions have low Q^2 values. These large fluxes of quasi-real photons have made HERA an outstanding laboratory in which to study photon physics. Here I survey some of the recent results reported by the experiments H1 and ZEUS, concentrating on the hard interactions which reveal the partonic structure of the photon.

Throughout, it will be helpful to have in mind three characteristic modes by which a photon can interact with a proton at high energies. These are illustrated in fig. 1.[1] The simplest is when the entire photon couples in a pointlike manner to a high transverse energy (E_T) $q\bar{q}$ pair; such diagrams are referred to as "direct" photon processes. The photon may also couple non-perturbatively to a hadronic state which then interacts with the proton. Both the hadronic state and the proton can be sources of partons which undergo hard QCD scattering; such diagrams are known as "resolved" photon processes.

These two classes of process are fully separable only in lowest order (LO) QCD. A more complex situation occurs when the photon couples in a pointlike way to a medium-E_T $q\bar{q}$ pair, one of which then undergoes a hard scatter. Here there are three perturbative couplings and the process is no longer LO. Such processes are referred to as "anomalous". Both pointlike and hadronic photon couplings are also present, of course, in higher order diagrams.

With the above theoretical ideas in the background, in practice we wish to make experimental observations. The principal measurable quantities available in hard photon reactions are high E_T jets, high E_T photons ("prompt" photons), various single particle spectra and the remnants of the incoming photon and proton. The quantities we would like to study include the na-

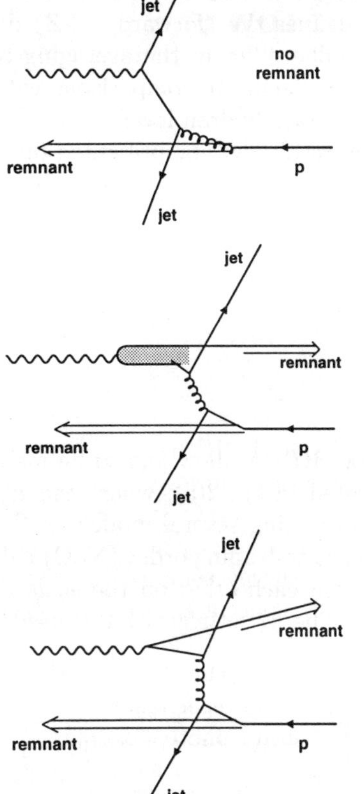

(a) Pointlike photon coupling to a high-E_T quark pair. There is no photon remnant. In the LO diagram illustrated, $x_\gamma = 1$.

(b) Hadronic photon coupling. The photon acts as a source of partons which can scatter off partons in the proton. There is a photon remnant as well as a proton remnant.

(c) Anomalous photon coupling. The photon couples perturbatively to a medium-E_T quark pair. There is a photon remnant at non-zero transverse momentum.

Figure 1. Schematic illustrations of photoproduction processes.

ture of the outgoing quarks and gluons, and the hadronic structure of the photon, in addition to the challenge of verifying that we have an adequate understanding of the basic QCD mechanisms that are operating.

A basic parton-level concept is that of x_γ, the fraction of the photon energy that takes part in the main hard QCD subprocess. (Above LO this may not always be defined simply.) In practice, what is wanted is a measurable quantity which correlates with x_γ in a chosen theoretical description of the process. Fro dijet final states, two such experimental estimators are

$$x_\gamma^{obs} = \frac{\Sigma_{jets} E_T^{jet} e^{-\eta^{jet}}}{2E_\gamma} \quad \text{and} \quad x_\gamma^{jets} = \frac{\Sigma_{jets}(E^{jet} - p_z^{jet})}{\Sigma_{hadr}(E^{hadr} - p_z^{hadr})}$$

which have been used by ZEUS and H1. Here as elsewhere, η denotes laboratory pseudorapidity and the proton beam defines the "forward" $(+Z)$ direction. $(E_T e^{-\eta} \equiv (E - p_z)$; the above formulae differ in the averaging of the parameters over the particles in the jet.) To facilitate comparisons with theory, both estimators are defined at the final-state hadron level.

In the following sections, photoproduction results will be presented on:
(i) dijet final states
(ii) prompt photons
(iii) the parton content of the photon
(iv) the photon remnant, and
(v) properties of the final-state jet system.

2   Dijet final states.

The integrated luminosities now available from HERA allow jets to be measured out to high E_T values. This is illustrated in fig. 2(a), where the E_T spectrum seen in ZEUS[2] is compared to theory using several models of the photon parton density, making use of recent next-to-leading order (NLO) calculations [3] whose predictions do not differ from each other on the scale of the figure. Agreement is excellent in all cases. The k_T jet algorithm is used,[4]

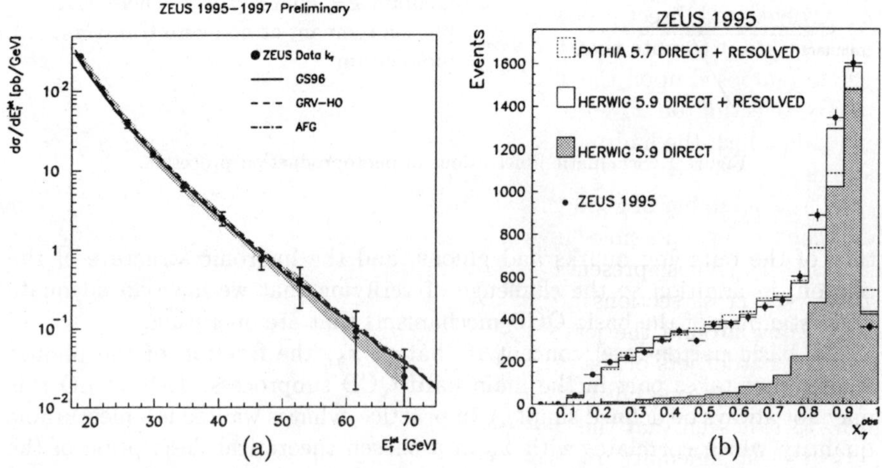

Figure 2. (a) Distribution of transverse energy of photoproduced jets in ZEUS in the pseudorapidity range $-0.75 < \eta < 2.5$. (b) Distribution of x_γ^{obs} in dijet events.

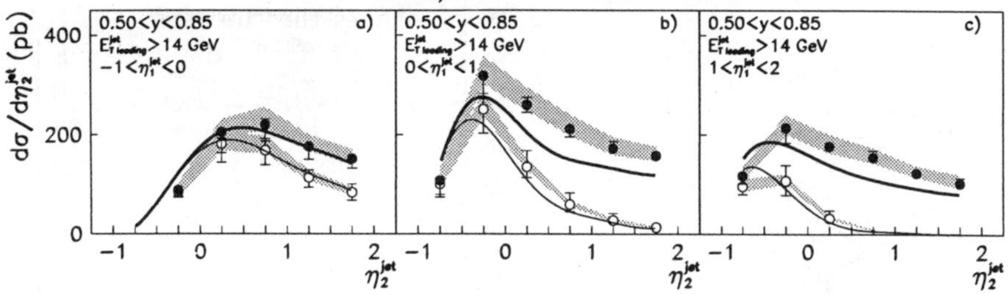

Figure 3. Cross sections for photoproduced jets, in dijet events with minimum jet E_T values of 14 GeV. The incident photon energy, as a fraction y of the e beam energy, is restricted to high values. Full circles = all x_γ^{obs}; open circles = $x_\gamma^{obs} > 0.75$. The pseudorapidity of one jet is plotted for given intervals of that of the other. The shaded bands represent experimental uncertainty due to the simulation of the response of the ZEUS calorimeter. Uncertainties of up to 15% due to the QCD scale, and 10% due to effects of hadronisation, should be allowed on the theoretical predictions.

in order to avoid the uncertainties concerning seed definition and jet merging present with the cone algorithm approach.[5] Figure 2(b) shows the x_γ^{obs} spectrum obtained by ZEUS using high E_T jet pairs. The shape of the distribution can be well fitted using PYTHIA and HERWIG LO simulations of the resolved and direct processes, provided that each contribution is rescaled. It is important to note that the shape of the distribution depends strongly on the cuts imposed upon the jet parameters, as will be seen below.

By selecting on $x_\gamma^{obs} < 0.75$ or $x_\gamma^{obs} > 0.75$, event samples can be obtained in which the hadronic (resolved) or pointlike (direct) photon diagrams dominate.

The availability of good jet statistics at high E_T removes most of the problems concerning a possible "underlying event", due to multi-parton interaction effects (MI),[6] whose presence has been postulated at lower jet energies where the low-x_γ cross sections are often higher than expected.[5,7] Studies of the energy flow around the jets have shown that MI effects appear to be small with the harder jets of the present study, giving stronger confidence that discrepancies between the data and theory are not due to this cause. Indeed in fig. 3 it is seen that the agreement between experiment and theory is not perfect in all kinematic regions. Discrepancies are particularly evident with higher incoming photon energies and when both emerging jets are at forward η values. The existing models are unable to account for the data, and more theoretical input would appear invited.

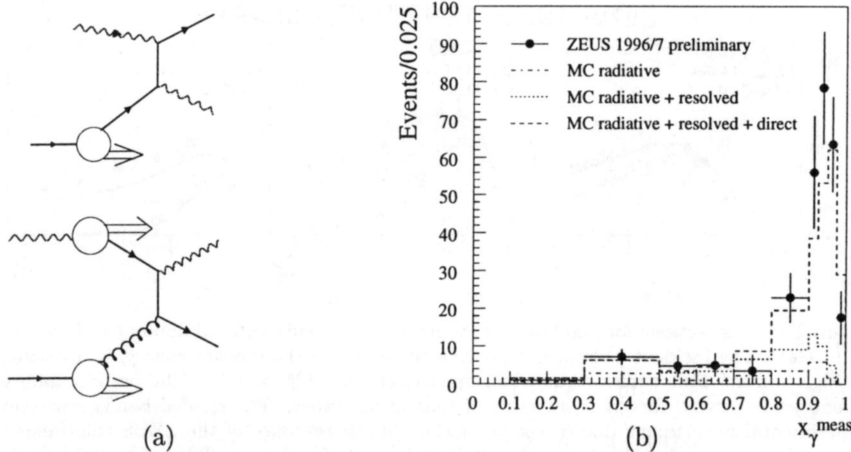

Figure 4. (a) Examples of direct and resolved LO diagrams in prompt photon photoproduction. (b) Distribution of x_γ^{meas}, defined at the detector level analogously to x_γ^{jets}, in photoproduced events having an isolated prompt photon and a jet. The photon satisfies $5 < E_T < 10$ GeV and $-0.7 < \eta < 0.9$; the jet satisfies $E_T > 4.5$ GeV and is required within a central η range. The GRV [8] photon structure is used.

3   Prompt photon measurements.

As an alternative to jets, one can seek to measure the production of high-E_T photons, known as "prompt photons". Since QCD processes of this type have a different set of diagrams from those of dijet processes (fig. 4(a)), they provide a complementary perspective on the physics. The photon is already a partonic object; so there are fewer complications associated with hadronisation (though a recoil jet is still present). There is less dependence on jet finding, and a photon is typically better measured than a jet. On the other hand, an important background from hard neutral mesons (mainly π^0, η) must be subtracted, and allowance must be made for photons radiated from high-E_T quarks. It is helpful to insist that the photon be isolated. ZEUS impose an isolation condition such that within a cone of unit radius in (η, ϕ) around a photon of transverse energy E_T, no more than a further $0.1E_T$ of transverse energy may be present.

When a jet is also observed, a value of x_γ may be measured from the jet and the photon. Figure 4(b) shows that the distribution is very different from that found in dijet photoproduction. The peak near unity, associated with

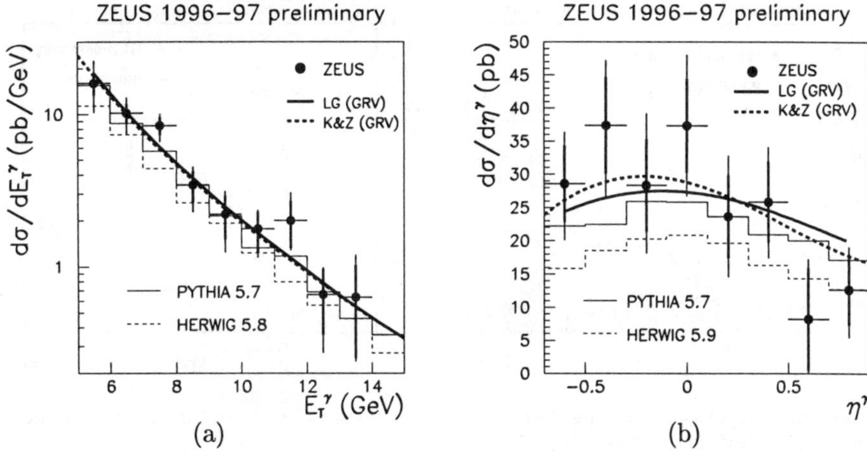

Figure 5. (a) E_T distribution of inclusive prompt photons with $-0.7 < \eta < 0.9$. (b) η distribution of inclusive prompt photons with $5 < E_T < 10$ GeV. Predictions are shown for PYTHIA 5.7, HERWIG 5.9 and two NLO calculations.[10,11]

the direct diagrams, is considerably more pronounced than with dijets. The distribution is quite well described using PYTHIA.

Following an earlier paper on the observation of photoproduced prompt photon processes,[9] ZEUS have presented further preliminary results. The inclusive cross section as a function of E_T is also well described by PYTHIA (fig. 5(a)) and by NLO calculations, although HERWIG appears low. Given the large errors at present, the available models describe the rapidity distributions satisfactorily (fig. 5(b)).

4 Gluon content of the photon.

The hadronic coupling of the photon has to involve charged components in the intermediate hadronic state, namely quarks – but quarks should be accompanied by a gluon component. In most models, the quarks tend to take more of the photon momentum than the gluons, so that a search for gluons in the photon should aim at a sensitivity at low x_γ values.

This is illustrated in fig. 6, which shows x_γ distributions from dijet events in H1. The shape of the distribution is changed radically when jets with lower E_T values are accepted, even while the minimum dijet mass remains at the same value. The Monte Carlo based predictions comprise components from direct processes, and from resolved processes initiated by quarks and

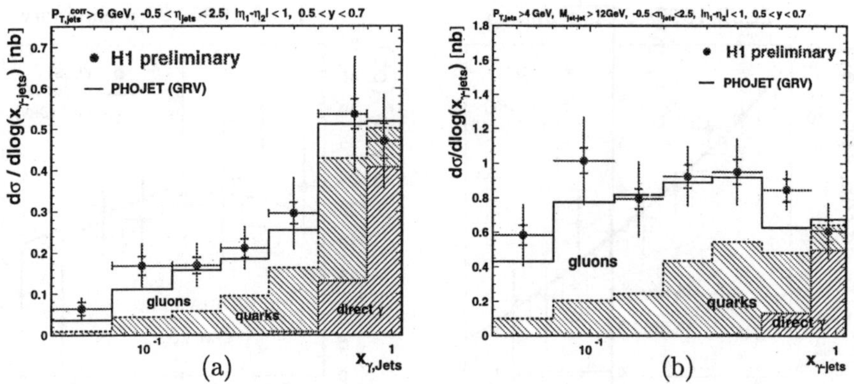

Figure 6. Plots of x_γ (H1) using dijets of mass greater than 12 GeV. In the (a), the minimum jet E_T is 6 GeV, in (b) it is 4 GeV.

by gluons within the photon. This partition of the resolved component is illustrated using the GRV photon parton densities,[8] which have a generally good record and here too give a good fit to the data. Note that although the direct peak is no longer visible as such in (b), this is entirely a consequence of the much higher rates now obtained at lower x_γ values. Through accepting the lower-E_T jets, one has greater acceptance at low x_γ giving a greater sensitivity to the gluon. Indeed, if the quark part of the resolved photon is as modelled above, the level and shape of the cross sections already imply a quantitative observation of gluons in the photon.

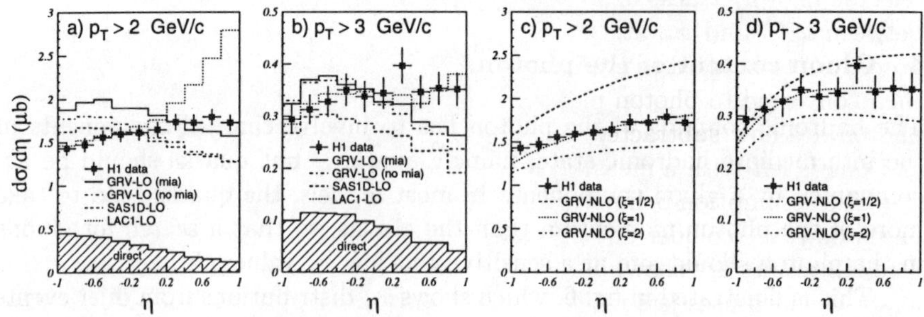

Figure 7. Distribution of inclusive charged particles measure by H1. The data are compared with (a,b) LO Monte Carlo distributions and (c,d) NLO curves.

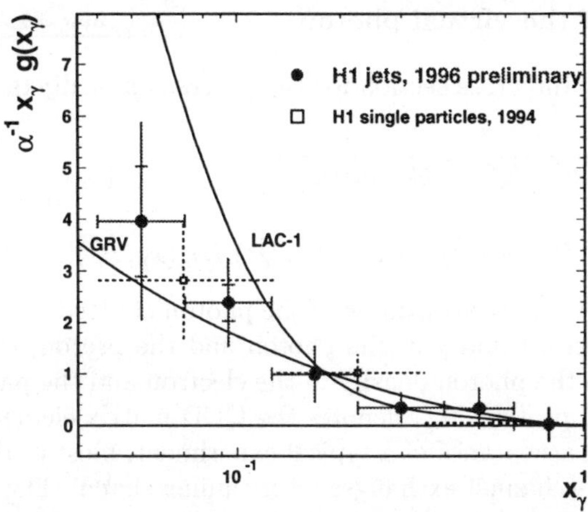

Figure 8. Gluon density in photon, as measured by H1 using inclusive tracks and dijet events. The error bars include uncertainties in the subtraction of the quark component.

A further study has now been performed by means of inclusive charged particles measured in H1.[12] Above transverse momenta of 2–3 GeV, such particles are fairly well associated with high-p_T partons; they are also well measured in the apparatus. Distributions are shown in fig. 7, where it is demonstrated that the 3 GeV data can be described by means of LO theory using a suitably chosen photon structure. NLO calculations are also successful, but there exists a large QCD scale uncertainty.

An x_γ estimator $x_\gamma^{rec} = \sum p_T e^{-\eta}/E_\gamma$ is now evaluated by summing over selected high-p_T tracks in an event. Using PYTHIA, good correlation is found between x_γ^{rec} and x_γ at the LO parton level, which permits an unfolding procedure to be used to evaluate cross sections as a function of x_γ. These are then converted to photon parton densities. Calculated quark densities in the photon are now subtracted off (with a small amount of model dependence), and one is left with a measurement of the gluon densities.

The assumption behind this procedure is that the existing photon models, based on e^+e^- collider data, have relatively well-determined quark densities. The H1 results are shown in fig. 8, together with those from a similar recent analysis which uses jets rather than charged particles. The two sets of points are consistent, and agree well with the GRV model of the photon gluon density. As is commonly the case, LAC1 [13] is too high.

5 Studies of the virtual photon

The LO differential cross section for the process $ep \to$ dijets can be written

$$\frac{d^5\sigma}{dy\, dx_\gamma\, dx_p\, d\cos\theta^*\, dQ^2} = \frac{1}{32\pi s_{ep}\, yx_\gamma x_p} \times$$

$$\times \sum f^k_{\gamma/e}(y,Q^2)\, f^k_{i/\gamma}(x_\gamma, P_t^2, Q^2)\, f_{j/p}(x_p, P_t^2)|M_{i,j}(\cos\theta^*)|^2. \qquad (1)$$

The sum is over the polarisation k of the photon emitted by the electron, and the parton species i and j in the photon and the proton, respectively. The terms f denote the photon density in the electron and the parton densities in the photon and proton; $M_{i,j}$ denotes the QCD matrix element.

Within the acceptance of a typical experiment, most of the $2 \to 2$ parton processes are t–channel exchanges with similar shape. The Single Effective Subprocess approximation [14] replaces all the $M_{i,j}$ terms by a common expression $M_{SES}(\cos\theta^*)$, and forms combinations $\tilde{f}$ of the parton densities. These are known as Effective Parton Densities (EPDs). Taking into account the colour charge on the gluon, one obtains for the proton:

$$\tilde{f}_p = \sum \left(f_{q/p}(x_p, P_t^2) + f_{\bar{q}/p}(x_p, P_t^2)\right) + \tfrac{9}{4} f_{g/p}(x_p, P_t^2),$$

summing over quark flavours. Only transverse polarised photons are retained. Then the sum in (1)can be replaced by the single effective term:

$$f^T_{\gamma/e}(y,Q^2)\, \tilde{f}_\gamma(x_\gamma, P_t^2, Q^2)\, \tilde{f}_p(x_p, P_t^2)|M_{SES}(\cos\theta^*)|^2.$$

The SES and EPD approximations provide an experimental procedure for comparing the properties of the photon in different situations. On this basis, H1 have measured the EPD $\tilde{f}_\gamma$ in the photon over a range of small to medium values of Q^2, using photoproduced jet pairs of transverse momentum P_t.[15] The aim is to examine the variation of $\tilde{f}_\gamma$ with x_γ, Q^2 and P_t^2 and compare with QCD predictions.

This is a two-scale situation. It is first found that the x_γ^{jets} distributions show a clear trend, as Q^2 and P_t^2 separately increase, to be more and more dominated by the "direct peak" at $x_\gamma^{jets} \approx 1$. In other words, the hadronic properties of the photon decrease as the process becomes "harder". Such a situation is not new and has commonly involved the employment of meson-like form factors. However the HO QCD diagrams add further complexity and interest to the present type of interaction.

Figure 9(a) shows the variation of $\tilde{f}_\gamma$ with Q^2 at two x_γ values. (The variation with P_t^2 is found to be weak.) Also plotted are predictions from a pure VMD ρ-pole form factor, and modern QCD-based calculations from

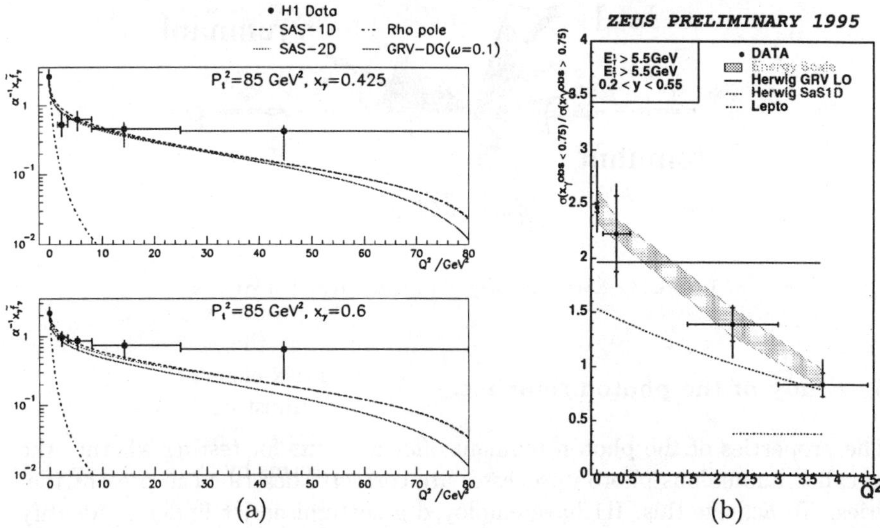

Figure 9. Q^2 dependence of (a) effective parton densities in photon (H1) and (b) ratio of low-x_γ^{obs} to high-x_γ^{obs} cross sections (ZEUS).

Schuler and Sjöstrand.[16] The latter are valid only in the kinematic range $\Lambda_{QCD}^2 \ll Q^2 \ll P_t^2$. Two conclusions follow; one is that as now expected, a ρ pole form factor alone is quite insufficient. The second is that the QCD approach gives good results in the region where its validity is claimed, namely for $0.1 \ll Q^2 \ll 85 \text{GeV}^2$ but perhaps fails elsewhere. Altogether, given the approximations made, this would seem to represent a considerable success, and confirms the H1 collaborations earlier results in terms of inclusive jets.[17]

Over a finer scale of low Q^2 values, however, ZEUS have shown that at present there still may be problems in describing the behaviour of the photon. Fig. 9(b) shows the cross section ratio $\sigma(x_\gamma^{obs} < 0.75)/\sigma(x_\gamma^{obs} > 0.75)$ for a kinematic range of centrally produced jet pairs. At $Q^2 \approx 0$ the lower x_γ^{obs} range is dominated by hadronic ("resolved") diagrams. The ratio falls with Q^2, as predicted by SaS, but the latter theory does not reproduce the magnitude of the effect. At $Q^2 \approx 0$, the data lie above the GRV prediction (which is not modelled to fall with Q^2), indicating the possible presence of multiparton interactions under these conditions. The Lepto line represents a model with no hadronic component to the photon, and NLO effects confined to parton showers; it fails completely at these Q^2 values.

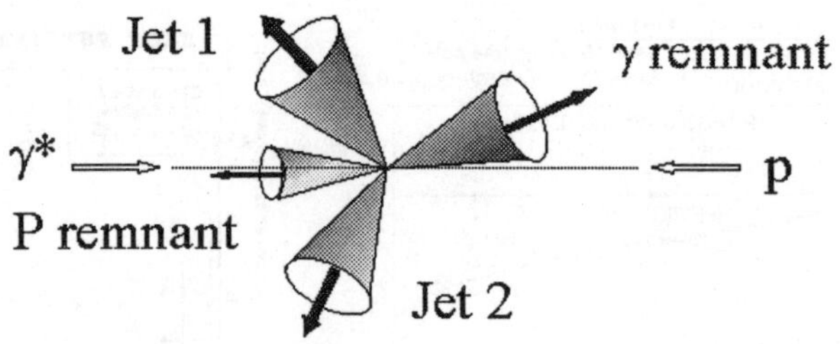

Figure 10. Event topology for photon remnant studies.

6 Study of the photon remnant.

The properties of the photon remnant offer a means for testing whether the so-called anomalous photon processes are correctly described in present theories. To achieve this, H1 have employed a customised jet finder to identify events with two high-E_T jets accompanied by a proton remnant and a photon remnant. The latter is then treated as a jet-like object whose properties can be studied. Fig. 11 shows the mean p_T of the photon remnant relative to the beam, plotted as a function of (a) the virtuality of the incident photon, and (b) the mean E_T of the two hard jets. In this way, theory can be tested over a range of kinematic parameters.

The results show that HERWIG describes the behaviour of the photon remnant quite well, apart perhaps from the case of the softest "hard" jets. An intrinsic k_T of partons within the photon of 0.66 GeV was taken. Within HERWIG, most of the transverse momentum observed at $Q^2 \approx 0$ arises from the effects of hadronisation: to study the parton level properties of the anomalous photon coupling it may be best to go to Q^2 values of more than a few GeV^2. On the other hand, the RAPGAP Monte Carlo is not successful in this context.

7 Three-jet events in photoproduction.

With the aim of confirming our understanding of the QCD processes that are operative in the photoproduction of jets, ZEUS have performed an investigation of three-jet events in photoproduction.[18] The most important experimental requirement was for the three-jet mass to exceed 50 GeV. The kinematics of the final state are best studied in the three-jet centre-of-mass frame, as illustrated in fig. 12. The angle ψ_3 is defined between the plane of the three

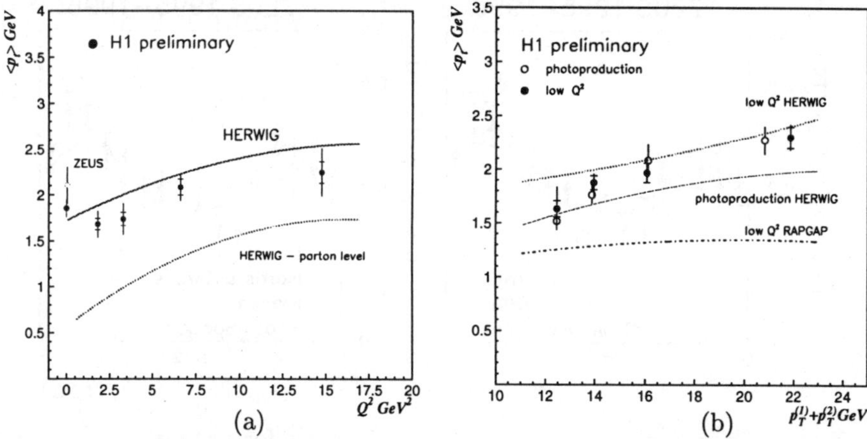

Figure 11. Variation of the mean p_T of the photon remnant (a) with the virtuality of the photon (b) with the mean transverse energy of the hard jets.

jets, and the plane containing the incident proton and photon directions and the highest-E_T jet, labelled "3".

The distribution of ψ_3 is sensitive to a number of kinematic and dynamical properties of the process (fig. 12). It should first be noted that owing to the kinematic requirements on the jets, the distribution is depleted near zero and 180°; what appears as a two-peaked structure would otherwise be a steep-sided valley. Likewise, the flat ψ_3 distribution that would be had if the three jets (or high-p_T partons) were produced according to phase space now becomes centrally peaked. The latter shape in no way resembles the data, confirming that dynamical mechanisms are shaping the three-jet production.

Higher-order calculations describe the shape of the observed distribution well, as indeed do HERWIG and PYTHIA. From further investigations it was found that the main contribution to the three-jet process comes from initial-state gluon radiation from the proton and photon. (The higher peak near 180° comes from the higher amount of initial-state radiation from the proton.) Final-state radiation is relatively suppressed since the third jet then needs to become substantially separated from its parent jet. It is also possible to switch off the colour coherence structures within PYTHIA. The result is a flatter ψ_3 distribution whose shape does not agree well with the data. With further statistics, one may hope to distinguish between the PYTHIA and HERWIG models and possibly some of the higher order calculations.

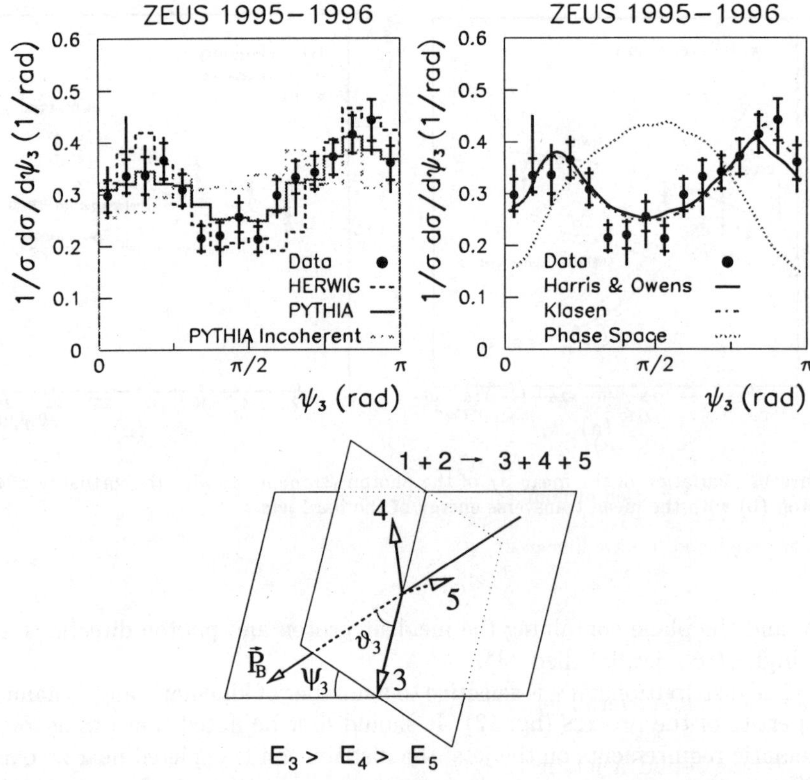

Figure 12. Definition and distribution of ψ_3 in three-jet events in photoproduction.

8 SubJets

A further useful tool for studying the parton structure of the photon would be a knowledge of the nature of a given hard jet. When a jet is due to a heavy quark, this may often be determined on a jet-by-jet basis; however the majority of jets in photoproduction are from light quarks and gluons. Studies made by ZEUS of jet widths [19] have now been supplemented by studies on the numbers of "subjets" within a jet. The concept of a subjet arises in terms of clustering jet algorithms such as the k_T algorithm, where a cut y_{cut} on the clustering parameter determines whether two jet candidates shall be merged. The algorithm is iterated until all remaining pairs of jet candidates have clustering parameters above this value, which is chosen to correspond

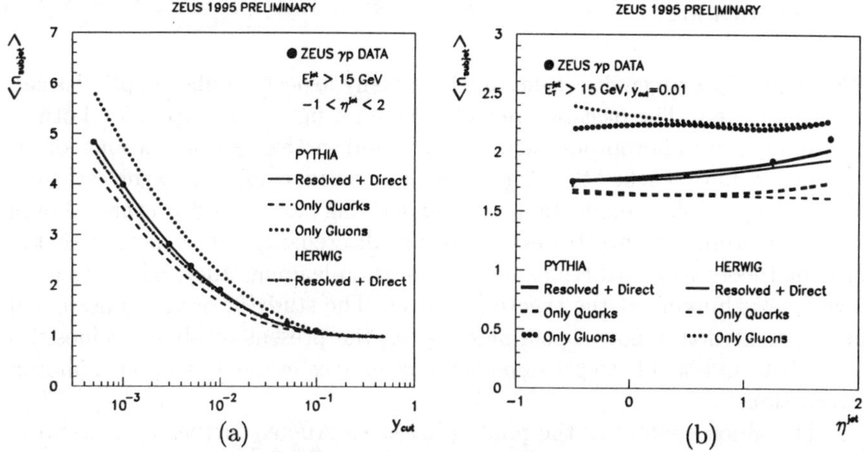

Figure 13. (a) Variation of number of subjets with cut parameter for photoproduced jets with $E_T > 15$ GeV at HERA (b) variation with pseudorapidity for fixed $y_{cut} = 0.01$. Gluon jets are predicted to have higher $<n_{subjet}>$ than quark jets.

to an acceptable overall jet radius. A large value of y_{cut} gives more merging, and fewer final jets.

Having found a jet, it is possible to rerun the algorithm using varying smaller values of y_{cut}; a number of smaller jets ("subjets") may now be obtained in place of the original one. This number, at given y_{cut}, is an indicator of the internal structure of the jet.

Fig. 13(a) shows the variation of the mean number of subjets $<n_{subjet}>$ with y_{cut} for photoproduced high-E_T jets in ZEUS. The form of the variation can be simulated using HERWIG and PYTHIA, and is intermediate between the expectations for pure quark-initiated and pure gluon-initiated jets, corresponding well to an appropriate mixture of direct and resolved processes.

With a chosen fixed y_{cut} value, it is then possible to study the variation of $<n_{subjet}>$ with, for example, the pseudorapidity of the primary jet, as shown in fig. 13(b). It is apparent that the jets are predominantly quarklike at negative η values but predominantly gluonlike for positive η. The variation of the data is reasonably well reproduced using the GRV photon structure. It is too early to say if this technique can prove fruitful in distinguishing between different photon models, but it is clearly a potentially useful addition to the physicist's repertoire in an area where investigations are notoriously difficult.

9 Conclusions

HERA provides an environment in which many aspects of photon physics can be investigated. The past two years have seen a significant expansion both in the numbers of photoproduction analyses and in their scope. Studies of jet production have reached the stage where kinematic regions can be investigated where the possible complications from underlying parton-parton events should not be a problem; nevertheless, there are discrepancies even here with the present theoretical predictions, even at next-to-leading order, which seem to invite new thinking at the theoretical level. The study of prompt photons in photoproduction is now a possibility, given the present possible luminosities at HERA, and is able to provide new perspectives in the area of hard photon interactions.

The gluon content of the photon has been studied further by the extension of existing techniques. These results from H1 are consistent with the best available photon models, but the errors remain fairly large. Both H1 and ZEUS have begun to study the transition from the quasi-real photon, with its extensive hadronic properties, to the virtual state, still at moderate Q^2 values, where "anomalous" photon coupling becomes a more dominant effect, providing an perturbative QCD element in low-x_γ photon physics. New studies of the photon remnant provide further insight into the "anomalous" photon.

The actual QCD mechanisms remain an important area of study. Here, multi-jet final states and the teazing apart of jets into "subjets" have begun to give improved confirmation of our understanding of detailed aspects of the hadronic interactions initiated by a photon.

To conclude, photon physics at HERA has entered a new and remarkably productive stage. The entire area of heavy flavour production was omitted here, being part of another speaker's remit! There are still many avenues to investigate and many questions to answer. The future at HERA will hopefully provide us with the opportunity to unravel further the properties of a particle, the photon, which proves to have a richer and more complex nature the more it is examined.

Acknowledgments

The author wishes to thank members of the H1 and ZEUS collaborations for much helpful information in the preparation of this talk, and also the Royal Society (London) for financial assistance in attending the Conference.

References

1. G A Schuler and T Sjöstrand, Nucl. Phys. **B 407**, 539 (1993);
   G A Schuler and T Sjöstrand, *Phys. Lett.* B **300**, 169 (1993).
2. ZEUS Collab., J. Breitweg et al., DESY 99-057, subm. to Eur. Phys. J. C.
3. P Aurenche et al., proceedings, *Future Physics at HERA,* eds. G. Ingelman et al., DESY (1996) 570;
   S Frixione and G Ridolfi, Nucl. Phys. **B 507**, 315 (1997);
   B Harris and J Owens, *Phys. Rev.* D **57**, 5555 (1998);
   M Klasen, T Kleinwort and G Kramer, Eur. Phys. J. **C1**, 1 (1998).
4. S. Catani et al., Nucl. Phys. **B 406**, 1993 (187).
5. ZEUS Collab., J. Breitweg et al., Eur. Phys. J. **C1**, 109 (1998).
6. J. M. Butterworth et al., J. Phys. **G 22**, 883 (1996).
7. H1 Collab., S. Aid et al., Z. Phys. **C 70**, 17 (1996).
8. M. Glück, E. Reya and A. Vogt, Phys. Rev. **D46**, 1973 (1992).
9. ZEUS Collab., J. Breitweg et al., *Phys. Lett.* B **413**, 201 (1997).
10. M. Krawczyk and A. Zembrzuski, proc. *ICHEP'98*, Vancouver, eds. A. Astbury, D. Axen and J. Robinson, (World Scientific, Singapore, 1999), 895 and hep-ph/9810253.
11. L. E. Gordon, Phys. Rev. **D57**, 235 (1998) and private communication;
    L. E. Gordon, proceedings, *Photon 97*, Egmond aan Zee, eds. A. Buijs and F. Erné (World Scientific, Singapore, 1998), 173 and hep-ph/9706355.
12. H1 Collab., C. Adloff et al., DESY 98-148, to appear in Eur. Phys. J. C.
13. H. Abramowicz, K. Charchuła and A. Levy, PLB **269**, 458 (1991).
14. B. V. Combridge and C. J. Maxwell, Nucl. Phys. **B 239**, 429 (1984).
15. H1 Collab., C. Adloff et al., EJP **C 1**, 97 (1998);
    H1 Collab., C. Adloff et al., DESY 98-205, to appear in Eur. Phys. J. C.
16. G. A. Schuler and T Sjöstrand, *Phys. Lett.* B **376**, 1973 (1996).
17. C. Adloff et al., PLB **415**, 418 (1997).
18. ZEUS Collab., J. Breitweg et al., *Phys. Lett.* B **443**, 394 (1998).
19. ZEUS Collab., J. Breitweg et al., Eur. Phys. J. **C 8**, 367 (1999).

B PRODUCTION and ONIUM PRODUCTION

Andrzej Zieminski

Indiana University, Bloomington, IN 47405, USA

email: zieminsk@indiana.edu

Abstract

We review recent results on b-quark and onia production by various processes, as measured at the Tevatron collider, fixed target experiments, HERA and LEP. Majority of the reported b-quark cross sections exceed predictions based on the next-to-leading order QCD calculations. We also discuss recent theoretical developments in this area. Next, we compare predictions of several models on onia production with results from a wide range of experiments. We point out the difficulties of the color octet model to account for the absence of the J/ψ and $\psi(2s)$ polarization in hadronic collisions and to describe inelastic J/ψ electroproduction. Finally, we report on the latest precision measurements of nuclear effects in the J/ψ hadroproduction.

Introduction

The production of b quarks in hadronic collisions is a subject of persistent theoretical and experimental interest. The comparison of the measured b quark production cross section with QCD calculations reveals the underlying dynamics. The next-to-leading order (NLO) calculations have been available for a decade [1]. In addition to the lowest order $b\bar{b}$ production processes (s-channel gluon diagrams or t-channel quark exchange), sizable contributions come from $\mathcal{O}(\alpha_s^3)$ diagrams including the t-channel gluon exchanges with subsequent gluon splitting, and flavor excitation in which a gluon–$b\bar{b}$ virtual fluctuation is put on-mass shell by an interaction.

The NLO QCD predictions show a large dependence on the choice of the renormalization and factorization scale, μ. This scale is usually chosen as $\mu = \mu_o \equiv \sqrt{m_b^2 + (p_T^b)^2}$, where p_T^b is the transverse momentum of the b-quark and m_b is its mass. The theoretical uncertainties are obtained by varying μ between $\mu_0/2$ and $2\mu_0$. The theoretical predictions are less affected by the choice of m_b, with a central value usually set to 4.75 GeV/c^2, and varied between 4.5 and 5.0 GeV/c^2.

It has been known for several years that the measured inclusive b-quark cross sections at the Tevatron collider exceed the NLO QCD predictions by a factor of ~ 2.5. There is an indication that the disagreement is even larger for the forward b-quark hadroproduction. In this paper we summarize the latest DØ and CDF measurements of the inclusive b-quark cross sections and $b\bar{b}$ correlations and review some of the recent theoretical attempts to explain the disagreement.

There have been several experiments at the fixed target energies that attempted to measure b-quark production with pion and proton beams incident on nuclear targets. The largest statistics measurement from the CERN WA92 Collaboration is consistent with the NLO QCD predictions, whereas some of the earlier measurements, and the recent result from Fermilab E771 , suggest a discrepancy similar to the one observed at the Tevatron collider.

First preliminary results on $b\bar{b}$ photoproduction at HERA became available last winter. Both experiments, H1 and ZEUS, report a large excess of events over the LO QCD predictions.

Finally, there are two recent b-quark production related results from LEP. The preliminary measurement of the inclusive $b\bar{b}$ production at 189 GeV by the L3 Collaboration indicates again a factor of ~ 2 data excess above the QCD predictions. On the other hand the gluon splitting into $b\bar{b}$ results from three LEP experiments agree very well with the predictions based on leading and next-to-leading logarithmic terms resummed.

The renewed interest in the charmonium and bottomonium production originated with the observation by the CDF Collaboration that the direct J/ψ and $\psi(2s)$ exceeds expectations based on the Color Singlet Model of charmonium production by a factor of ~ 50. This observation led to a development of the Color Octet Model (COM), nonperturbative parameters of which were fitted to match the CDF data. However, the COM is not without predictive power. The model predicts that onium states are produced transversely polarized. The polarization prediction is the most likely effect to help distinguish the COM model from the multiple soft gluon exchanges model, a resurrected version of the old color evaporation model.

The latest results on onia production, subject to this review, include a recent DØ measurement of the J/ψ production in previously unexplored kinematical regions of small J/ψ transverse momenta and large rapidities, and preliminary polarization measurements of the directly produced J/ψ s and $\psi(2s)$ from CDF. New CDF data on the Υ production cross sections, fraction of directly produced $\Upsilon(1s)$ states, and their polarization have also become available. The Color Octet Model predictions are also verified at the fixed target energies, for the electroproduction of J/ψ s, and for the direct J/ψ production in Z boson decays.

Finally, we discuss the new, precise data on nuclear effects in the hadroproduction of J/ψ, reported by the Fermilab E866 Collaboration.

Central b-Quark HadroProduction

Both CDF and DØ have extracted the inclusive b quark cross sections for rapidity $|y| < 1$ using several different final state channels [2,3]. Most of the available DØ results on the inclusive b-quark production were derived from the inclusive muon and dimuon transverse momentum spectra. The transition from the muon (dimuon) spectrum to the inclusive b-quark cross sections involves b-quark decay tables, which have been updated since the completion of some of the analyses. A compilation of the revised and the latest [4] DØ measurements together with the CDF results is shown in Fig. 1. These measurements agree in shape with the NLO QCD predictions [1] over the studied transverse momentum range, $6 < p_T^b < 30$ GeV/c, but are larger by a factor ~ 3.0 than the central value [5].

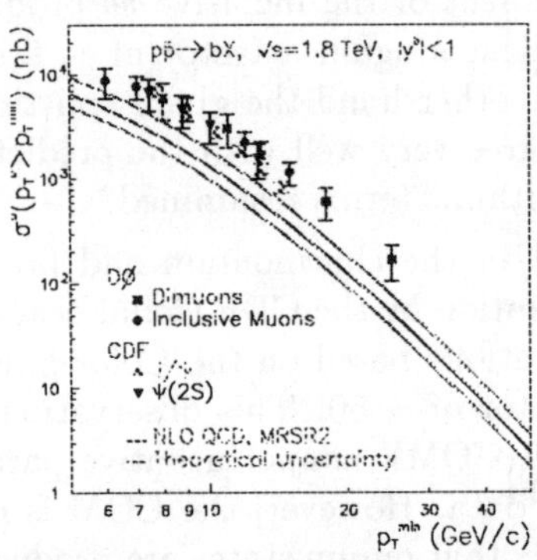

Figure 1: Compilation of the intergrated inclusive b-quark production cross sections as a function of the transverse momentum cutoff p_T^{min}, compared with the NLO QCD predictions. The theoretical uncertainty shows the uncertainty associated with the factorization and renormalization scale and b-quark mass, as described in the text.

Forward μ and b-Quark Production

In the DØ analysis of the b-quark production in the forward region muons are selected in the rapidity range $2.4 < |y^\mu| < 3.2$, with momentum $p^\mu < 150$ GeV/c and transverse momentum $p_T^\mu > 2$ GeV/c. After subtraction of the π/K contribution, the muon cross section is multiplied by the the ratio of the muon yield due to b-quark decays to the total muon yield from b- and c-quark decays, f_b. This ratio is determined using NLO QCD predictions for the ratio of b- and c-quark production cross sections. These predictions are verified for events containing a muon associated with a jet ($E_T^{jet} > 10$ GeV). The f_b fraction for this subsample of events (7%) is determined by fitting distributions of the muon transverse momentum with respect to the jet direction, p_T^{rel}, to the expected shapes from $b, c,$ and π/K decay.

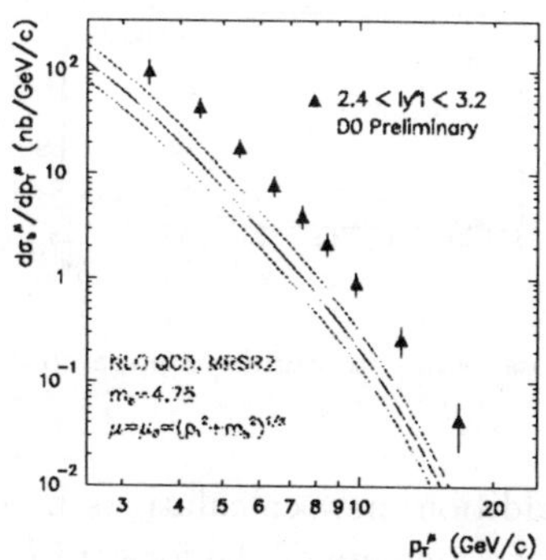

Figure 2: p_T spectrum of forward muons from b decays compared to NLO QCD prediction, generated with $m_b = 4.75$ GeV/c, $\mu = \mu_0$, and the MRSR2 [13] parton distribution functions. The theoretical uncertainty is determined by varying the parameters m_b and μ from 4.5 and 5.0 GeV/c^2 and from $\mu_0/2$ to $2\mu_0$, respectively.

The differential muon cross section from b-quark production and decay is presented in Fig. 2. The data are compared to the NLO QCD prediction obtained using the Monte Carlo simulation that relies on MNR calculations [1] for the production of the $b/\bar{b}$ partons, and ISAJET for the fragmentation to b hadrons and the decay of these hadrons to muons. The predictions of this model match the measured cross section fairly well in shape, but are approximately a factor of four lower than the data.

These measurements, combined with previous DØ results [3] obtained in the central region ($| y^\mu |< 0.8$), permit to study the rapidity dependence of the b-produced muon cross section. In Fig. 3 the differential muon cross section from b decay as a function of rapidity is compared to the NLO QCD predictions for two p_T^μ ranges, $p_T^\mu > 5$ GeV/c and $p_T^\mu > 8$ GeV/c. For $p_T^\mu > 5$ GeV/c, the ratio data/theory is equal to 2.5 ± 0.5 in the central region and increases to 3.6 ± 0.8 in the forward region. For $p_T^\mu > 8$ GeV/c, these numbers become 3.0 ± 0.6 and 4.6 ± 1.3 in the central and forward regions, respectively.

There have been several theoretical attempts to account for the observed increase in the data/theory discrepancy with increasing b quark rapidity [9]. A good agreement with the data requires a reduction of the renormalization

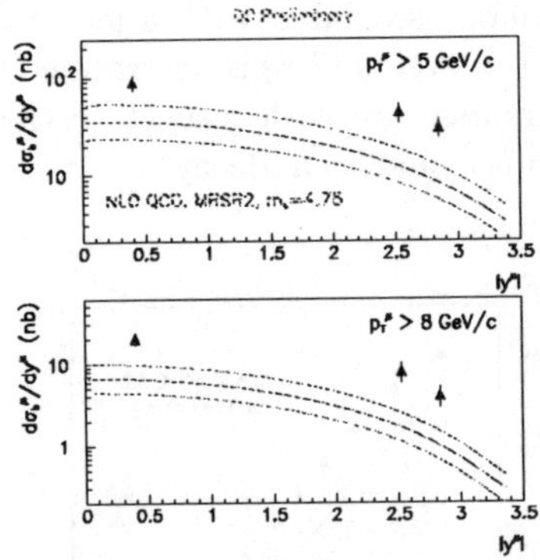

Figure 3: b-produced muon cross section *vs.* rapidity compared to the NLO QCD prediction.

scale to $\sim 0.25\ \mu_0$. In addition, new calculations based on a variable flavor number scheme [10] predict an increase of the forward b-quark production cross section by 30 to 50% with respect to the standard NLO QCD calculations. An increase in the B-meson production cross section of 50% in the forward region and 30% in the central region can also be obtained by using a stiffer b-quark fragmentation function than the usual Peterson one [11]. More data are needed, in particular at larger transverse momenta, in order to better understand the discrepancy.

$b\bar{b}$ Rapidity Correlations

Studies of $b\bar{b}$ correlations provide an even more stringent test of the NLO QCD calculations. Several previous studies of the $b\bar{b}$ correlations performed by CDF [6] and DØ [4] focused on the difference in azimuthal angle between the two muons coming from $b\bar{b}$ decays. The distribution of the opening angle between the two muons is found to be consistent with the theory within the uncertainties (see Fig. 4). However, another CDF analysis [7] of the correlations between the b and $\bar{b}$ quarks in μ–jet events, exploring production of b and $\bar{b}$ quarks with relatively large transverse momenta, shows discrepancies between the predicted and observed distribution of the azimuthal opening angle between the two quarks.

Recently, CDF has extended its correlation studies to the forward region [8], measuring the $b\bar{b}$ production cross section when one quark is produced in the forward region ($1.8 < |\eta^b| < 2.6$) and the other in the central pseudorapidity range ($|\eta^b| < 1.5$). This measured cross section was found to be higher than

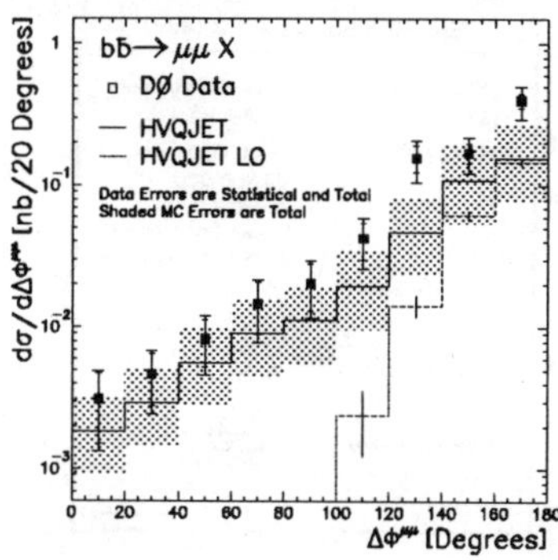

Figure 4: Azimuthal correlations between muons from b and $\bar{b}$ decays.

the NLO QCD prediction by a factor $2.4^{+0.6}_{-0.5}$, with the uncertainty reflecting the experimental error only.

In order to reduce the systematic uncertainties on both the data and the theory, CDF now measures [8] a ratio of correlated cross sections for a central-central and central-forward $b\bar{b}$ production. The event selection requires a central b-quark jet with $E_T > 26$ GeV and $| \eta^{\text{jet}} | < 1.5$, accompanied by a second b-quark giving rise to a muon and a jet. This jet must have an $E_T > 15$ GeV and the events are classified as central-central or forward-central depending upon the muon pseudorapidity range, $| \eta^{\mu} | < 0.6$ or $2.0 <| \eta^{\mu} | < 2.6$, respectively. For the μ-jet combination, the b-quark decay is identified on the basis of the muon momentum transverse to the μ-jet direction, p_T^{rel}. The central b-quark jet is identified through a secondary vertex tag. The final data sample, corresponding to an integrated luminosity of 80 pb^{-1}, contains 382 forward-central and 7544 central-central events.

To extract the signal fraction in each sample, the p_T^{rel} of the muon and the pseudo-$c\tau$ of the b-jet are fitted simultaneously using a binned maximum likelihood method. The template histograms for the signal and the dominant background sources ($c\bar{c}$ production and $b\bar{b}$ production where one b tag is a fake) are obtained either from Monte Carlo simulation or directly from the data. The $b\bar{b}$ signal fraction is about 75% in the forward-central events and 60% in the central-central events.

The ratio of the cross sections is given by:

$$R_{\text{data}} = \frac{\sigma(p\bar{p} \to b_1 b_2 X;\ 2.0 <| y_{b_1} |< 2.6)}{\sigma(p\bar{p} \to b_1 b_2 X;\ | y_{b_1} |< 0.6)} \tag{1}$$

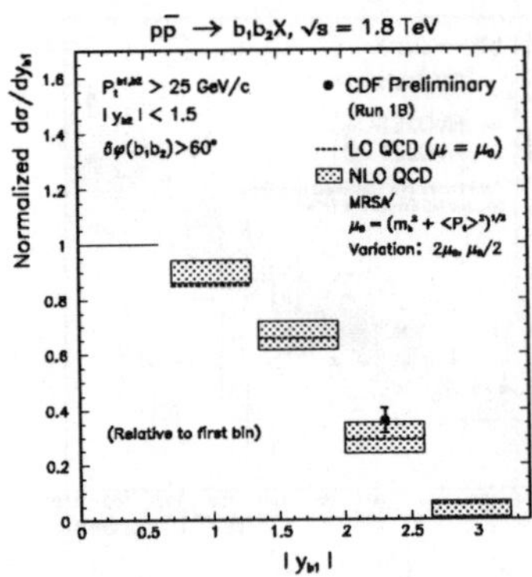

Figure 5: Evolution with rapidity of the cross section ratio.

where $p_T(b_1, b_2) > 25$ GeV/c, $|y_{b_2}| < 1.5$, and, $\delta\phi(b\bar{b}) > 60°$ for both cross sections. It is calculated using the number of signal events in both samples and the ratio of the total efficiency for central-central and forward-central events. Due to the much smaller kinematic acceptance of the forward toroids relative to the central detector, this ratio is equal to 5.4. The result,

$$R_{\mathrm{data}} = 0.361 \pm 0.033(\mathrm{stat})^{+0.015}_{-0.031}(\mathrm{syst}),$$

is in good agreement with the NLO QCD prediction obtained with MRSA′[13]:

$$R_{\mathrm{theory}} = 0.338^{+0.014}_{-0.097}.$$

The predicted evolution of the cross section ratio with the rapidity range of the muon coming from the b-quark semileptonic decay is shown in Fig. 5. The measurement is in agreement with both the LO and NLO QCD predictions.

With the cross section definition given in (1), the forward-central events arise from collisions between one parton with an average $x \approx 0.025$ and a second parton at $x \approx 0.25$. The cross section ratio should thus be sensitive to the gluon distribution in the proton at high x values, i.e. in a region where this gluon distribution is not very well known[14].

Figure 6 shows a comparison of the R_{data} measurement with R_{theory} obtained using the parton distribution sets MRSR1(2)[12] and CTEQ4HJ[15]. The data point and theory curves are normalized to MRSA′[13] and are presented as a function of the rapidity of the b-quark decaying to μ-jet. The data point is in good agreement with the MRS sets and slightly disfavors the CTEQ4HJ distribution. Since the CDF result is dominated by the statistical uncertainty,

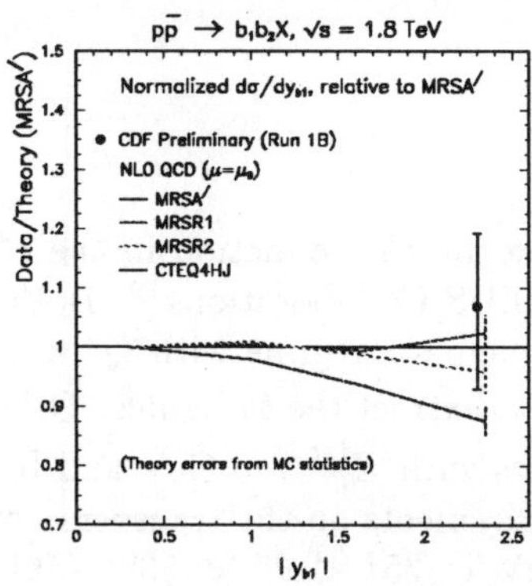

Figure 6: Comparison of the cross section ratio with various parton distribution functions.

this comparison illustrates the potential resolving power of a high statistics forward b production measurement.

$b\bar{b}$ Hadroproduction at Fixed Target Energies

There have been several experiments at fixed target energies dedicated to extracting b-quark production signal. The measurement is particularly difficult with only one $b\bar{b}$ pair produced per $10^6 - 10^7$ inelastic collisions. Therefore the largest $b\bar{b}$ signal observed by these experiments consist of 26 events, reconstructed among 10^8 collected triggers [16]. The CERN WA92 experiment used 350 GeV/c π^- beam incident on Cu and W targets to measure $\sigma(\pi^- N \to b\bar{b}) = 5.7^{+1.5}_{-1.2}$ nb/nucleon.

The measured cross section is well within the acceptable band of the NLO QCD predictions, but somewhat lower than earlier Fermilab measurements at higher energies. The WA92 analyses also include measurements of various kinematic distributions of reconstructed B mesons as well as an observation of azimuthal $b\bar{b}$ correlations (for 13 events with both Bs reconstructed). The strength of the correlations is consistent with an intrinsic transverse momentum of colliding partons, $< k_T^2 > \approx 0.5 - 1.0 \ (GeV/c)^2$.

The Fermilab E771 published the cross section for $b\bar{b}$ production in proton nucleus collisions at 800 GeV, based on 6 $b\bar{b}$ candidate events observed through either semileptonic or J/ψ b-quark decays [17]. The cross section of 42^{+20}_{-13} nb is approximately 3 times higher than the NLO QCD predictions and 2.4 standard deviations larger than the E789 result at the same energy [18]. Unfortunately, the size of the experimental uncertainties precludes drawing any strong conclusions.

$b\bar{b}$ production at Hera

The preliminary results on the $b\bar{b}$ production in the $e^{+}p$ collisions have been reported by the H1 and ZEUS Collaborations[19]. Both analyses are restricted to the photoproduction kinematic regime with $Q^2 < 1$ GeV^2 and employ fits to the $p_{T,rel}^{\mu(e)}$ distributions to extract the $b\bar{b}$ signals. Other requirements include presence of at least two jets with $E_T^{jet} > 6$ GeV and $|\eta^{jet}| < 2.5$ (**H1** - similar cuts for **ZEUS**). The requirements on the transverse momenta of the charged lepton are: $p_T^{\mu} > 2.0$ GeV/c; $35^o < \theta^{\mu} < 130^o$ (**H1**) or $p_T^e > 1.6$ GeV/c; $|\eta^e| < 1.1$ (**ZEUS**). The H1 result for the visible cross section is $\sigma_{b\bar{b}}^{vis} = 0.93\pm 0.08_{-0.12}^{+0.21}$ nb , whereas the AROMA 2.2 (LO MC) [44] predicts $\sigma_{b\bar{b}}^{vis} = 0.191$ nb. To relate these numbers to the NLO QCD predictions the authors note that the AROMA prediction for full kinematic range is 3.8 nb (m_b=4/75 GeV, MRS(G) pdf), to be compared with (4.7 - 10) nb expected range from the NLO QCD calculations.

The ZEUS result, $\sigma_{b\bar{b}}^{vis} = 39\pm11_{-16}^{+23}$ pb, is ~ 3.7 times above the HERWIG 5.9, LO Monte Carlo prediction, using CTEQ4D p and GRV-LO γ parton densities, m_c=1.55 GeV and m_b= 4.95 GeV.

It is fair to conclude that the preliminary $b\bar{b}$ photoproduction results exhibit similar level of discrepancy between the data and theory as observed for $b\bar{b}$ hadroproduction at the Tevatron collider.

Gluon Splitting Rate into $b\bar{b}$ Pairs in Hadronic Z Decays (LEP)

The rate of gluons splitting to a $b\bar{b}$ quark pair per hadronic Z^o decay, $g_{b\bar{b}}$, and the rate of final states with two $b\bar{b}$ quark pairs, g_{4b}, were recently measured by the OPAL Collaboration[21]. Events containing four jets were selected where two of the jets that are close in phase-space contain secondary decay vertices. In addition, four-jet events were selected that have secondary decay vertices in at least three of the four jets. Information from the event topology was combined in a likelihood analysis to extract the values of $g_{b\bar{b}}$ and g_{4b}. The values of $g_{b\bar{b}}$ were previously extracted by the DELPHI and ALEPH collaborations[20]. The results from the three experiments are summarized below:

$$g_{b\bar{b}} = (2.1 \pm1.1\ (stat) \pm 0.9\ (syst)) \times 10^{-3} \quad \text{DELPHI},$$
$$g_{b\bar{b}} = (2.77\pm0.42(stat) \pm 0.57(syst)) \times 10^{-3} \quad \text{ALEPH},$$
$$g_{b\bar{b}} = (2.15\pm0.43(stat) \pm 0.80(syst)) \times 10^{-3} \quad \text{OPAL},$$

$g_{4b} = (0.53\pm0.20(stat) \pm 0.23(syst)) \times 10^{-3}$ OPAL.

The QCD predictions are based on calculations including leading and next-to-leading logarithmic terms resummed[22]:

$\Lambda^5_{\overline{MS}}$	$\alpha_s(M_Z)$	$g_{c\bar{c}}$ (%)	$g_{b\bar{b}}$ (%)
150 MeV	0.112	$1.35^{+0.48}_{-0.30}$	0.20 ± 0.02
300 MeV	0.125	$1.85^{+0.69}_{-0.44}$	0.26 ± 0.03

There is a good agreement between experimental results and theoretical predictions within rather large uncertainties. Also the OPAL result for g_{4b} is compatible with the naive expectation $g_{4b} \approx R_b \times g_{b\bar{b}}$, where R_b stands for the fraction of $Z \to b\bar{b}$ decays.

At this conference the L3 Collaboration reported on a preliminary measurement of the $b\bar{b}$ production cross section at $\sqrt{s} = 189$ GeV [23], $\sigma_{b\bar{b}} = 9.9 \pm 2.9(stat) \pm 3.8(syst)$ pb, that, again turns out to be a factor ~ 2 above predictions [24].

Review of the onia production at the Tevatron collider

In high energy $p\bar{p}$ collisions J/ψ's are produced directly, from decays of higher mass charmonium states [χ and $\psi(2S)$], and from b-quark decays. Existing experimental results in the central rapidity region from UA1 [25] at $\sqrt{s} = 0.63$ TeV, and from CDF [26] and DØ [27] at $\sqrt{s} = 1.8$ TeV demonstrate that the measured inclusive J/ψ transverse momentum distribution cannot be described solely by contributions from b quark decays and prompt production predicted by the color singlet model [28]. In the color singlet model the charmonium meson retains the quantum numbers of the produced $c\bar{c}$ pair and thus each J/ψ state can only be directly produced via the corresponding hard scattering color singlet subprocess. The model predicts direct J/ψ and $\psi(2S)$ production rates fifty times smaller than those observed by CDF [26]. To explain this discrepancy, a color octet model was introduced [29,30]. The color octet mechanism extends the color singlet approach by taking into account the production of $c\bar{c}$ pairs in a color octet configuration accompanied by a gluon. The color octet state evolves into a color singlet state via emission of a soft gluon. The parameters of the model were derived from a fit to CDF data for direct J/ψ and $\psi(2S)$ production at central rapidity. The inclusion of the color octet mechanism leads to a prediction that directly produced ψ charmonia will be increasingly transversely polarized at high p_T. This is due to the dominance of gluon fragmentation into the color octet ($^3S_1^{(8)}$ $c\bar{c}$ states), and the preservation of the gluon's transverse polarization as the $c\bar{c}$ pair evolves into a bound ψ state.

Other recent analyses of J/ψ photo- and hadro-production data[31,32] resurrect the twenty year old color evaporation approach. This approach differs from Ref. [29,30] in the way the $c\bar{c}$ pair exchanges color with the underlying event. It assumes multiple soft gluon exchanges, whereas such interactions are suppressed in the color octet model by powers of v, the relative velocity of the heavy quarks within the $c\bar{c}$ system. These multiple soft gluon exchanges destroy the initial polarization of the heavy quark pair. Therefore, polarization measurements may provide the best tool to distinguish between these phenomenological models.

Small Angle J/ψ production

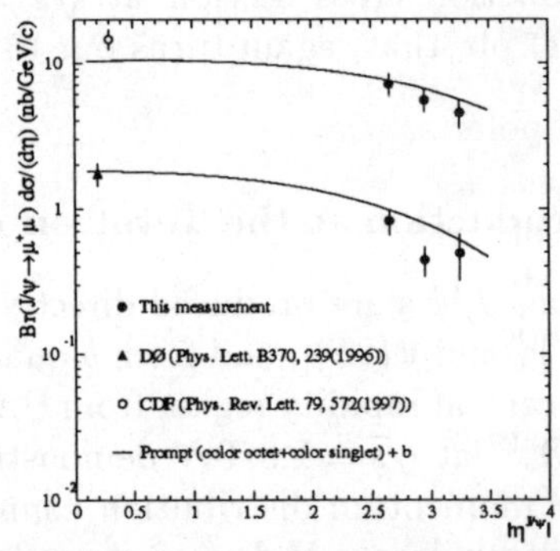

Figure 7: The pseudorapidity dependence of the J/ψ production cross section with $p_T >$ 5 GeV/c (upper points and curve) and $p_T > 8$ GeV/c (lower points and curve). The error bars are statistical and systematic errors (polarization uncertainties not included) summed in quadrature.

Recently, the DØ collaboration utilized the large rapidity coverage of the DØ muon system to study the process $p\bar{p} \rightarrow J/\psi + X \rightarrow \mu^+\mu^- + X$ in previously unexplored kinematical regions of small J/ψ transverse momenta and large rapidities[33].

Figure 7 shows the pseudorapidity dependence of the measured J/ψ cross section for $p_T^{J/\psi} > 5$ and 8 GeV/c along with the corresponding central rapidity measurements of DØ[27] and CDF[26]. Within uncertainties, the color octet model plus b quark decays describe the η dependence of the inclusive J/ψ

production in the full rapidity region. Therefore, the new results are consistent with the published results for the central J/ψ production, once the expected effects due to the process kinematics and variations of the gluon structure functions are taken into account.

CDF Measurement of Prompt J/ψ and $\psi(2s)$ Polarization

Last summer CDF made public a measurement of the polarization of promptly produced J/ψ mesons using the full Run I data set [34]. The polarization parameter α is measured over the kinematic range 4 GeV/c $< p_T^{J/\psi} <$ 20 GeV/c and $|y^{J/\psi}| <$ 0.6. The contribution of J/ψ mesons from B-hadron decay is separated using lifetime information, however feeddown from heavier charmonia is not separated.

The polarization is measured using the distribution of the decay angle θ, which is the angle between the μ^+ direction in the J/ψ rest frame and the J/ψ direction in the laboratory frame. The distribution has the form: $\omega(\theta) \sim (1 + \alpha cos^2\theta)$, where α is known as the polarization parameter. Fully transverse polarization corresponds to $\alpha = +1$, fully longitudinal to $\alpha = -1$, and $\alpha = 0$ indicates no polarization.

The data sample contains a signal of approximately 180,000 $J/\psi \rightarrow \mu^+\mu^-$ decays. The separation of the promply produced J/ψ mesons from those originating from B-hadron decays is achieved by splitting the event sample into two sets using the "pseudo"-proper lifetime $c\tau$ of the decay: $-0.1 < c\tau < 0.01$ cm and $c\tau > 0.01$ cm.

The "short-lived" sample contains mostly prompt J/ψ mesons and the "long-lived" sample contains mostly B-hadron decays. As the fraction of prompt J/ψ mesons depends on p_T, such fits are done separately in p_T bins. The polarization parameter α is measured for prompt and B-decay production using template fits to the $cos\theta$ distributions. The fit results for seven $p_T^{J/\psi}$ bins, with statistical and systematic uncertainties are shown in Fig. 8.

The parameter α for J/ψ mesons from B-decay is generally consistent with zero. The polarization of the promptly produced J/ψs is positive at intermediate $p_T^{J/\psi}$ but does not appear to continue to rise significantly at high $p_T^{J/\psi}$. The measured α values for this sample include feeddown from χ_c ($\approx$ 30%) and $\psi(2s)$ ($\approx$ 10%).

A similar analysis of the 1776$\pm$62 $\psi(2s)$ candidates with $p_T^{\psi(2s)} > 5$ GeV/c led to results shown in Fig. 9.

The polarization for prompt $\psi(2s)$ candidates is extracted to be $-0.76 \pm 0.49(stat) \pm 0.06(syst)$ for $9 < p_T^{\psi(2s)} < 20$ GeV/c. This measurement has

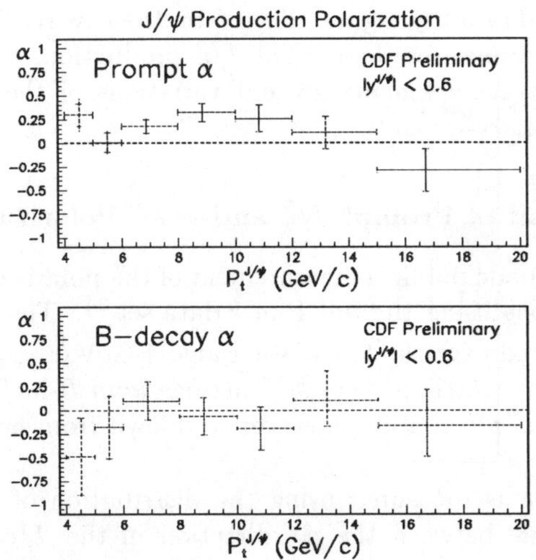

Figure 8: J/ψ polarization parameter α *vs.* transverse momentum for: (a) a prompt J/ψ production, (b) J/ψ s from B-hadron decays.

large statistical uncertainties, but it appears to not support the Color Octet prediction that the $\psi(2s)$ is transversely polarized at high p_T. The prompt $\psi(2s)$ candidates are expected to be directly produced.

New CDF Results on Υ Production.

CDF has also completed the analysis of the production of Υ states using ~ 77 pb^{-1} of data collected during Run 1b[34]. The new preliminary results include:

- measurements of the differential p_T cross sections for the $\Upsilon(1S)$, $\Upsilon(2S)$, and $\Upsilon(3S)$ states for values of Υ rapidity $|y|$ <0.4 in the p_T interval 0 $< p_T^\Upsilon < 20$ GeV/c.

The improved precision over the published Run Ia results will allow a more meaningful comparison with the theory predictions.

- a measurement of $\Upsilon(1S)$ polarization.

A χ^2 fit of the templates to the $\cos\theta$ decay distributions for $\Upsilon(1S)$ yield a longitudinally polarized fraction of Γ_L/Γ=0.37±0.04, therefore $\Upsilon(1S)$ are unpolarized within the error of the measurements.

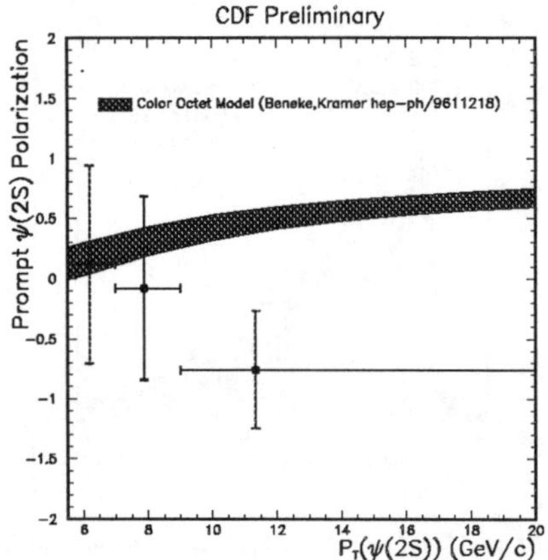

Figure 9: $\psi(2s)$ polarization parameter α *vs.* transverse momentum for a prompt $\psi(2s)$ production

- determination of a fraction of $\Upsilon(1s)$ mesons from χ_b decays.

This analysis is based on the $L = 90$ pb^{-1} sample of Run I data. The contributing χ_b states are separated in mass from $\Upsilon(1s)$ by: $\Delta M = M(\mu^+\mu^-\gamma) - M(\mu^+\mu^-) = 443\,\mathrm{MeV}/c^2$ $(\chi_b(1P) \to \Upsilon(1s)\gamma)$ and $\Delta M = 802\,\mathrm{MeV}/c^2$ $(\chi_b(2P) \to \Upsilon(1s)\gamma)$. The resulting difference between the dimuon-photon and the dimuon mass, ΔM, is shown in Fig. 10 for $p_T^{\mu\mu} > 8.0$ GeV/c and $E_T^\gamma > 700$ MeV. There are two well separated peaks corresponding to the 1P and 2P states, consistent with expectations based on detector simulations. The background mass spectrum is estimated using data and decaying charged particles as if they were π^o or K^o mesons.

The measured fractions of $\Upsilon(1s)$ with $p_T^\Upsilon > 8.0$ GeV/c are:

$(26.7 \pm 6.9(stat) \pm 4.3(syst))\%$ from $\chi_b(1P)$

$(10.8 \pm 4.4(stat) \pm 1.3(syst))\%$ from $\chi_b(2P)$

Taking into account feeddown from higer Υ states, one finds that the direct $\Upsilon(1s)$ production is responsible for $(51.8 \pm 8.2(stat)^{+9.0}_{-6.7}(syst))\%$ of observed $\Upsilon(1s)$ mesons.

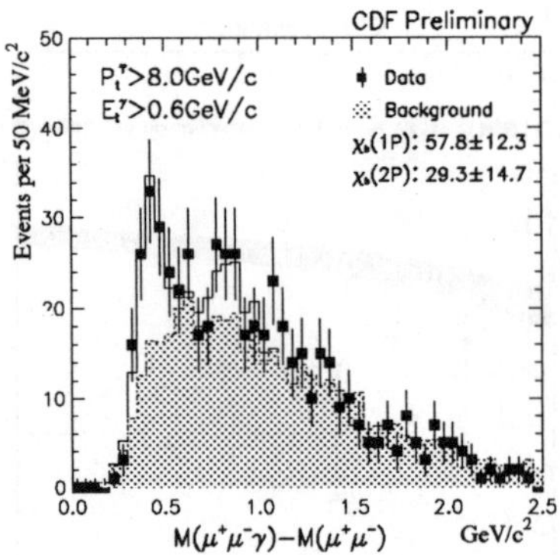

Figure 10: $\Delta M = M(\mu^+\mu^-\gamma) - M(\mu^+\mu^-)$ mass distribution.

J/ψ production at fixed target energies

Gupta and Sridhar[35] made a detailed comparison of the color-octet model predictions with measured x_F distributions from several fixed-target experiments. All values of the required non-perturbative matrix elements were derived from other experiments, primarily from charmonium photoproduction data and from J/ψ production data from the Tevatron. In general, the parameter-free predictions of the model for x_F - distributions agree with the proton and pion induced data both in magnitude and shape.

On the other hand, most of the recent results are consistent with unpolarized J/ψ production. The E771 Collaboration reported $\alpha = $ -0.09±0.12[36] in pSi interactions at 800 GeV/c. The E672 measurement of α for J/ψ production from the high-statistics $\pi^- Be$ data at 515 GeV/c gave $\alpha = $ -0.01±0.08[37]. These results confirm some of the earlier observations which reported small values of α[38], and are consistent with expectations from the color evaporation model [31]. They are in disagreement with expectations from the color octet model[35], which predicts a sizeable J/ψ transverse polarization, with mild energy dependence (e.g. $0.31< \alpha < 0.63$ at $\sqrt{s} = 15.3$ GeV).

Inelastic charmonium production at HERA

The precision of the HERA data on inelastic J/ψ production has substantially improved since last year due to the increased statistics, corresponding to integrated luminosity greater than 26 pb^{-1} for each experiment.

The ZEUS Collab. studied **inelastic photoproduction** of J/ψ mesons with kinematic constraints of $Q^2 < 1$ GeV^2, 50$<$ W$<$ 180 GeV and $p_T^{J/\psi} >$ 1 GeV/c [42]. The selection of inelastic events was based on requiring the observed energy in the forward region of the detector to exceed 1 GeV. This requirement corresponds roughly to $M_X > 4$ GeV and $0.4 < z < 0.9$, where M_X is the mass of the hadronic system excluding the J/ψ and z is defined as $z = p^{J/\psi} \cdot q/p^{J/\psi} \cdot P$, and is equal to the ratio of the J/ψ energy to that of the initial photon, both calculated in the proton rest frame.

The total cross section as a function of W (invariant mass of the γp system) is correctly reproduced by the NLO calculations within the Color Singlet Model (CSM) [39]. Equally good agreement between the data and the CSM predictions is observed for differentian cross sections $d\sigma/dz$. The inclusion of the Color Octet Model contributions results in a strong enhancement of cross sections at large values of z, that is not observed in the data. An improvement can be achieved by including intrinsic k_T effects on the level of 1 GeV/c [40], however even with this modification the agreement in the shape of the distribution remains poor.

The H1 Collaboration studied **inelastic J/ψ electroproduction** requesting 2$< Q^2 < 80$ GeV^2, 40$< W < 180$ GeV and $z <$0.2 [41]. An additional cut requiring $M_X > 10$ GeV suppresses diffractive events and is favored by the theorists for comparison with the Color Octet model predictions [43]. Nevertheless, the theory calculations including LO Color Singlet and Color Octet contributions disagree in both shapes and normalization for most of the differential cross sections (versus Q^2, $p_T^{J\psi}$, W, y*, z - not shown). The Color Octet Model, leading order NRQCD matrix elements were taken from the fits to the CDF data. The comparisons may indicate the need to decrease the size of the color octet long distance matrix elements or to change the relative importance of the different color octet contributions, and/or to include higher orders in the NRQCD perturbative expansions. The color singlet contribution alone is below the data by factors 2 - 3. The model of Soft Color Interactions, a non-perturbative phenomenological approach to the description of inclusive J/ψ production [32], describes the shape of the differential distributions reasonably well, but fails to reproduce the normalization and the z - dependence.

The H1 Collab. has also measured the J/ψ polarization. The values of paprameter α are $0.77^{+0.44}_{-0.38}$ for the inelastic sample selected with the $M_X > 10$ GeV cut. The expectation for the color singlet model is $\alpha \sim 0.5$ whereas the Color Octet Model predicts $|\alpha| <$0.5. The uncertainties are too large to draw strong conclusions here. For a more comprehensive review of the data see Ref. [45].

Z → prompt J/ψ + X **decays at LEP**

L3 reported measurements of the inclusive production of heavy quarkonium states in Z decays based on the analysis of 3.6 million hadronic events [46]. Processes contributing here include heavy quark fragmentaton, gluon fragmentation and gluon radiation, with only the two latter processes leading to a prompt onium production. The analysis employed two different isolation cuts to separate prompt J/ψ mesons form J/ψ mesons produced in the decays of B-hadrons. The cuts were based on the energy in a 30° cone around the J/ψ direction, excluding the J/ψ energy, and on the angle between the J/ψ and the nearest jet. The results, together with earlier measurements by other LEP experiments [47] are:

$$\mathrm{BR}(Z \to prompt\ J/\psi + X) = (1.9 \pm 0.7 \pm 0.5 \pm 0.5) \times 10^{-4} \qquad \text{OPAL}$$
$$\mathrm{BR}(Z \to prompt\ J/\psi + X) = (3.0 \pm 0.8 \pm 0.3 \pm 0.1) \times 10^{-4} \qquad \text{ALEPH}$$
$$\mathrm{BR}(Z \to prompt\ J/\psi + X) = (4.4^{+3.6}_{-3.0}) \times 10^{-4} \qquad \text{DELPHI}$$
$$\mathrm{BR}(Z \to prompt\ J/\psi + X) = (2.1 \pm 0.6 \pm 0.4^{+0.4}_{-0.2}) \times 10^{-4} \qquad \text{L3}$$

These results can be compared with recent theoretical work including $\log(z)$ resummations within the COM [48]. The predictions can be approximated by a formula:

$$BR(Z \to prompt\ J/\psi + X) = (1.47\ \frac{M^\psi(^3S_1^{(8)})}{0.014\ GeV^3} + 0.47\ \frac{M^\psi(^3S_1^{(1)})}{1.45\ GeV^3}) \times 10^{-4}$$

The combined LEP data give an extracted effective octet matrix element $M^\psi(^3S_1^{(8)}) = (0.019 \pm 0.005(stat) \pm 0.010(theo))$ GeV^3 to be compared to the effective fit to the CDF data $M^\psi(^3S_1^{(8)}) = (0.014 \pm 0.002(stat) \pm large(theo))$ GeV^3.

In this case the COM contribution is 3 times that of CSM and the data seem to favor its inclusion.

Nuclear dependence of onia production

Studies of the charmonium hadroproduction from nuclei are necessary to understand J/ψ suppression in heavy ion collisions, one of expected signatures for the formation of the of quark-qluon plasma. These studies are also indirectly related to the color octet model investigations discussed in this paper. One expects the color octet $c\bar{c}$ state propagating through nuclear matter to be absorbed stronger than the corresponding color singlet state. To study nuclear effects the charmonium production cross sections are parametrized as $\sigma_A =$

$\sigma_N A^\alpha$, where σ_N and σ_A are cross sections on nuclear and nucleon target, respectively, A is the atomic number, and the exponent α measures the level of suppression. $\alpha = 1$ indicates no suppression. The values of α are expected to vary with kinematic variables like p_T, Feynman-x (x_F) etc.

Several generations of experiments at the Fermilab Meson East Spectrometr (E772,E789,E866) have provided ever improving data on nuclear effects in onia production. The latest results from E866/NuSea[49] are shown in Fig. 11, where we plot the dependence of the parameter α on x_F.

There are at least three mechanisms proposed to account for the observed J/ψ suppression[49]. They manifet themselves at somewhat different regions of the J/ψ Feynman-x variable.

- nuclear and "co-mover" absorption

The absorption is understood here as a dissociation of the nascent $c\bar{c}$ by interactions with the nucleus or with co-movers into a separate charmed quark and anti-quark which would eventually hadronize into D mesons. A slow (small and negative x_F) $c\bar{c}$ pair may become a physical J/ψ (or $\psi(2s)$) meson and have an absorption related to the radius of the particle. Therefore one expects $\psi(2s)$ with radius of 0.56 fm to be more strongly absorbed than J/ψ mesons, whose radius is only 0.29 fm. The upsilon (radius = 0.13 fm) production is expected to be suppressed even less. The data exhibit, as expected, a stronger suppression of $\psi(2s)$ than J/ψ mesons at small values of x_F (see Fig. 11). Comparison of these results with the lack of suppression seen for the production of open charm D mesons (at $x_F \approx 0$), where this absorption mechanism is not available, supports this general picture.

Fast (large x_F) ψ mesons are formed from $c\bar{c}$ pairs that hadronize well outside nucleus. The fast $c\bar{c}$ pairs experience similar absorption, the strength of which depends critically on the production mechanism, with much stronger effects expected for the color octet mechanism than for the color-singlet one. It was suggested that the observed falloff in α with x_F could be explained by the short octet state lifetime[51]. However, this explanation is difficult to reconcile with the lack of scaling of the exponent α at large laboratory momenta, $p_{LAB}^{J/\psi}$ (not shown).

- gluon shadowing

Gluon fusion dominates charmonium production for $x_F < 0.65$. Any effects of reduced gluon density inside nuclei are expected to show up as a suppression of the J/ψ meson production again at relatively small values of x_F. However, the expected size to the effect is not well known. In addition, a lack of scaling

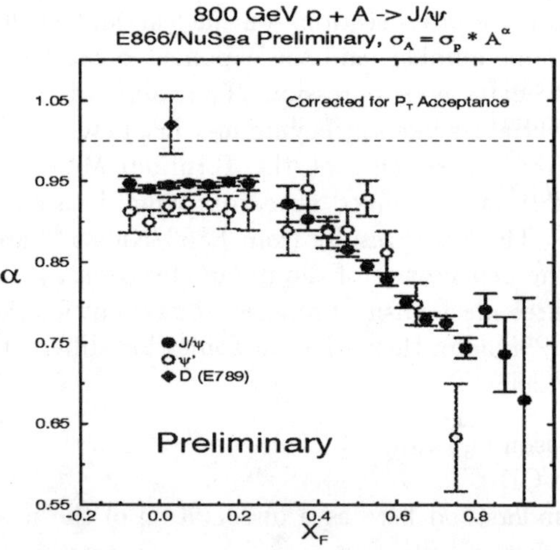

Figure 11: Exponent α dependence on x_F for J/ψ, $\psi(2s)$ and D meson production in p - nucleus interactions at 800 GeV/c.

with the x_2 variable (a fraction of the target nucleon momentum carried by an initial parton), appears to rule out large contribution from shadowing.

- energy losses of the initial and final partons

Energy losses, suffered by initial and final partons propagating through nuclear matter, shift x_F distribution toward smaller values. This shift results in values of the exponent α less than 1 in the large x_F region.

Drell-Yan pair production is influenced by energy loss in the initial, but not in the final, state. The experimentally observed values of α (for Drell-Yan pairs) close to one [50] indicate that the energy losses of initial partons are relatively modest. Although the incoming gluon's energy loss is expected to be larger by a color factor of 9/4 and the additional energy loss of the outgoing $c\bar{c}$ pair may be as large as that of a gluon, the overall contribution of energy loss is expected to be small.

The authors of Ref. [51] succeeded in describing most of the available data on the nuclear J/ψ suppression, including the latest E866/NuSea results summarized in Fig.11, using the standard color neutralization and state expansion scenario with a common set of parameters. However, the octet lifetime determined from the data is too short to make the proposed scenario realistic. They conclude that, most likely, the suppression effects observed at large x_F are not yet understood.

Summary

There has been an impressive increase in the number and quality of experimental results on the b-quark and onia productions since last year. The new data confirm the discrepancy between the measured inclusive b-quark cross sections and the NLO QCD predictions. There is no clear explanation for this observation, most likely there are several effects contributing to the excess of events over the expectations.

The Color Octet Model for onia production is being severely tested by the latest CDF data on onia polarization and by the HERA data on the J/ψ photo- and electroproduction.

1. M. Mangano, P.Nason and G. Ridolfi, *Nucl. Phys.* B **373**, 295 (1992).
2. F. Abe *et al.*, The CDF Collaboration, *Phys. Rev. Lett.* **71**, 500, 2396, 2537 (1993); *Phys. Rev. Lett.* **75**, 1451 (1995); *Phys. Rev. Lett.* **79**, 572 (1997).
3. S. Abachi *et al.*, The DØ Collaboration, *Phys. Rev. Lett.* **74**, 3548 (1995).
4. B. Abbott *et al.* The DØ Collab., FERMILAB-Pub-99/144-E (1999).
5. D. Zieminska, in: Proceedings of the XXXIIIrd Rencontres de Moriond '98, QCD and Hadronic Interactions, p. 79.
6. F. Abe *et al.*, The CDF Collaboration, *Phys. Rev.* D **55**, 2546 (1997).
7. F. Abe *et al.*, The CDF Collaboration, *Phys. Rev.* D **53**, 1051 (1996).
8. F. Abe *et al.*, The CDF Collaboration, preprint FERMILAB-Pub-98/392-E, *Phys. Rev.* D - in press.
9. G. Ridolfi, invited talk at the Tevatron Run 2 b-physics workshop, Fermilab, September 1999.
10. F.I. Olness, R.J. Scalise, and Wu-Ki Tung, *Phys. Rev.* D **59**, 14506 (1999).
11. M.L. Mangano, `hep-ph/9711337` ; CERN preprint TH-97-328 (unpublished).
12. A.D. Martin, R.G. Roberts and W.J. Stirling, *Phys. Lett.* B **387**, 419 (1996).
13. A.D. Martin, R.G. Roberts and W.J. Stirling, *Phys. Rev.* D **47**, 867 (1993).
14. J. Huston *et al.*, The CTEQ Collaboration, *Phys. Rev.* D **58**, 4034 (1998).
15. H. L. Lai *et al.*, The CTEQ Collaboration, *Phys. Rev.* D **55**, 1280 (1997).
16. M. Adamovich *et al.*, The CERN WA92 Collaboration, *Nucl. Phys.* B **519**, 19 (1998); Y. Alexandrov *et al.*, *Phys. Lett.* B **433**, 217 (1998); Y. Alexandrov *et al.*, CERN-EP-99-071, submitted to *Phys. Lett. B*.

17. T. Alexopoulos *et al.*, The Fermilab E771 Collaboration, *Phys. Rev. Lett.* **82**, 41 (1999).

18. D. M. Jansen *et al.*, The Fermilab E789 Collaboration, *Phys. Rev. Lett.* **74**, 3118 (1995).

19. H1 Collab., paper 575 submitted to ICHEP98 Vancouver, July 1998; A. Bertolin at Les Recontres de Physique de la Vallee d'Aoste, March 1999; M. Hayes at Rencontres de Moriond, March 1999, hep-ex/9905033.

20. P. Abreu *et al.*, The DELPHI Collaboration, *Phys. Lett.* B **405**, 1947 (1997); R. Barate *et al.*, The ALEPH Collaboration, *Phys. Lett.* B **434**, 437 (1998).

21. The OPAL Collaboration, Physics Note PN 383 (1999).

22. S. Frixione et al. CERN-TH-97-16, to be published in "Heavy Flavors II", eds. A. J. Buras and M. Linder (World Scientific).

23. Maneesh Wadhwa at this Conference.

24. M. Drees *et al.*, *Phys. Lett.* B **306**, 371 (1993).

25. C. Albajar *et al.*, The UA1 Collaboration, *Phys. Lett.* B **256**, 112 (1991).

26. F. Abe *et al.* The CDF Collaboration, *Phys. Rev. Lett.* **69**, 3704 (1992); *ibid* **79**, 572 (1997); *ibid* **79**, 578 (1997).

27. S. Abachi *et al.* The DØ Collaboration, *Phys. Lett.* B **370**, 239 (1996).

28. R. Baier and R. Ruckl, *Z. Phys.* C **19**, 251 (1983).

29. E. Braaten and S. Fleming, *Phys. Rev. Lett.* **74**, 3327 (1995); P. Cho and M. Wise, *Phys. Lett.* B **346**, 129 (1995).

30. P. Cho and A.K. Leibovich, *Phys. Rev.* D **53**, 6203 (1996).

31. J. F. Amundson *et al.*, *Phys. Lett.* B **390**, 323 (1997).

32. A. Edin, G. Ingelman and J. Rathsman *Z. Phys.* C **75**, 57 (1997).

33. B. Abbott *et al.*, The DØ Collaboration, *Phys. Rev. Lett.* **82**, 35 (1999).

34. The CDF Collaboration, www-cdf.fnal.gov/physics/ new/ bottom/ bottom. html. quarkonia - August 1999.

35. S. Gupta and K. Sridhar, *Phys. Rev.* D **54**, 5545 (1996); **55**, 2650 (1997); M. Beneke and I. Z. Rothstein, *Phys. Rev.* D **54**, 2005 (1996); **54**, 7082 (E) (1996).

36. T. Alexopoulos *et al.*, The Fermilab E771 Collaboration, *Phys. Rev.* D **55**, 3927 (1997).

37. A. Gribushin *et al.*, The Fermilab E672/E706 Collaboration, *Phys. Rev.* D **53**, 4723 (1996); Fermilab-Pub-99/243-E (1999), submitted to Phys. Rev. D.

38. C. Akerlof *et al.*, *Phys. Rev.* D **48**, 5057 (1993) and references therein.

39. M. Kramer *et al.*, *Phys. Lett.* B **348**, 657 (1995); *Nucl. Phys.* B **459**, 3 (1996).

40. M. Cacciari *et al.*, *Phys. Rev. Lett.* **76**, 4128 (1996); K. Sridhar, A. D. Martin and W. J. Stirling, *Phys. Lett.* B **438**, 211 (1998).
41. C. Adloff *et al.*, The H1 Collaboration, preprint DESY-99-026 (hep-ex/9903008) - submitted to Eur. Phys. J. C (1999).
42. J. Breitweg *et al.*, The ZEUS Collaboration, *Z. Phys.* C **76**, 599 (1997).
43. S. Fleming and T. Mehen, *Phys. Rev.* D **57**, 1846 (1998).
44. G. Ingelman, J. Rathsman and G. A. Schuler, *Comput. Phys. Commun.* **101**, 1135 (1997).
45. B. Naroska "Inelastic Charmonium Production at HERA", invited talk at the PHOTON'99 Conf., Freiburg, May 1999.
46. M. Acciarri *et al.*, The L3 Collaboration, *Phys. Lett.* B **453**, 94 (1999).
47. D. Buskulic *et al.*, The ALEPH Collaboration, *Phys. Lett.* B **295**, 396 (1992); P. Abreu *et al.*, The DELPHI Collaboration, *Phys. Lett.* B **341**, 109 (1994); G. Alexander *et al.*, The OPAL Collaboration, *Z. Phys.* C **70**, 197 (1996).
48. C. G. Boyd, A. K. Leibovich and I. Z. Rothstein, *Phys. Rev.* D **59**, 054016 (1999).
49. M. J. Leitch *et al.*, The Fermilab E866/NuSea Collaboration, preprint NUCL-EX/9909007 and LA-UR-99-5007, submitted to *Phys. Rev. Lett.* and references therein; M. J. Leitch, invited talk at the QM'99 Conf., May 1999.
50. M. A. Vasiliev *et al.*, The Fermilab E866/NuSea Collaboration, *Phys. Rev. Lett.* **83**, 2304 (1999).
51. F. Arleo *et al.*, HEP-PH/9907286.

REVIEW OF PARTON DISTRIBUTIONS AND IMPLICATIONS FOR THE TEVATRON AND LHC (PARTONS IN COLLISION AT PHYSICS IN COLLISION)

J. HUSTON

Department of Physics and Astronomy, Michigan State University, East Lansing, MI 48824, USA

E-mail: huston@pa.msu.edu

This talk is intended to serve as a pedagogical guide on the determination of, the proper use of, and the uncertainties of parton distribution functions and their impact on physics cross sections at the Tevatron and LHC. A longer writeup of this talk is available at http://www.pa.msu.edu./~huston/lhc/lhc_pdfnote.ps.

1 Introduction

The calculation of the production cross sections at the Tevatron and LHC, for both interesting physics processes and their backgrounds, relies upon a knowledge of the distribution of the momentum fraction x of the partons in a proton in the relevant kinematic range. These parton distribution functions (pdf's) are determined by global fits to data from deep inelastic scattering (DIS), Drell-Yan (DY), and jet and direct photon production at current energy ranges. Two major groups, CTEQ and MRS, provide semi-regular updates to the parton distributions when new data and/or theoretical developments become available. The newest pdf's, in most cases, provide the most accurate description of the world's data, and should be utlilized in preference to older pdf sets. The newest sets from the two groups are CTEQ5[1] and MRST[2].

2 Processes Involved in Global Analysis Fits

Lepton-lepton, lepton-hadron and hadron-hadron interactions probe complementary aspects of perturbative QCD (pQCD). Lepton-lepton processes provide clean measurements of $\alpha_s(Q^2)$ and of the fragmentation functions of partons into hadrons. Measurements of deep-inelastic scattering (DIS) structure functions (F_2, F_3) in lepton-hadron scattering and of lepton pair production cross sections in hadron-hadron collisions provide the main source on quark distributions $f^a(x, Q)$ inside hadrons. At leading order, the gluon distribution function g(x,Q) enters directly in hadron-hadron scattering processes with direct photon and jet final states. Modern global parton distribution fits are carried out to next-to-leading (NLO) order which allows $\alpha_s(Q^2), q^a(x, Q)$ and

$g(x, Q)$ to all mix and contribute in the theoretical formulae for all processes. Nevertheless, the broad picture described above still holds to some degree in global pdf analyses.

In pQCD, the gluon distribution is always accompanied by a factor of α_s, in both the hard scattering cross sections and in the evolution equations for parton distributions. Thus, determination of α_s and the gluon distribution is, in general, a strongly coupled problem. One can determine α_s separately from e^+e^- or determine α_s and $g(x, Q)$ jointly in a global pdf analysis. In the latter case, though, the coupling of α_s and the gluon distribution may not lead to a unique solution for either. (See for example the discussion in the CTEQ4 paper where good fits were obtained to a global analysis data set, including the inclusive jet data, for a wide range of α_s values.[3] Adjustments to the gluon distribution compensated for the changes in α_s.)

Currently, the world average value of $\alpha_s(M_Z)$ is on the order of 0.118-0.119[4] The average value from LEP is .121 while the DIS experiments prefer a somewhat smaller value (of the order of 0.116-0.117). Since global pdf analyses are dominated by the high statistics DIS data, they would tend to favor the values of α_s closer to the lower DIS values. The more logical approach is to adopt the world average value of $\alpha_s(M_Z)$ and concentrate on the determination of the pdf's. This is what both CTEQ and MRS currently do.[a]

The data from DIS, DY, direct photon and jet processes utilized in pdf fits cover a wide range in x and Q. The kinematic 'map' in the (x,Q) plane of the data points used in a recent parton distribution function analyses is shown in Figure 1. The HERA data (H1+ZEUS) are predominantly at low x, while the fixed target DIS and DY data are at higher x. There is considerable overlap, however, with the degree of overlap increasing with time as the statistics of the HERA experiments increase. DGLAP-based NLO pQCD should provide an accurate description of the data (and of the evolution of the parton distributions) over the entire kinematic range shown. At very low x and Q, DGLAP evolution is believed to be no longer applicable and a BFKL description must be used. No clear evidence of BFKL physics is seen in the current range of data; thus all global analyses use conventional DGLAP evolution of pdf's.

There is a remarkable consistency between the data in the pdf fits and the NLO QCD theory fit to them. Over 1300 data points are shown in Figure

---

[a]One can either quote a value of $\alpha_s(M_Z)$ or the value of $\Lambda^{\overline{MS}}$. In the latter case, however, the number of flavors has to be clearly specified, since the value of α_s (and not $\Lambda^{\overline{MS}}$) has to be continuous across flavor thresholds. The range for $\alpha_s(M_Z)$ from .105 to .122 corresponds to $100 < \Lambda_5^{\overline{MS}} < 280 MeV$ and $155 < \Lambda_4^{\overline{MS}} < 395 MeV$.

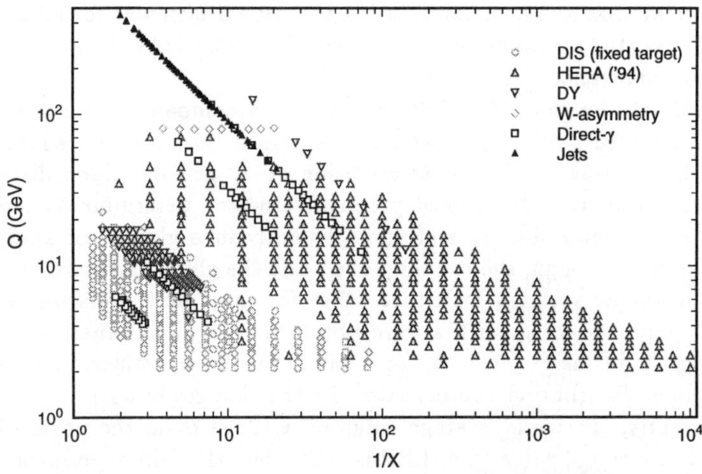

Figure1. The kinematic map in the (x,Q) plane of data points used in the CTEQ5 analysis.

1 and the χ^2/DOF for the fit of theory to data is on the order of 1.

Parton distributions determined at a given x and Q^2 'feed-down' to lower x values at higher Q^2 values. The accuracy of the extrapolation to higher Q^2 depends both on the accuracy of the original measurement and any uncertainty on $\alpha_s(Q^2)$.[b] For the structure function F_2, the typical measurement uncertainty at medium to large x is on the order of $\pm 3\%$. At large x, the DGLAP equation for F_2 can be approximated as $\frac{\partial F_2}{\partial \log Q^2} = \alpha_s(Q^2) P^{qq} \otimes F_2$. There is an extrapolation uncertainty of $\pm 5\%$ in F_2 from low to high Q^2 (10^5 GeV^2) from the given uncertainty in α_s (0.1175 ± 0.005).

Evolution is susceptible to a feed-down on an anomalously large contri-

---

[b]The evolution can be carried out in either moment space or configuration space. Improvements have been made in the CTEQ and MRST evolution programs so that both now agree with the 'DESY standard' evolution prescription. The CTEQ and MRST packages should be able to carry out the evolution using NLO DGLAP to an accuracy of a few percent over the LHC kinematic range, except perhaps at very large and very small x. Note that the theoretical predictions for the W and Z total cross sections at the LHC may have uncertainties of less than 5%. [?] This puts a great demand for the pdf evolution to have accuracies of better than a few percent, since any error on a pdf gets doubled in the cross section calculation. Evolution programs at NNLO may be available at the time of the LHC turnon, but the advantages over NLO evolution should be minimal.

bution to F_2 near x values of 1. Such a contribution may not be evident in fixed target measurements at low x and low Q^2, but may influence higher Q^2 measurements. Such an example was considered in Reference[5].

3 Parameterizations and Schemes

A global pdf analysis carried out at next-to-leading order needs to be performed in a specific renormalization and factorization scheme. The evolution kernels are in a specific scheme and to maintain consistency, any hard scattering cross section calculations used for the input processes or utilizing the resulting pdf's need to also have been implemented in that same renormalization scheme. Almost universally, the $\overline{MS}$ scheme is used; pdf's are also available in the DIS scheme, a fixed flavor scheme (a la GRV[6]) and several schemes that differ in their specific treatment of the charm quark mass.

It is also possible to use only leading-order matrix element calculations in the global fits which results in leading-order parton distribution functions. Such pdf's are preferred when leading order matrix element calculations (such as Monte Carlo programs like Herwig[7] and Pythia[8]) are used. The differences between LO and NLO pdf's, though, are formally NLO; thus, the additional error introduced by using a NLO pdf with Herwig rather than a LO pdf, for example, should not be significant, in principle, and NLO pdf's can be used when no LO alternatives are available.

All global analyses use a generic form for the parameterization of both the quark and gluon distributions at some reference value Q_o: $F(x, Q_o) = A_o x^{A_1}(1 - x)^{A_2} P(x; A_3, ...)$. The reference value Q_o is usually chosen in the range of 1-2 GeV. The parameter A_1 is associated with small-x Regge behavior while A_2 is associated with large-x valence counting rules. In some pdf fits, A_1^{gluon} has been tied to $A_1^{seaquark}$; in more recent fits like CTEQ4, CTEQ5 and MRST, the two small x exponents are allowed to vary independently. The current statistical power of the low x and Q^2 DIS data from HERA warrants the separation.

The first two factors, in general, are not sufficient to describe either quark or gluon distributions. The term $P(x; A_3, ...)$ is a suitably chosen smooth function, depending on one or more parameters, that adds more flexibility to the pdf parameterization. In general, both the number of free parameters and the functional form can have an influence on the global fit. For example, the MRS group traditionally uses $P_{MRS}(x; A_3, A_4) = 1 + A_3\sqrt{x} + A_4 x$. The CTEQ3 pdf used $P_{CTEQ3} = 1 + A_3 x$ while CTEQ2, CTEQ4 and CTEQ5 all use the more general form $P_{CTEQ2,4,5} = 1 + A_3 x^{A_4}$. The flexibility in the latter form, for example, makes possible the larger gluon at high x observed

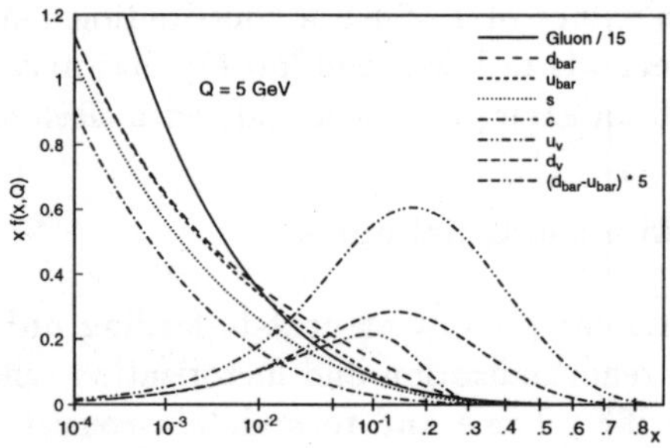

Figure2. The parton distributions from the CTEQ5 set plotted at a Q value of 5 GeV.

in the CTEQ4HJ pdf.

Although the pdf's determined from global analyses should, in principle, be universal, in practice they could depend on the choice of data sets, and in particular on the choice of Q_{cut} values that specify the minimum hard physical scale $(Q, p_T, ..)$ required for data points to be included in the fit.

The parton distributions from the recent CTEQ pdf release are plotted in Figure 2 at a Q value of 5 GeV. The gluon distribution is dominant at x values of less than .01 with the valence quark distributions dominant at higher x.

4 Evolution in time and Q^2

As discussed in the introduction, the MRS and CTEQ groups provide semi-regular updates to their parton distributions as new data and/or theory becomes available. The latest parton distributions are the most accurate and should be used in preference to previous pdf's. However, in some cases calculations using older pdf's are necessary; for example, none of the more recent pdf's are implemented in Pythia, and most comparisons in the ATLAS TDR are made with the CTEQ2L pdf (the default pdf in Pythia).

A comparison of the CTEQ1M[9], CTEQ2M[10], CTEQ3M[11] and CTEQ4M[3] parton distributions (in particular the up sea quark distribution) are shown in Figure 3, at a Q value of 5 GeV. The CTEQ2-4 up quark sea distributions are substantially steeper than that of CTEQ1, reflecting the influence of the HERA data. A similar effect is seen with the gluon distribution while, as expected, there is little change in the valence up quark distribution

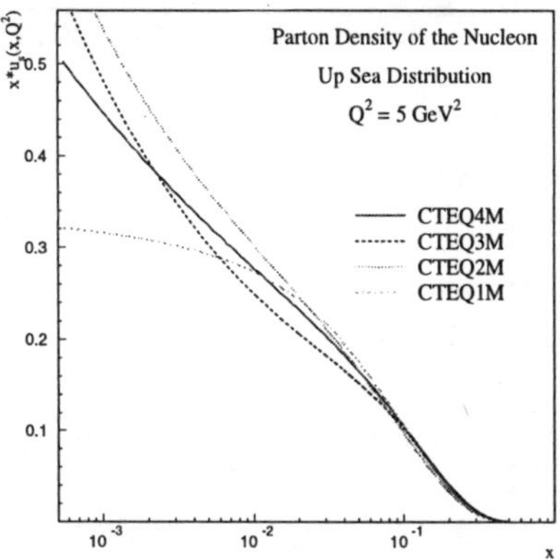

Figure3. The up sea quark parton distributions from the CTEQ1-4 sets plotted at a Q^2 value of 5 GeV^2.

The up sea quark distributions is shown in Figures, 4 at a larger Q^2 value of 10^4 GeV^2. Evolution has evened out many of the differences observed at lower Q^2 values. A Q^2 value of 10^4 GeV^2 corresponds to a mass scale at the LHC of about 100 GeV.

5 NLO and LO pdf's

As mentioned previously, pdf's are also available from leading order fits to the same data sets used in the NLO fits. For many hard matrix elements for processes used in the global analysis, there exist K factors (NLO/LO) signficantly different from unity. Thus, one expects there to be comparable differences of the LO parton distributions from the NLO ones.

The differences between NLO and LO parton distributions are not that great for many pdf's in many regions of x, however, and again tend to shrink at higher Q^2.

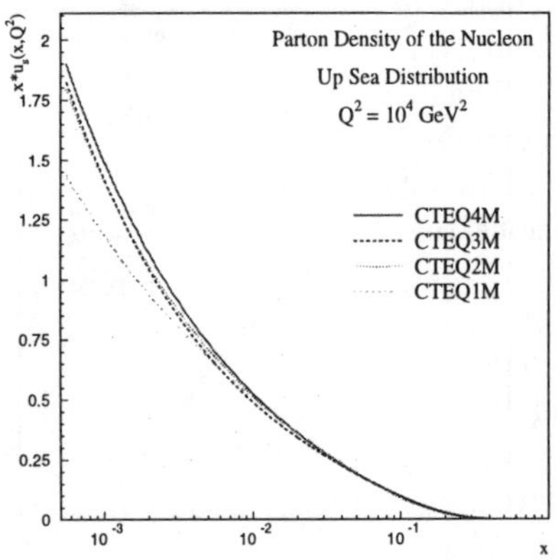

Figure4. The up sea quark parton distributions from the CTEQ1-4 sets plotted at a Q^2 value of $10^4\ GeV^2$.

6 Uncertainties on pdf's

In addition to having the best estimates for the values of the pdf's in a given kinematic range, it is also important to understand the allowed range of variation of the pdf's, i.e. their uncertainties. The conventional method of estimating parton distribution uncertainties is to compare different published parton distributions. This is unreliable since most published sets of parton distributions (for example from CTEQ and MRS) adopt similar assumptions and the differences between the sets do not fully explore the uncertainties that actually exist. Ideally, one might hope to perform a full error analysis and provide an error correlation matrix for all the parton distributions. (See for example, Ref.[12].) This goal is an admirable one but is difficult to carry out for two reasons. Experimentally, only a subset of the experiments usually involved in global analyses provide correlation information on their data sets in a way suitable for the analysis. Even more important, there is no established way of quantifying the theoretical uncertainties for the diverse physical processes that are used and uncertainties due to specific choices of parameterizations. Both of these are highly correlated.

The sum of the quark distributions $(\Sigma(q(x) + \overline{q}(x))$ is, in general, well-determined over a wide range of x and Q^2. As stated above, the quark distributions are predominantly determined by the DIS and DY data sets which have large statistics, and systematic errors in the few percent range ($\pm$3% for $10^{-4} < x < 0.75$). Thus the sum of the quark distributions is basically known to a similar accuracy. The individual quark flavors, though, may have a greater uncertainty than the sum. This can be important, for example, in predicting distributions that depend on specific quark flavors, like the W asymmetry distribution [13] and the W rapidity distribution.

Information on the $\bar{d}$ and $\bar{u}$ distributions comes, at small x, from HERA and at medium from fixed target DY production on H_2 and D_2 targets. It is now well-established [14,15] that the $\bar{d}$ and $\bar{u}$ distributions are not the same. The difference between the CTEQ4M and CTEQ5M pdf's is due primarily to the influence of the data from the E866 experiment. It is worth noting that our detailed knowledge of $\bar{d}/\bar{u}$ is limited primarily to the x region (.03-.35) covered by E866.

The strange quark sea is determined from dimuon production in ν DIS (CCFR[16]), with the strange quark distribution ($s + \bar{s}$) being approximately $1/2$ ($\bar{u} + \bar{d}$). The charm and bottom quark distributions are calculated perturbatively from gluon splitting for given masses of m_c and m_b. (See also the previous discussion on schemes.)

Current information on d/u at large x comes from fixed target DY productin on H_2 and D_2 and the lepton asymmetry in W production at the Tevatron. In the CTEQ5 and MRST fits, the NMC D_2/H_2 data are used to constrain the large x d quark distribution in this way. Bodek and Yang have argued that the D_2 data need to be corrected for nuclear binding effects, which would lead to a larger d/u ratio at large x (and thus a larger d quark distribution as the u quark distribution is well-determined from DIS). [17] The larger d quark distribution would lead to an increase in the high E_T Tevatron jet cross section of about 10%. A similar excess would be expected for high E_T jet production at the LHC.

The largest uncertainty of any parton distribution, however, is that on the gluon distribution. The gluon distribution can be determined indirectly at low x by measuring the scaling violations in the quark distributions ($\partial F_2/\partial log Q^2$), but a direct measurement is necessary at moderate to high x. Direct photon production has long been regarded as potentially the most useful source of information on the gluon distribution with fixed target direct photon data, especially from the experiment WA70[18], being used in a number of global analyses. However, as will be discussed in Section VII, there are a number of theoretical complications with the use of direct photon data.

The LHC is essentially a gluon-gluon collider and many hadron-collider signatures of physics both within and beyond that Standard Model involve gluons in the initial state. Thus, it is important to estimate the theoretical uncertainty due to the uncertainty in the gluon distribution.

The momentum fraction of the proton carried by quarks is determined very well from DIS data; at a Q_o value of 1.6 GeV, in the CTEQ4 analysis for example, the momentum fraction carried by quarks is 58% with an uncertainty of $\pm 2\%$. Thus, the momentum fraction carried by gluons is 42% with a similar uncertainty. This constraint is important; if the gluon distribution increases in one x range, momentum conservation forces it to decrease in another x range. Thus, if the gluon flux in the x range from 0.01 to 0.3 were to decrease by 20%, the gluon flux would have to increase by a fairly dramatic amount in the other x ranges to compensate. For example, if this compensation were to come in the high x region, the gluon distribution would have to double.

An alternative approach, to those described above, for estimating the uncertainty on the gluon distribution is to systematically vary the gluon parameters in a global analysis and then look for incompatibilities with the data sets that make up the global analysis database. This study has been recently carried out by CTEQ using only DIS and Drell-Yan data where the theoretical and experimental systematic errors are under good control.[19] [c] Except at larger values of $x(x > 0.2 - 0.3)$, the variation in the gluon distributions is less than 15% at low values of Q, decreasing to less than 10% at high values.[d] Note that the DIS and DY datasets used in this analysis do not provide any strong constraints on the gluon distribution at high values of x. This study used the CTEQ4 value of α_s (0.1116). If α_s is varied in the range from 0.113 to 0.122, the gluon distribution varies by 3% for $x < 0.15$.

In order to assess the range of predictions on physics cross sections, it is more important to know the uncertainties on the gluon-gluon and gluon-quark luminosity functions at the appropriate kinematic region (in $\tau = x_1 x_2 = \hat{s}/s$) rather than on the parton distributions themselves. Therefore it is useful to define the relevant integrated parton-parton luminosity functions. The gluon-gluon luminosity function can be defined as: $\tau dL/d\tau = \int_\tau^1 G(x, Q^2) G(\tau/x, Q^2) dx/x$.

This quantity is directly proportional to the cross section for s-channel production of a single particle and it also gives a good estimate for more complicated production mechanisms. Here, Q^2 is taken to be τs, which naturally takes the Q^2 dependence of the gluon distribution into account as one changes $\sqrt{\tau}$. Above an x value of 0.1, the allowed variation grows dramatically; this

---

[c]All of the parton distribution functions used in the calculation of the gluon uncertainty are available from the CTEQ web site, http://www.phys.psu.edu/+cteq/.

[d]As noted earlier, evolution is the great equalizer for parton distributions.

Table1. The parton-parton luminosity uncertainty as a function of $\sqrt{\tau}$.

$\sqrt{\tau}$ range	gluon-gluon	gluon-quark
< 0.1	±10%	±10%
0.1 − 0.2	±20%	±10%
0.2 − 0.3	±30%	±15%
0.3 − 0.4	±60%	±20%

indicates the need for more information about the gluon distribution at large x than provided by the DIS and DY data sets used in this analysis.

In analogy with the discussion of gluon-gluon luminosities, one can also study the gluon-quark luminosity (again normalized to the CTEQ4M result). The uncertainties on the parton-parton luminosities, as a function of $\sqrt{\tau}$, are summarized in Table 1. Note that the region of production of a 100-140 GeV Higgs at the LHC lies in the region where the range of variation in the gg luminosity is ±10%.

7 Direct Photons and Jets in Global Fits

7.1 Direct Photons

As mentioned previously, direct photon production has long been viewed as an ideal vehicle for measuring the gluon distribution in the proton. The quark-gluon Compton scattering subprocess ($gq \rightarrow \gamma q$) dominates photon production in all kinematic regions of pp scattering, as well as for low to moderate values of parton momentum fraction x in $\bar{p}p$ scattering. As mentioned in the previous section, the gluon distribution is relatively well constrained at low $x(x < 0.1)$ by DIS and DY data, but less so at higher x. Consequently, fixed target direct photon data have been incorporated in several modern global parton distribution function analyses with the hope of providing a major constraint on the gluon distribution at moderate to high x.

A pattern of deviations of direct photon data from NLO predictions has been observed[20], however, with the deviations being particularly striking for the E706 experiment. The origin of the deviations lies in the effects of initial state soft gluon radiation, or k_T.[e] Direct evidence of this k_T has long been evident from Drell-Yan, diphoton and heavy quark measurements. The values of $< k_T >$/parton for these processes vary from 1 GeV/c at fixed target energies to 3-4 GeV/c at the Tevatron Collider. The growth is approximately

---

[e]I should add that this view is not universally held; see for example Reference[21].

140

logarithmic with center of mass energy. (The value expected at the LHC for relatively low mass states (30-40 GeV/c^2) is in the range of 6.5-7.0 GeV/c.

Perturbative QCD corrections are insufficient to explain the size of the observed k_T and full resummation calculations are required to explain Drell-Yan, W/Z and diphoton distributions.[22] These resummation calculations qualitatively describe the growth of the $< k_T >$ with center-of-mass energy. Currently, however, there is no rigorous k_T resummation calculation available for single photon production. The calculation is quite challenging in that the final state parton takes part in soft gluon emission and in color exchange with initial state partons, in contrast with the Drell-Yan and diphoton cases. Also, the calculation is complicated by the fact that several overlapping power-suppressed corrections can contribute and, at high x, threshold effects are important. Nevertheless, there has been recent theoretical progress.[23,24,25]

In lieu of a rigorous calculation of the resummed direct photon p_T distribution, the effects of soft gluon radiation can be approximated by a convolution of the NLO cross section with a Gaussian k_T smearing function. The value of $< k_T >$ to be used for each kinematic regime should be taken directly from relevant experimental observables, given the lack of a rigorous formalism, rather than from a theoretical prediction. The behavior of the k_T smearing correction is quite different for the Tevatron collider and for fixed target experiments. For the Tevatron, there are two points to note: (1) the agreement with the data is improved if the k_T correction is taken into account and (2) the k_T smearing effects fall off roughly as $1/p_T^2$.[26] The latter behavior is the expectation for such a power-suppressed type of effect and is the behavior expected at the LHC, where the effects of the k_T smearing should not be important past p_T values of 30 GeV/c.^{f}

The k_T correction obtained for E706 at a center-of- mass energy of 31.6 GeV is shown in Figure 5. The value of $< k_T >$ of 1.2 GeV/c was obtained from measurements of several kinematic observables in the experiment.[26] The k_T smearing effect is much larger here then observed at the collider and does not have the $1/p_T^2$ falloff. Also shown are the k_T corrrections using values of $< k_T >$ of 1.0 and 1.4 GeV/c (a reasonable estimate of the range of experimental uncertainty in the $< k_T >$ determination). Also shown is is k_T correction for the E706 data used in the recent MRST pdf's. The MRST k_T correction is larger which leads to a smaller gluon distribution in the relevant x range. (Both the CTEQ4 and MRST pdf's, with their respective k_T corrections, lead to good agreement with the E706 direct photon cross sections.)

---

^{f}Similar k_T smearing effects should be present in all hard scattering cross sections, as for example, jet production at the Tevatron. The size of the experimental and theoretical systematic errors in the low E_T region make such a confirmation difficult.

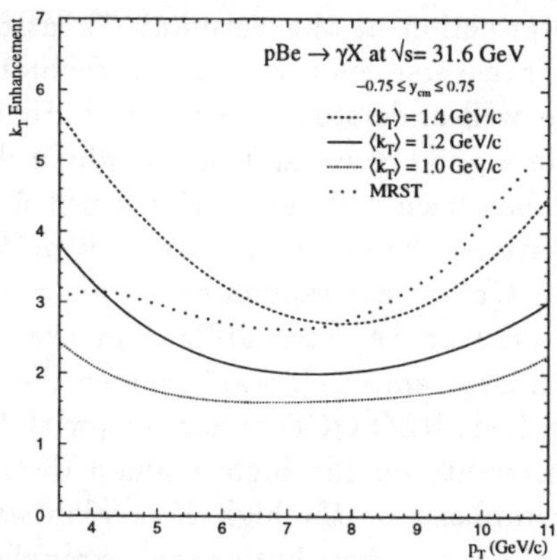

Figure5. The variation of k_T enhancements (ratio of cross sections with and without the k_T corrections) relevant to E706 direct photon data at 31.6 GeV, for different values of average k_T. In addition, the k_T correction for E706 used in the recent MRST fit is indicated.

The differences between the k_T correction from Reference[26] and that from the MRST pdf's can be taken as an indication of the uncertainty in the value of this correction. Good agreement with the E706 direct photon and π^o cross sections at $\sqrt{s}$ is observed when the nominal k_T correction of 1.2 GeV/c is used; however, the allowed range of variation of $\langle k_T \rangle$(1.0-1.4 GeV/c) makes quantitative comparisons, and thus an extraction of the gluon distribution, difficult.[g] Since the high p_T E706 data agrees well with CTEQ4M, it would thus disfavor the CTEQ4HJ pdf; as stated before, however, a definitive conclusion must await a more rigorous theoretical treatment.

7.2 Influence of Jets

An important process that is sensitive to the gluon distribution is jet production in hadron-hadron collisions. Processes responsible for jet production include gluon-gluon, gluon-quark and quark-quark(or anti-quark) scattering.

---

[g]NLO QCD predictions for fixed-target direct photon production also contain a non-negligible renormalization and factorization scale dependence; for example, changing the scale from $Q = p_T/2$ to $Q = p_T$ typically results in a decrease in the cross section of about 30%.

Precise data on jet production at the Fermilab Tevatron are now available over a wide range of transverse energy, and the theoretical uncertainties in most of this range are well-understood. Thus, it is to be expected that jet production can provide a good constraint on the gluon distribution

The jet data that has been utilized in global pdf fits has been from the CDF and D0 collaborations. [h] The data cover a wide kinematic range (E_T values from 15 to 450 GeV/c corresponding to an x range of .02 to 0.5). The CDF jet data from Run IA were utilized in the CTEQ4HJ pdf fit.[27] In the CTEQ4HJ fit, a large emphasis was given to the high E_T data points which show a deviation from NLO QCD predictions with "conventional" pdf's. Given the lack of constraints on the high x gluon distribution discussed in Section VI, the extra emphasis on the high E_T region was enough to cause a significant increase in the gluon distribution; for example, the gluon distribution at an x value of 0.5 (Q=100 GeV) increases by a factor of 2. Since the dominant jet subprocess in this region is $\bar{q}q$ scattering the increase in the gluon distribution of a factor of 2 causes only a 20% increase in the jet cross section. This is sufficient to pass through the bottom of the CDF high E_T jet error bars. The preliminary jet cross sections from Run 1B(90 pb^{-1}) from both the CDF and D0 experiments were used in the CTEQ4M fits, but with statistical errors only and only for the E_T range of 50-200 GeV/c. The E_T points lower than 50 GeV/c have substantial systematic errors on both the theoretical and experimental sides and the E_T points higher than 200 GeV/ccontain the CDF excess at high E_T. The inclusion of the jet data serves to considerably constrain the gluon distribution over the x range of 0.1 to 0.2. The resulting gluon (CTEQ4M) does not decrease the excess observed in CDF at high E_T.

The published D0 jet cross section[28] along with the (soon-to-be published) CDF jet cross section[29] from Run 1B were used in the recently released CTEQ5 parton distributions. The fits use the full E_T range for the cross sections and use the correlation information on the systematic errors as contained in the covariance matrices for both experiments. The two experiments are in agreement with each other except for a slight normalization shift[i]; the two highest E_T data points for CDF are above those for D0, but both experiments have large statistical errors in this region. As can be seen in Figure 6 the NLO QCD prediction with the CTEQ5M pdf is in good agreement with the CDF data. The conclusions are exactly the same for the D0 jet data, as well. The CTEQ5M gluon is very similar to CTEQ4M, except

---

[h]The experimental and theoretical errors associated with the UA2 jet cross section make its use in pdf fits difficult.

[i]A shift on the order of 3% is expected since the two experiments use values for the total inelastic cross section that differ by that amount.

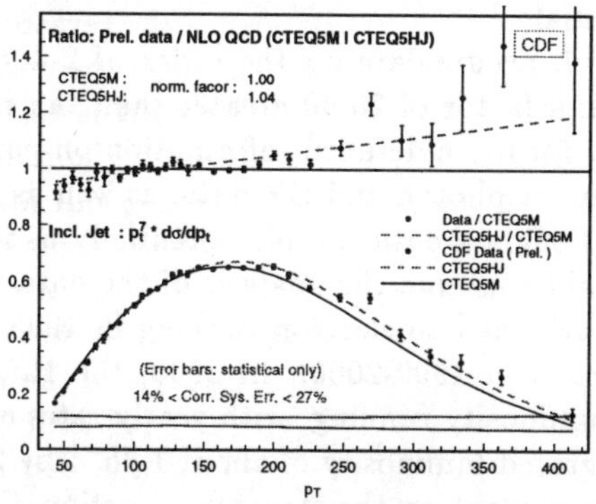

Figure6. A comparison of the Run 1B CDF inclusive jet cross section to the CTEQ5 fits. The bottom plot shows the measured cross section multiplied by p_T^7 in order to allow a linear display. The top plot shows the ratio of the measured cross section to that calculated with CTEQ5M, as well as the ratios of CTEQ5HJ to CTEQ5M.

perhaps at very high x. The CTEQ4HJ pdf has been updated to complement the new CTEQ5M pdf. The CTEQ5HJ pdf gives almost as good a global fit as CTEQ5M to the full set of data on DIS and DY processes, and has the feature that the gluon distribution is significantly enhanced in the high x region, resulting in improved agreement with the observed trend of jet data at high E_T in both the CDF and D0 experiments.

8 Progress Before the LHC Turns on

DGLAP-based perturbative QCD calculations have been extremely successful in describing data in DIS, DY and jet production, as well as describing the evolution of parton distributions over a wide range of x and Q^2. From the pdf point-of-view, the primary problem lies in the calculation of fixed target direct photon cross sections; they can serve as a primary probe of the gluon distribution at high x. However, rigorous theoretical treatment of soft gluon effects (perhaps requiring both k_T and Sudakov resummation) will be required before the data can be used with confidence in pdf fits.

Differential dijet data from the Tevatron explore a wider kinematic range than the inclusive jet cross section. Both CDF and D0 have dijet cross section measurements from Run I which may also serve probe the high x gluon distribution, in regions where new physics is not expected but where any parton distribution shifts should be observable. The ability to perform such cross-checks is essential.

CDF and D0 will accumulate on the order of 2-4 fb^{-1} of data in Run II (from 2000-2003), a factor of 20-40 greater than the current sample. This sample should allow for more detailed information on parton distributions to be extracted from direct photon and DY data, as well as from jet production. Run III (2003-2007) offers a data sample potentially as large as $30\,fb^{-1}$.

H1 and ZEUS will continue the analysis of the data taken with positrons in 1991-97. HERA switched to electron running in 1998 and plans to deliver approximately 60 pb^{-1} in 1999-2000. In 2000, the HERA machine will be upgraded for high luminosity running, with yearly rates of 150 pb^{-1} expected, allowing for an integrated luminosity of about 1 fb^{-1} by 2005. This will allow for an error of a few percent on the structure function F_2 for Q^2 scales up to $10^4 GeV^2$. The gluon density, derived from scaling violations of F_2, should be known to an accuracy of less than 3% in the kinematic range $10^{-4} < x < 10^{-1}$.

9 Physics cross sections at the LHC and the role of LHC data in pdf determination

ATLAS measurements of DY (including W and Z), direct photon, jet and top production will be extremely useful in determining pdf's relevant for the LHC. The data can be input to the global fitting programs, where it will serve to confirm/constrain the pdf's in the LHC range. Again, DY production will provide information on the quark (and anti-quark) distributions while direct photon, jet and top production will provide, in addition, information on the gluon distribution.

Another possibility that has been suggested is to directly determine parton-parton luminosities (and not the parton distributions per se) by measuring well-known processes such as W/Z production.[30] This technique would not only determine the product of parton distributions in the relevant kinematic range but would also eliminate the difficult measurement of the proton-proton luminosity. It may be more pragmatic, though, to continue to separate out the measurements of parton pdf's (through global analyses which may contain LHC data) and of the proton-proton luminosity. The measurement of the latter quantity can be pegged to well-known cross sections, such as that of the W/Z, as has been suggested for the Tevatron.

References

1. H.L. Lai, J. Huston et al., **hep-ph/9903282**.

2. A.D. Martin, R.G. Roberts, W.J. Stirling and R. Thorne, *Eur. Phys. J.* **C4**, 463 (1998).

3. H.L. Lai, J. Huston et al., *Phys. Rev.* **D55**, 1280 (1997), **hep-ph/9606399**.

4. See, for example, the plenary talk given by Y. Dokshitzer at the ICHEP conference in Vancouver, **hep-ph/9812252**.

5. S. Kuhlmann, W.-K. Tung and H. L. Lai, *Phys. Lett.* **409B**, 271 (1997)

6. M. Gluck, E. Reya, A. Vogt, *Eur. Phys. J.* **C5**, 461 (1998).

7. G. Marchesini et al., **hep-ph/9607393**.

8. T. Sjostrand, **hep-ph/9508391**.

9. J. Botts et al., *Phys. Lett.* **304B**, 159 (1993).

10. CTEQ internal report (unpublished).

11. H.L. Lai, J. Huston et al., *Phys. Rev.* **D51**, 4763 (1995).

12. W.T. Giele, Stephane Keller, *Phys. Rev.* **D58**, (1998), **hep-ph/9803393**. Science.

13. CDF Collaboration, F. Abe et al., *Phys. Rev. Lett.* **81**, 5754 (1998), **hep-ex/9809001**.

14. A. Baldit et al., *Phys. Lett.* **332B**, 244 (1994).

15. E.A. Hawker et al., *Phys. Rev. Lett.*80, 3715 (1998), **hep-ex/9803011**.

16. W. G. Seligman et al., *Phys. Rev. Lett.* **79**, 1213 (1997).

17. U.K. Yang, A. Bodek, **hep-ph/9809480**, submitted to *Phys. Rev. Lett.*

18. M. Bonesini et al., *Z. Phys.* **C38**, 371(1988); *ibid.* **C37**,535(1988); *ibid.* **C37**, 39.

19. J. Huston et al., *Phys. Rev.* **D58** (1998), **hep-ph/9801444**.

20. J. Huston et al., *Phys. Rev.* **D51**, 6139 (1995), **hep-ph/9501230**.

21. P. Aurenche et al., **hep-ph/9811382**.

22. C. Balazs, C.-P. Yuan, *Phys. Rev.* **D56**, 5558 (1997) and references therein.

23. E. Laenen, G. Oderda, G. Sterman, *Phys. Lett.* **438B**, 173(1998), **hep-ph/9806467**.

24. S. Catani, M. Mangano, P. Nason, **hep-ph/9806487**.

25. H-n. Li, hep-ph/9812363; H-n. Li, **hep-ph/9811340**.

26. L. Apanasevich, J. Huston et al., *Phys. Rev. Lett.* **59** (1999), **hep-ph/9808467**.

27. J.Huston et al., *Phys.Rev.Let.* **77**, 444(1996).

28. B. Abbott et al., **hep-ex/9807018**.

29. F. Bedeschi, talk at 1999 Hadron Collider Physics Conference, Bombay, India, January, 1999.

30. M. Dittmar, F. Pauss, D. Zuercher, *Phys. Rev.* **D56**, 7284; **hep-ex/9705004**.

SPIN STRUCTURE OF THE NUCLEON

N.C.R. MAKINS

*University of Illinois at Urbana-Champaign, 1110 W Green St, Urbana IL 61801,
USA*
E-mail: makins@uiuc.edu

This report summarizes recent progress in the physics of nucleon spin structure,
including results from experiments at DESY, CERN, and SLAC. First, the global
analysis of the inclusive spin-structure functions $g_1^{p(n)}(x)$ by the SMC collaboration
is described. Next, new measurements of the flavour-separated quark polarizations
from HERMES are presented, based on the measurement of semi-inclusive spin
asymmetries. Finally, novel measurements are presented from HERMES and SMC,
of single-spin asymmetries in semi-inclusive deep inelastic scattering, and of the
double spin asymmetry in diffractive vector meson production.

1 Introduction

Deep-inelastic scattering of polarized leptons from polarized targets has been
used successfully for more than a decade to yield information about the spin
structure of the nucleon. The central question at issue is the manner in which
the partonic components of the nucleon manage to produce its overall spin of
1/2. The expression

$$\frac{1}{2} = \frac{1}{2}\Delta\Sigma + \Delta g + L_q + L_g \tag{1}$$

illustrates that the total spin of the nucleon must arise from a combination of
three sources: the spin distribution of the quarks ($\Delta\Sigma$), the spin distribution
of the gluons (Δg), and the angular momentum of quarks and gluons (L_q, L_g).
In a naive SU(3)-symmetric parton picture of the nucleon, the spins of the
valence quarks provide the only non-zero contributions, with $\Delta u = +4/3$ and
$\Delta d = -1/3$. However, in 1988 the EMC experiment at CERN measured the
spin-dependent inclusive structure function g_1^p of the proton, and found that
when taken together with information from hyperon decay and interpreted in
an SU(3)-symmetric context, their measurement yielded a total quark spin
contribution $\Delta\Sigma$ that was consistent with zero. Since then, this "spin crisis"
has been explored by further experiments at CERN, SLAC, and DESY. Analyses of these measurements conclude that the contribution of the quark spins
is non-zero but accounts for less than half of the nucleon spin, at scales of
$Q^2 \simeq 1 \text{ GeV}^2$.

2 Inclusive Measurements

The principal goal of the experiments described above is to determine the longitudinal polarization of the quarks in the nucleon. In polarized DIS, a longitudinally polarized lepton beam scatters from quarks in the target nucleons via the exchange of polarized virtual photons. Because of helicity conservation, only quarks whose spins are aligned *against* that of the virtual photon may absorb the photon. Thus, by aligning the polarization of the target nucleons alternately along and against the direction of the polarized beam, one may "image" the momentum distributions of quarks with a particular helicity. These distributions are the polarized counterpart of the familiar parton distribution functions, and are denoted $\Delta q(x) \equiv q^\uparrow(x) + \bar{q}^\uparrow(x) - q^\downarrow(x) - \bar{q}^\downarrow(x)$. Here, x is the Bjorken scaling variable, representing the fraction of the nucleon's momentum carried by the quark, and $q^\uparrow(q^\downarrow)$ indicates quarks whose spins are parallel (anti-parallel) to the spin of the nucleon.

In this section, we describe the status of inclusive measurements, where only the scattered beam lepton is detected. By flipping the orientations of the beam and/or target spins, one obtains the experimental asymmetry $A_\parallel$:

$$A_\parallel = \frac{N^{\uparrow\downarrow} L^{\uparrow\uparrow} - N^{\uparrow\uparrow} L^{\uparrow\downarrow}}{N^{\uparrow\downarrow} L_P^{\uparrow\uparrow} + N^{\uparrow\uparrow} L_P^{\uparrow\downarrow}} \qquad (2)$$

The pairs of arrows in this expression refer to the relative orientations of the beam and target polarizations. For each spin arrangement, N denotes the number of events collected, L is the integrated luminosity, and L_P is the integrated luminosity weighted by the product of beam and target polarizations.

The structure function ratio g_1/F_1 is related to $A_\parallel$ via the relation

$$\frac{g_1}{F_1} = \frac{1}{1+\gamma^2}\left[\frac{A_\parallel}{D} + (\gamma - \eta)A_2\right]. \qquad (3)$$

Here, F_1 is the familiar unpolarized structure function, D is the virtual photon depolarization factor, and γ and η are kinematical factors [1]. The magnitude of the transverse virtual photon absorption asymmetry A_2^p has been measured previously to be negligible [2]. In leading order QCD and the quark-parton model, the spin structure function g_1 is given by the charge weighted sum over the polarised quark spin distributions Δq_f for flavour f:

$$g_1(x) = \frac{1}{2}\sum_f e_f^2 \Delta q_f(x). \qquad (4)$$

A summary of the published world data on $g_1(x)$ is presented in Figure 1a. This figure does not include the precise but as yet unpublished data on g_1^p

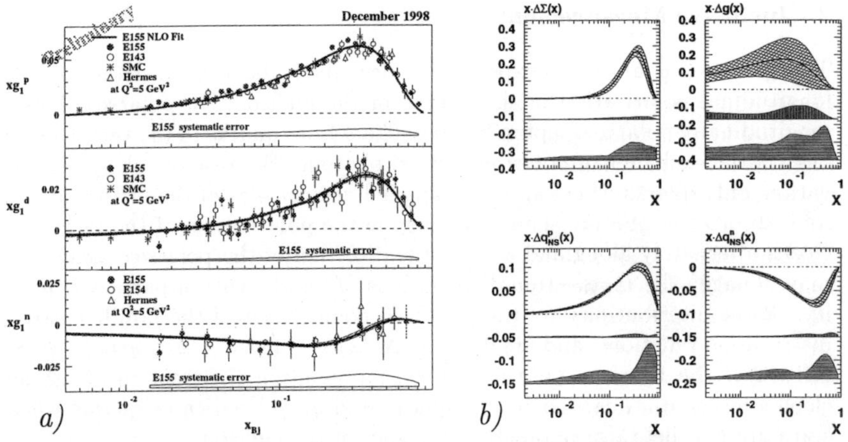

Figure 1. (a) Summary of published world data on the spin structure function $g_1(x)$. The data on g_1^p from the E155 experiment are as yet only preliminary. (b) Polarized parton distribution functions at $Q_i^2 = 1$ GeV^2 determined from the NLO analysis of [7].

from the SLAC experiment E155 [3].

The first analyses of EMC's original measurement of $g_1^p(x)$ concentrated on its first moment $\Gamma_1^p(Q^2)$ [4]:

$$\Gamma_1^{p(n)}(Q^2) \equiv \int_0^1 g_1^{p(n)}(x, Q^2)dx = \frac{1}{9}\left(C_{NS}\left[\pm\frac{3}{4}a_3 + \frac{1}{4}a_8\right] + C_S a_0\right) \quad (5)$$

Here, a_0 and a_3, a_8 are the singlet and nonsinglet nucleon matrix elements of the axial current, while the coefficient functions C_S and C_{NS} describe the QCD evolution of these quantities. The symbol Q^2 denotes the negative square of the 4-momentum of the virtual photon. The nonsinglet matrix elements are related to the quark helicity distributions in a straightforward manner, and assuming SU(3) flavour symmetry, can be obtained from the F and D parameters describing the weak decay of hyperons [6]:

$$a_3 = \Delta q_3 \equiv \Delta u - \Delta d = F + D = 1.2601 \pm 0.0025, \quad (6)$$

$$a_8 = \Delta q_8 \equiv \Delta u + \Delta d - 2\Delta s = 3F - D = 0.579 \pm 0.032. \quad (7)$$

The singlet matrix element a_0 may then be extracted from the first moment of $g_1(x)$. Now a_0 itself, like a_3 and a_8, is a physical, scheme-independent quantity. However, its interpretation in terms of the parton helicity distributions is not quite so straightforward. In the $\overline{\text{MS}}$ scheme, a_0 is simply identified with the fits moment of the singlet quark spin distribution:

$$a_0 = \Delta\Sigma \equiv \Delta u + \Delta d + \Delta s. \tag{8}$$

It should be noted that in this scheme, $\Delta\Sigma$ is a scale-dependent quantity. The well-known analysis of Ellis and Karliner [5] of the first moment of g_1 was performed in this scheme, and yielded $\Delta u = 0.83 \pm 0.03$, $\Delta d = -0.43 \pm 0.03$, $\Delta s = -0.10 \pm 0.03$, and so $\Delta\Sigma = 0.31 \pm 0.07$, at a renormalization scale of $Q^2 = 10$ GeV^2. The rather surprising conclusions were that the quark spins account for less than a third of the nucleon spin, and that the strange sea has a small but distinctly negative polarization, and are often collectively referred to as the "Spin Puzzle".

The Spin Muon Collaboration has recently published a global analysis of the world's data on $g_1(x, Q^2)$, in the framework of next-to-leading order (NLO) QCD [7]. This analysis seeks to obtain information about the polarizations of the individual quark flavours by exploiting the different Q^2 dependence of the various contributions to $g_1(x, Q^2)$:

$$g_1^{p(n)} = \frac{1}{9}\left(C_{NS} \otimes \left[\pm\frac{3}{4}\Delta q_3 + \frac{1}{4}\Delta q_8\right] + C_S \otimes \Delta\Sigma + 2N_f C_g \otimes \Delta g\right). \tag{9}$$

The symbols C_S, C_{NS}, and C_g denote the coefficient functions describing the evolution of the singlet, non-singlet, and gluon terms respectively. Note the appearance of the gluon spin distribution Δg in this expression at next-to-leading order. The SMC analysis is performed in both the $\overline{\text{MS}}$ and AB schemes. In the former scheme, the first moment of the gluon coefficient function C_g is equal to zero. This is not true in the AB scheme, where the gluon polarization Δg contributes directly to the integral $\int_0^1 g_1(x)dx$ via the axial anomaly. The integral of the singlet quark distribution $\Delta\Sigma$ is thus related differently to the singlet matrix element in the two schemes:

$$a_0 = \Delta\Sigma_{\overline{\text{MS}}} = \Delta\Sigma_{\text{AB}} - n_f \frac{\alpha_s(Q^2)}{2\pi}\Delta g(Q^2). \tag{10}$$

The non-singlet moments Δq_3 and Δq_8 are constrained to the known F and D values (Eq. (7)), under the assumption of SU(3) symmetry. The quantity n_f represents the number of active quark flavours.

Figure 1b shows the polarized parton distributions determined from the SMC analysis, at a scale of $Q_i^2 = 1$ GeV^2. This analysis is distinguished by a careful treatment of uncertainties. The upper curve and cross-hatched band in each panel of the figure represent the fitted distributions and accompanying statistical errors. The middle, vertically hatched bands represent the experimental systematic errors, while the lower, horizontally hatched bands indicate the theoretical uncertainty. The latter includes sensitivity to variations of factorization and renormalization scales, and uncertainties in the value of α_s, the funtional form of the initial parton distribution functions, the values of quark mass thresholds, and the value of g_A/g_V.

Some of the moments derived from this analysis are summarized in Table 1. The value of Δs has been obtained using the singlet moment $\Delta\Sigma$ and the value of Δq_8 from Eq. (7). The table shows that the basic conclusions of this analysis are similar to those of [5]: the quark spins account for a relatively small fraction of the total nucleon spin, and the strange sea would seem to have a negative polarization. Further, the gluon polarization Δg, which was extracted only in the AB scheme, is clearly only poorly constrained by the inclusive data.

Table 1. First moments determined from the analysis of [7], evaluated at $Q^2 = 1$ GeV^2.

	$\overline{\mathrm{MS}}$ scheme	AB scheme
a_0	$0.19 \pm 0.05 \pm 0.04$	$0.23 \pm 0.07 \pm 0.19$
$\Delta\Sigma$	$0.19 \pm 0.05 \pm 0.04$	$0.38^{+0.03+0.03+0.03}_{-0.03-0.02-0.05}$
Δg	-	$0.99^{+1.17+0.42+1.43}_{-0.31-0.22-0.45}$
Δs	-0.13 ± 0.02	-0.07 ± 0.02

3 Quark Polarizations from Semi-Inclusive Measurements

To proceed further in our knowledge of the polarized parton distributions, experimenters have turned to *semi-inclusive* asymmetry measurements. In these measurements, a hadron is detected in coincidence with the scattered beam lepton. Through the agency of the fragmentation functions, a probabilistic relation exists between the flavour of the struck quark and the flavour content of the hadrons generated in the final state. By measuring the semi-inclusive production of hadrons of various types, one has the possibility to separate the parton distributions by flavour, with a sensitivity beyond that which is possible with inclusive measurements alone.

Hadron production in DIS is described by the absorption of a virtual photon by a point-like quark and the subsequent fragmentation into a hadronic final state. The two processes can be characterized by the quark distribution function $q_f(x, Q^2)$, and the fragmentation function $D_f^h(z, Q^2)$. Here, z represents the energy fraction E_h/ν carried by the detected hadron. The semi-inclusive DIS cross section $\sigma^h(x, Q^2, z)$ to produce a hadron of type h

with energy fraction $z = E_h/\nu$ is given by:

$$\sigma^h(x, Q^2, z) \propto \sum_f e_f^2 q_f(x, Q^2) D_f^h(z, Q^2). \tag{11}$$

In this case, the sum is over quark and antiquark flavours $f = (u, \overline{u}, d, \overline{d}, s, \overline{s})$. One obtains the following expression for the longitudinal spin asymmetry:

$$A_1^h(x, Q^2, z) = \frac{\sum_f e_f^2 \Delta q_f(x, Q^2) D_f^h(z, Q^2)}{\sum_f e_f^2 q_f(x, Q^2) D_f^h(z, Q^2)} \frac{(1 + R(x, Q^2))}{(1 + \gamma^2)} \tag{12}$$

Semi-inclusive asymmetries for positive and negative hadron production have been measured and analysed by both the HERMES [8] and SMC experiments [9]. We describe here the method employed by HERMES to extract the polarized quark distributions, based on measurements from both p and ^{3}He targets. Eq. (12) can be written as

$$A_1^h(x) = \sum_f P_f^h(x) \frac{\Delta q_f(x)}{q_f(x)} \frac{(1 + R(x))}{(1 + \gamma^2)} \tag{13}$$

where $P_f^h(x)$ are the integrated purities defined as

$$P_f^h(x) = \frac{e_f^2 q_f(x) \int_{0.2}^1 D_f^h(z)\, dz}{\sum_{f'} e_{f'}^2 q_{f'}(x) \int_{0.2}^1 D_{f'}^h(z')\, dz'}. \tag{14}$$

The inclusive asymmetry A_1 can be expressed in a similar way. Combining the expressions for the various asymmetries in matrix form, one can write

$$\vec{A}(x) = \mathcal{P}(x) \cdot \vec{Q}(x) \tag{15}$$

where the vector $\vec{A} = (A_{1p}, A_{1p}^{h^+}, A_{1p}^{h^-}, A_{1He}, A_{1He}^{h^+}, A_{1He}^{h^-})$ contains as elements the measured asymmetries. The vector $\vec{Q}(x)$ contains the quark and antiquark polarizations.

The matrix $\mathcal{P}$ contains the integrated purities for the two targets, as well as the $(1 + R)/(1 + \gamma^2)$ factor and corrections for nuclear effects in ^{3}He. Physically, these purities describe the conditional probability that the virtual photon was absorbed by a quark of flavor f when a hadron of type h is detected in the experiment. Note that the purities depend only on *spin-independent* functions ($q_f(x)$ and $D_f^h(z)$). They further include the effects of the acceptance of the experiment, and have been determined with a Monte Carlo simulation based on the LUND string fragmentation model. The LUND fragmentation parameters were tuned to fit the hadron multiplicities measured at HERMES.

Eq. (15) can be solved for $\vec{Q}(x)$ by a χ^2-minimization procedure, taking into account the correlated uncertainties between the different components of the asymmetry vector $\vec{A}$. In the fit procedure, constraints were imposed relating the various flavour components of the sea polarization to improve statistical significance. Two alternative were considered. As a first possibility it was assumed that the polarization $\Delta q_s(x)/q_s(x)$ of sea quarks is independent of flavor:

$$\frac{\Delta u_s(x)}{u_s(x)} = \frac{\Delta d_s(x)}{d_s(x)} = \frac{\Delta s(x)}{s(x)} = \frac{\Delta \bar{u}(x)}{\bar{u}(x)} = \frac{\Delta \bar{d}(x)}{\bar{d}(x)} = \frac{\Delta \bar{s}(x)}{\bar{s}(x)}. \tag{16}$$

This approach is used for all calculations unless otherwise stated. As a second approach, a pure singlet spin distribution of the sea is considered:

$$\Delta u_s(x) = \Delta d_s(x) = \Delta s(x) = \Delta \bar{u}(x) = \Delta \bar{d}(x) = \Delta \bar{s}(x). \tag{17}$$

The flavor decomposition is obtained by solving Eq. (15) for a vector $\vec{Q}$, which contains the sum of quarks and antiquarks

$$\vec{Q} = \left(\frac{\Delta u(x) + \Delta \bar{u}(x)}{u(x) + \bar{u}(x)}, \frac{\Delta d(x) + \Delta \bar{d}(x)}{d(x) + \bar{d}(x)}, \frac{\Delta s(x) + \Delta \bar{s}(x)}{s(x) + \bar{s}(x)} \right), \tag{18}$$

where due to the assumption of Eq. (16) the polarizations of the strange quarks and of the total sea are equal: $(\Delta s(x) + \Delta \bar{s}(x))/(s(x) + \bar{s}(x)) = \Delta q_s(x)/q_s(x)$.

Figure 2 shows the HERMES and SMC results for the extracted valence $(\Delta u_v, \Delta d_v)$ and sea (Δq_s) quark polarizations. All data points are evolved to a common Q^2 of 2.5 GeV/c^2. The solid lines indicate the positivity limit $|\Delta q(x)| \leq q(x)$, while the dashed lines show the 'Gluon A' leading order parametrization of Gehrmann and Stirling [10]. Finally, Table 2 presents the integrals of the distributions extracted by HERMES. The results are compared with the the SU(3)-symmetric analysis of inclusive data by Ellis and Karliner [5]. The results of the two analyses are clearly in agreement.

HERMES is presently collecting data using a deuterium target, with an operational RICH detector in the spectrometer. The addition of the new data

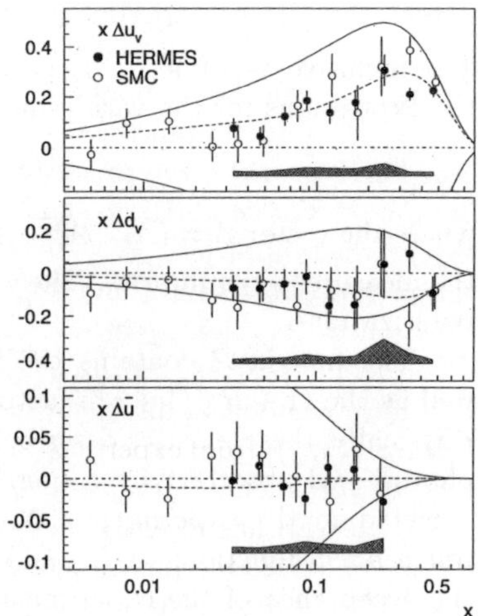

Figure 2. Polarized valence and sea quark distributions extracted from semi-inclusive data.

will significantly improve the uncertainties on the d-valence and sea quark polarizations. In addition, the presence of the RICH detector enables the measurement of charged kaon asymmetries, and it is anticipated that these asymmetries will yield a first separation of the strange sea polarization that is not subject to the assumptions of SU(3) symmetry made by the analyses of inclusive data.

Table 2. First moments of quark polarizations determined from HERMES semi-inclusive measurements. The prediction column refers to the SU(3)-symmetric analysis [5] of inclusive data. All values are evolved to a scale of $Q^2 = 2.5$ GeV/c^2.

	total integral	SU(3) prediction
$\Delta u + \Delta \overline{u}$	0.56 ± 0.04	0.66 ± 0.03
$\Delta d + \Delta \overline{d}$	-0.25 ± 0.08	-0.35 ± 0.03
$\Delta s + \Delta \overline{s}$	-0.02 ± 0.05	-0.08 ± 0.02

4 Gluon Polarization from Pairs of High-p_T Hadrons

As described earlier, the gluon polarization is only poorly constrained by existing data. One way to measure $\Delta g(x)$ directly is via the photon gluon fusion process, where a gluon from the nucleon participates in the hard scattering vertex. Two useful experimental signatures of this process are charm production and the production of jets with high transverse momentum p_T. In the former case, the very small charm content of the nucleon and the supression of charm production in the fragmentation process (due the the large mass of the charm quark) serve to suppress the single-quark scattering diagram that dominates the production of light hadrons. A similar argument applies to the production of jets: the transverse momentum produced in the fragmentation process is small, and two back-to-back jets with sufficiently high p_T thus reflect the high p_T of the quark and anti-quark produced in the photon gluon fusion process. Both charm production and high-p_T jet production in DIS have resulted in direct measurements of the unpolarized gluon structure function $g(x_g)$, from experiments H1, ZEUS, and NMC. At lower energy fixed target experiments, high-p_T hadrons must serve in place of jets [11].

HERMES has recently made a first measurement of the spin asymmetry in the photoproduction of pairs of high-p_T hadrons. Events were selected that contained at least one positively charged hadron h^+ and at least one negatively charged hadron h^-. The observation of the scattered positron was not required, in order to include the very low Q^2 region which dominates the cross section. The highest momentum hadron of each charge was required to

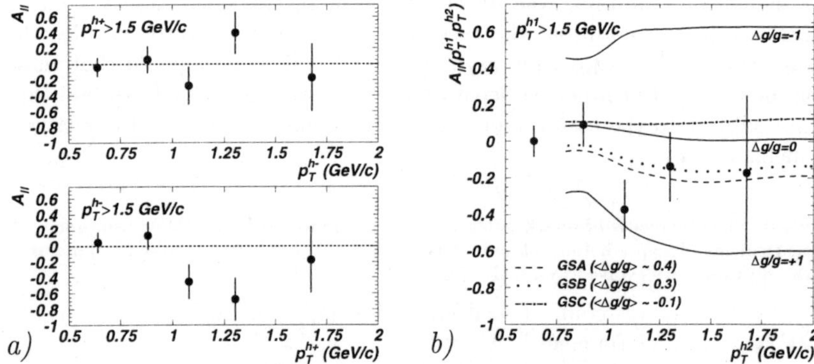

Figure 3. (a) HERMES measurement of $A_{||}(p_T^{h^+}, p_T^{h^-})$ for $p_T^{h^+} > 1.5$ GeV/c (top) and for $p_T^{h^-} > 1.5$ GeV/c (bottom). Note that the rightmost data point is identical in both panels. (b) $A_{||}$ for high-p_T hadron production measured at HERMES compared with Monte Carlo predictions for $\Delta g/g = \pm 1$ (lower/upper solid curves), $\Delta g/g = 0$ (middle solid curve), and the phenomenological LO QCD fits of [10] (dashed, dotted, and dot-dashed curves).

have a momentum above 4.5 GeV/c and a transverse momentum p_T above 0.5 GeV/c. Here p_T is defined as the momentum transverse to the positron beam direction, which is approximately equal to the momentum transverse to the photon direction when $Q^2 \approx 0$. To suppress contributions from vector meson resonances in the data sample, a minimum value of the invariant mass of the two hadrons (assuming both hadrons to be pions) $M(2\pi) > 1.0$ GeV/c^2 was imposed.

Figure 3a presents the measured $A_{||}$ for the highest values of transverse momenta accessible at HERMES; in the top (bottom) panel the positive (negative) hadron was required to have a p_T greater than 1.5 GeV/c and the asymmetry $A_{||}$ is then plotted as a function of the p_T of the hadron of opposite charge. The data suggest a more negative asymmetry when the transverse momentum of the negative hadron is higher than that of the positive hadron. Ignoring this charge asymmetry and averaging over the five bins satisfying the requirement $p_T^{h_1} > 1.5$ GeV/c and $p_T^{h_2} > 1.0$ GeV/c, a negative asymmetry $A_{||} = -0.28 \pm 0.12$ (stat.) ± 0.02 (syst.) is observed. (The symbol h_1 signifies the hadron with the higher p_T.) When the requirement $p_T^{h_1} > 1.5$ GeV/c is not enforced, the asymmetry is consistent with zero. The observed negative asymmetry is in contrast to the positive asymmetries typically measured in deep inelastic scattering from protons.

The measured asymmetry was interpreted using the PYTHIA Monte Carlo generator at leading order. In this model, several different processes contribute to the two-hadron cross section: lowest order deep inelastic scattering (containing no hard QCD vertex), interaction via the hadronic structure of the photon – described by the vector meson dominance model (VMD) and by non-resonant hadronic "anomalous" photon structure, and the two first-order QCD processes (termed "direct") that describe the interaction of a pointlike photon. These are photon gluon fusion (PGF) and the QCD Compton effect (QCDC).

The contribution from lowest order DIS is suppressed by the requirement of high p_T, and was confirmed to be negligible by a simulation based on the LEPTO Monte Carlo generator. Contributions from VMD were assumed to have a negligible spin asymmetry, and were thus treated as a dilution of the other asymmetries. Finally, we neglect possible contributions from anomalous photon structure, where the photon fluctuates into a non-resonant $q\bar{q}$ pair that interacts via hard processes with the partons inside the nucleon. This assumption is supported by the model of [12].

Under the assumptions described above, only the two direct processes contribute significantly to the measured asymmetry:

$$A_{||} \approx \left(\hat{a}_{\mathrm{PGF}} \frac{\Delta g}{g} f_{\mathrm{PGF}} + \hat{a}_{\mathrm{QCDC}} \frac{\Delta q}{q} f_{\mathrm{QCDC}} \right) D. \tag{19}$$

(The kinematic dependences have been suppressed for brevity.) The subprocess asymmetries $\hat{a}$ for the PGF and QCDC processes are directly calculable in LO QCD. The effective quark polarization $\Delta q/q$ is computed as a suitably weighted combination of $\Delta u/u$ and $\Delta d/d$, which are relatively well known from inclusive and semi-inclusive polarized DIS measurements (as described above). Finally, the fractional contributions f of each process to the total cross section were determined from the PYTHIA Monte Carlo simulation.

In the same region of phase space where a negative asymmetry is observed ($p_T^{h_1} > 1.5$ GeV/c and $p_T^{h_2} > 1.0$ GeV/c), the simulated cross section is dominated by photon gluon fusion. The consequent sensitivity of the measured asymmetry to the polarized gluon distribution is demonstrated in Figure 3b, where $A_{||}$ at high transverse momenta (i.e. the average of the two panels of Figure 3a) is compared with Monte Carlo predictions for different distributions of $\Delta g/g$.

For the four values of $A_{||}$ at $p_T^{h_2} > 0.8$ GeV/c presented in Figure 3b, $\langle \Delta g/g \rangle$ was extracted according to Eq. (19). Since these four measurements probed essentially the same range of x_g, the results for $\langle \Delta g/g \rangle$ were averaged. Using the assumptions and model parameters described above, a value for

$\langle \Delta g/g \rangle$, at an average $\langle x_g \rangle = 0.17$, was determined in LO QCD to be 0.41 ± 0.18 (stat.) ± 0.03 (syst.), where the systematic uncertainty represents the experimental contribution only.

It must be pointed out that this measurement is subject to theoretical uncertainties of an as-yet undetermined magnitude. At the kinematics of this measurement, no spin-dependent analyses of higher order QCD processes or contributions from anomalous photon structure are presently available; these processes have therefore been neglected in the model presented here. If such processes would be important but have no significant spin asymmetry, the extracted value of $\langle \Delta g/g \rangle$ would increase, but still differ from zero by 2.3σ. To alter the principal conclusion of this analysis, i.e., that $\langle \Delta g/g \rangle$ at $\langle x_g \rangle = 0.17$ is positive, a significant contribution from a neglected process with a large negative spin asymmetry would be needed.

5 Spin Azimuthal Asymmetries

As discussed above, a significant body of data has been collected on the spin structure function $g_1(x)$, which represents the distribution of quark helicities in a longitudinally polarized nucleon. Additional spin-distribution functions have been identified, but remain unmeasured. One of these, $h_1(x)$, is called transversity, and corresponds to the distribution $\delta q(x)$ of *transverse* quark spin in a nucleon polarized transverse to its (infinite) momentum [13]. In the absence of relativistic effects, $\delta q(x)$ is simply equal to the longitudinal helicity distribution $\Delta q(x)$. Transversity, and other related distribution functions, are predicted to be measurable via semi-inclusive single-spin asymmetries, where only the beam *or* target are polarized.

Mulders and Tangerman have performed a complete tree-level analysis of semi-inclusive deep-inelastic leptoproduction, including polarization of the beam, target hadron, and final state hadron [14]. As part of this work, they have identified and classified a complete set of nucleon structure functions, at the twist 2 and twist 3 levels. Figure 4a provides a summary of these functions at twist 2, and serves to illustrate graphically the operators represented by each of the functions. A completely analogous list of fragmentation functions also exists.

The familiar unpolarized and longitudinally polarized distribution functions appear as f_1 and g_{1L} respectively; the transversity distribution is denoted h_{1T}. The remaining functions in general represent a step in sophstication beyond the naive picture of collinear, non-interacting quarks that is commonly used in the description of nucleon structure. First of all, only the three distribution functions f_1, g_{1L}, and h_{1T} survive on integration over the

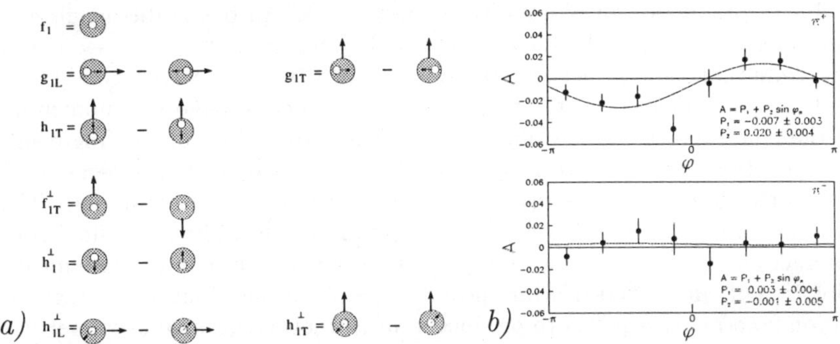

Figure 4. (a) Summary of the classification scheme of [14] for twist-2 distribution functions, and graphical representation of the operators associated with each function. The larger (shaded) circles and arrows represent the spin orientation of initial/final state hadrons, while the inner (open) circles represent the hadron's partonic constituents. The virtual photon momentum is understood to be pointing from left to right. (b) HERMES measurement of the single spin asymmetry of pion production from a longitudinally polarized proton target, plotted against the azimuthal angle ϕ of the pion relative to the lepton scattering plane.

transverse momentum component of the struck quark. The remaining functions are innately sensitive to intrisic k_t in the nucleon. All functions whose symbols contain the letter h are *chiral-odd*, and describe the transverse orientation of the quark spins. Since the strong interaction is chiral-even, DIS cross section terms involving a chiral-odd distribution function always involve a chiral-odd fragmentation function as well. The transversity distribution h_1 is such a function. One consequence of its chiral-odd nature is that, unlike in the case of g_1, the polarized gluon distribution does not mix with the polarized quark distributions. The scheme-dependent complications described earlier concerning the interpretation of $\Delta\Sigma$ are thus not present in the case of the transversity distribution. The distribution functions $f_{1T}^{\perp}$ and $h_1^{\perp}$ also have the unusual property of being odd under time-reversal, as a result of initial state interactions. As illustrated in Figure 4a, these functions relate transversely polarized partons (hadrons) to unpolarized hadrons (partons). Finally, the higher twist functions (not shown in Figure 4a) are sensitive to parton-parton correlations.

Transversity and related spin-distribution functions are as yet unmeasured because their unusual chiral-odd structure implies that they are not directly observable in inclusive lepton-nucleon scattering experiments. However, it has been suggested that the needed sensitivity can be provided by the semi-inclusive production of pions with modest $p_\perp$ [15]. An observable single-spin dependence is predicted to appear in the dependence of the cross section on the azimuthal angle ϕ_c between the spin axis of a *transversely* polarized target and the plane defined by the virtual photon momentum and the momentum of the pion (known as the Collins angle). The chiral-odd 'Collins' fragmentation function involved here is denoted $H_1^\perp$ in the classification scheme of [14]. (It is analogous to the distribution function $h_1^\perp$ depicted in Figure 4a). This function also has a time-reversal odd structure, resulting from the final-state interactions in the fragmentation process rather than from any fundamental violation of time-reversal invariance [15]. One T-odd distribution or fragmentation function is required to produce a measurable single-spin asymmetry in semi-inclusive DIS.

In the case of semi-inclusive pion production from a *longitudinally* polarized nucleon, chiral-odd quark spin-distribution functions closely related to transversity can be manifest. In such experiments, the Collins angle becomes the azimuthal angle ϕ of the pion around the virtual photon direction, with respect to the lepton scattering plane. Analyses such as those of [14] and [16] have shown how each chiral-odd spin-distribution function coupled with the Collins fragmentation function $H_1^\perp$ gives rise to a specific single-spin dependent moment of the pion yield distribution in ϕ.

HERMES has recently made the first measurement of single-spin asymmetries for semi-inclusive pion production in DIS, extracting the $\sin\phi$ and $\sin 2\phi$ moments of π^+ and π^- production using an unpolarized beam and a longitudinally polarized proton target. The ϕ dependence of the target asymmetry is shown in Figure 4b. The plotted quantity is defined as follows:

$$A(\phi) = \frac{N^+(\phi)L^+ - N^-(\phi)L^-}{N^+(\phi)L_P^+ + N^-(\phi)L_P^-},\tag{20}$$

where $N^\pm(\phi)$ are the numbers of pions collected in each ϕ bin for the two target spin states, and $L^\pm$ and $L_P^\pm$ are defined as in Eq. (2). A clear $\sin\phi$ dependence is observed for π^+, but not for π^-. The average $\sin\phi$ moment of the asymmetry for positive pions is $0.022 \pm 0.005\text{(stat.)} \pm 0.003\text{(sys.)}$.

When plotted vs x, the extracted $\sin\phi$ moment for π^+ is found to rise almost linearly from $x \simeq 0.05$ to $x \simeq 0.3$. Similarly, the moment increases with $p_\perp$ of the detected pion, from a value consistent with zero at low $p_\perp$ to a maximum value around $p_\perp = 1$ GeV/c. These kinematic dependences were anticipated by Collins [15]. He predicted that the asymmetry should be largest in the valence region ($x \gtrsim 0.3$), and that it should peak at moderate $p_\perp$ values, corresponding to typical k_T values of a few hundred MeV to about one GeV. Furthermore, he expected that the largest asymmetry from a proton target would be observed for the production of *positive* pions, because of the dominance of scattering from the positively polarized u quarks. In general, the appearance of this asymmetry can be interpreted as an effect of chiral-odd spin distribution functions coupled with a time-reversal-odd fragmentation function. The fact that it is measurably non-zero in scattering from a *longitudinally* polarized target is most encouraging for future measurements of transversity with a transversely polarized target, as are planned by both the HERMES and COMPASS experiments.

We point out that the SMC experiment has recently produced a measurement of the Collins angle asymmetry, using *transversely* polarized proton and deuteron targets [18]. However, the statistical precision of the measurement is able to resolve asymmetries only at the level of 0.1, and the extracted $\sin \phi_c$ azimuthal moments are consistent with zero.

6 Double-Spin Asymmetry in Diffractive Vector Meson Production

In the scattering of lepton beams from nuclear targets, a virtual photon is exchanged. In this process, there is the possibility that the photon fluctuates into a virtual vector meson state, carrying the same $J^{PC} = 1^{--}$ quantum numbers as the photon itself. (This hadronic fluctuation is often described by the Vector Meson Dominance model, termed VMD). The hadronic state may then be scattered *diffractively* from the target, becoming a real meson in the final state and leaving the target nucleon more-or-less intact.

The spin dependence of the diffractive scattering of vector mesons has to date been limited to the analysis of the spin-density matrix elements of the process. Broadly, these matrix elements describe the transition amplitudes between the various possible spin states of the virtual photon and those of the final state vector meson. One common hypothesis that is explored by these measurements is that of s-channel helicity conservation (SCHC), which postulates simply that the helicity of the virtual photon is transferred without change to that of the vector meson. So far, only a small deviation from this hypothesis has been observed [21].

Until this year, the *double-spin* asymmetry of diffractive vector meson production has remained unexplored. Successful analyses of unpolarized vector meson production in terms of Pomeron exchange suggest that this asymmetry should be zero: the Pomeron is a spinless quasi-particle with vacuum quantum numbers, and so processes involving its exchange may be expected to have no significant spin dependence.

However, HERMES has recently measured this asymmetry, for the diffractive production of ρ^0 mesons from a longitudinally polarized hydrogen target, and has obtained a non-zero result [19]. When averaged over the full kinematic range of the measurement, an asymmetry A_1^ρ of 0.30 ± 0.11(stat.)±0.04(sys.)±0.05(theor.) is found, which deviates from zero by 2.4 σ. No significant dependence of the asymmetry has been found on the kinematic variables Q^2, x, W^2, and t'. Here, W^2 is the square of the invariant mass of the photon-target system, and t' is a measure of the transverse momentum of the ρ meson.

The SMC experiment has also produced a recent measurement of A_1^ρ [20]. Unlike the HERMES measurement, the SMC results are consistent with an asymmetry of zero. Averaged over the kinematics of the measurement, and over the polarized proton and deuterium targets employed, an asymmetry of $A_1^\rho = -0.18 \pm 0.11$ is obtained.

As yet, there is little theoretical guidance on the possible interpretation of a non-zero spin asymmetry in the diffractive production of the ρ^0 meson. One interpretation may be in terms of off-forward parton distributions (OFPD) [22]. The HERMES measurement lies in the transitional kinematic region $4 \leq W^2 \leq 6$ GeV^2, where the exchange of meson-like Reggeons believed to be dominant at lower energies yields to the dominance of Pomeron exchange at higher energies. Interpreted in an OFPD framework, the exchange of quark (gluon) pairs is expected to dominate at low (high) energies [23]. The SMC measurement of a zero asymmetry by comparison lies in the higher-energy region $\langle W \rangle \simeq 15$ GeV.

7 Summary

The past year has seen significant advances in the experimental exploration of spin-dependent processes. New information from HERMES on semi-inclusive spin asymmetries in DIS has confirmed the primary conclusion from earlier analyses of inclusive data: the u quarks in the proton carry a large positive polarization, while the d quarks are negatively polarized. The semi-inclusive data on the sea quark polarization favour a value of zero, but are not yet of sufficient statistical precision to challenge the small negative strange-quark polarization indicated by the earlier analyses and by a recent NLO QCD analysis of the world's data on g_1 from the SMC experiment. Next, HERMES has made a first measurement of the asymmetry in photoproduction of pairs of hadrons at high transverse momentum. A negative asymmetry was found, which yields a positive value for the gluon polarization when intepreted in a LO QCD framework. HERMES and SMC have also explored single spin asymmetries in semi-inclusive hadron production, and extracted azimuthal moments of these asymmetries. These novel measurements are expected to be sensitive to the as-yet unmeasured transversity distribution $h_1(x)$, and to time-reversal-odd effects in the fragmentation process. HERMES has observed a distinctly non-zero effect, which is encouraging for future dedicated measurements of transversity planned by both the HERMES and COMPASS experiments. Finally, the first measurements of the double spin asymmetry of diffractive vector meson production have been made by HERMES and SMC. The lower-energy HERMES result is distinctly positive, while the higher-energy SMC measurement favours a value of zero. The HERMES result is presently without theoretical explanation.

References

1. HERMES Collab., A. Airapetian *et al.*, Phys. Lett. B 442 (1998) 484.
2. E143 Collab., K. Abe *et al.*, Phys. Rev. D 58 (1998) 112003.
3. G.S. Mitchell, hep-ex/9903055.
4. EMC, J. Ashman *et al.*, Nucl. Phys. B 328 (1989) 1.
5. J. Ellis and M. Karliner, Phys. Lett. B 341 (1995) 397.
6. F.E. Close and R.G. Roberts, Phys. Lett. B 316 (1993) 165.
7. SMC, B. Adeva *et al.*, Phys. Rev. D 58 (1998) 112002.
8. HERMES Collab., K. Ackerstaff *et al.*, Phys. Lett. B 464 (1999) 123.
9. SMC, B. Adeva *et al.*, Phys. Lett. B 420 (1998) 180.
10. T. Gehrmann and W.J. Stirling, Phys. Rev. D 53 (1996) 6100.
11. A. Bravar, D. von Harrach, A. Kotzinian, Phys. Lett. B 421 (1998) 349.
12. G. A. Schuler, T. Sjöstrand, hep-ph/9403393 and CERN-TH-7193-94;
    G. A. Schuler, T. Sjöstrand, Nucl. Phys. B 407 (1993) 539.
13. J.P. Ralston and P.E. Soper, Nucl. Phys. B 152, 109 (1979);
    R. Jaffe and X. Ji, Nucl. Phys. B 375, 527 (1992).
14. P.J. Mulders and R.D. Tangerman, Nucl. Phys. B 461, 197 (1996).
15. J.C. Collins, Nucl. Phys. B 396 (1993) 161.
16. A.M. Kotzinian, Nucl. Phys. B 441, 234 (1995).
17. HERMES Collab., A. Airapetian *et al.*, hep-ex/9910062 (1999).
18. A. Bravar, in: J.Blümlein, T.Riemann (Eds.), Proc. of the 7th International Workshop on Deep Inelastic Scattering and QCD, Zeuthen, Germany, Nucl. Phys. B (Proc. Suppl.) in press.
19. F. Meissner, *ibid.*
20. A. Tripet, *ibid.*
21. H1 Collab., C. Adloff *et al.*, hep-ex/99902019;
    ZEUS Collab., J. Breitweg *et al.*, hep-ex/9908026.
22. B. Lehmann-Dronke *et al.*, hep-ph/9901283.
23. M. Vanderhhaegen *et al.*, Phys. Rev. Lett. 80 (1998) 5064.

Recent QCD results at LEP

Stan Bentvelsen

CERN

CH-1211 Genève 23 Switserland

E-mail: stan.bentvelsen@cern.ch

In this note we discuss a selection of recent experimental results on QCD at LEP, as obtained by the Aleph, Delphi, L3 and Opal collaborations. First, we discuss QCD in e^+e^- annihilations at the highest available energies. This includes measurements of event shape distributions, charged particle related distributions, determination of the strong coupling constant and a discussion on power corrections. Secondly we discuss experimentally obtained differences in quark versus gluon initiated jets. Various methods to determine the colour factor ratio C_A/C_F are discussed. Some implications for a colour-reconnection model are given.

1 Introduction

LEP, the large electron positron storage ring at CERN near Geneva, is in operation for ten years now. During its first seven years, LEP operated at centre-of-mass (c.m.) energies around the Z^0 pole (LEP-1). In this period, about 4 million of multi-hadronic events have been recorded per experiment. This large amount of data allowed detailed tests (and confirmation) of many aspects of the Standard Model. Since 1995, the c.m. energy at LEP has increased. First, in a series of initial runs, the c.m. energy went up to 130-136 GeV. After that, from 1996, the c.m. energy increased to 161 GeV and higher (LEP-2), and the threshold for W^+W^--pair production was passed. In 1998 the c.m. energy reached 189 GeV, and a luminosity of about 190 pb^{-1} per experiment, corresponding to roughly 3000 multi-hadronic events, was recorded during that year. This provides the largest data sample at LEP-2 sofar[a].

Those samples of e^+e^- annihilation data offer unique possebility to study various aspects of the theory of strong interactions, QCD. In section 2 we discuss QCD at the highest energies, discuss event generators and event selection, determine α_s using the event shapes and present the particle multiplicity. In section 3 we give a short overview of power corrections.

Utilizing the large amount of data at the Z^0 scale at LEP-1, one can study the structure of jet-production with high precision, and probe extreme corners of phase space. In section 4 we will discuss properties of jets, initiated by quarks or by gluons. We will describe experimental techniques to select topologically equivalent quark and gluon jets, and make comparisons between

---

[a]At the time of writing, the LEP energy increased from 196 to 200 GeV.

them. We present an analysis of 'gluon-inclusive hemispheres', which allow direct and unambiguous comparisons between quarks and gluons. This offers the best technique so far to make detailed and useful comparison to theoretical calculations.

2 QCD at the highest energies

Many aspects of QCD analysis benefit from the large lever arm in energy between the LEP-1 and LEP-2 data. Due to non-perturbative long range effects, QCD cannot predict the absolute value for many observables but does predict their energy evolution behaviour. Experimentally one benifits from having the identical experimental apparatus in the wide energy range between LEP-1 and LEP-2 energies and hence a large fraction of systematic uncertainties cancel when studying evolution of QCD quantities.

The four experiments at LEP all perform the basic QCD tests that can be done with limited statistics [1,2]. This includes the determination of event shapes, tests of MC generator description, determination of the strong coupling α_s, and measurements of particle spectra and charged multiplicity. We will briefly describe these measurements in the following.

2.1 Event Generators

Various models that simulate the complete process of $e^+e^- \to$ hadrons are available [3]. All of them produce initial s-channel $q\bar{q}$-pairs according to the Electro-Weak theory. Emission of hard gluon radiation is described by the QCD matrix element, which is subsequently followed by a parton shower. The parton shower represents an approximate perturbative treatment of QCD dynamics at scales of momentum greater than some infra-red cut-off, and it includes coherence effects. Below this scale, a non-perturbative model of the hadronization process takes place.

The generators that are used are JETSET [4], which combines the parton shower with string fragmentation. In this model, separating partons moving in opposite directions loose energy to the colour field, which is supposed to collapse into a stringlike configuration between them. The string then breaks up in to hadron-sized pieces through spontaneous $q\bar{q}$ pair production in its intense color field. The HERWIG [5] event generator implements the cluster model to hadronize the partons. After the parton shower phase, colour singlet clusters are formed of neigbouring partons. This is brought about through non-perturbative splitting of gluons in $q\bar{q}$ pairs. The ARIADNE [6] model uses a dipole model for the perturbative phase of the parton shower. The string model is

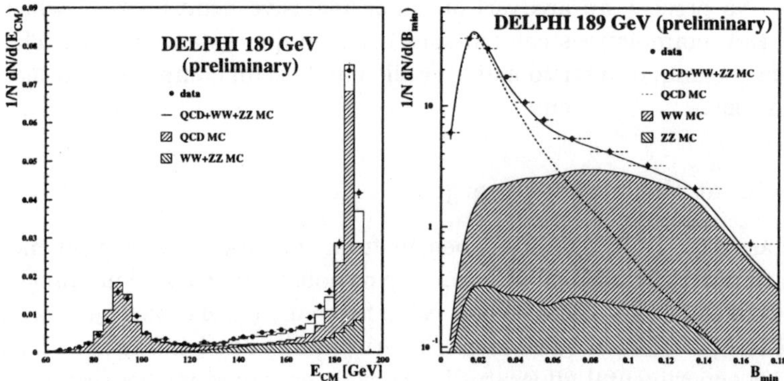

Figure 1: In (a) the c.m. energy of the hadronic system. In (b) the discriminant variable B_{min}, as used by DELPHI[2]. In both plots, the simulations are based on PYTHIA.

used for subsequent hadronization. The COJETS[7] generator is somewhat different as it has independent jet fragmentation and does not include coherence effects, although both aspects are prescribed by QCD.

All the event generators have a set of parameters to describe the event generation, which can be varied. Tuned parameter sets at $\sqrt{s} = M_{Z^0}$ provide good description of the large amount of hadronic decays recoreded at the Z^0 pole, for all the above mentioned generators.

2.2 Event Selection

The selection of events at LEP-2 of the type $e^+e^- \to (Z^0/\gamma)^* \to$ hadrons suffers from two main sorces of background. First, a fraction of events that are recorded have radiated hard photons in the initial state, such that the effective c.m. energy of the hadron system is reduced to $\sqrt{s} = M_{Z^0}$. Since the cross section at the pole $\sqrt{s} = M_{Z^0}$ is huge, this fraction is large and has to be removed from the sample. One generally search through the detector for isolated hard photons, signalling large initial state radiation. For the class of events for which the photon radiated down the beam-pipe, the larger fraction, one can determine the effective c.m. energy of the hadronic system using energy-momentum conservation, see e.g. Figure 1a.

The second class of events that are removed from the sample correspond to events of the type $e^+e^- \to$ 4-fermions (diagrams without intermediate gluons). These 4-fermion events with quarks and charged leptonic final states originate mainly from W^+W^- pair production, which is kinematically possible at c.m. energies above 161 GeV. For c.m. energies above 183 GeV, also $Z^0 Z^0$ pair production start to play a role. These type of events are rejected by requirements on variables that test the compability of events with either 4-fermion or '$(Z^0/\gamma)^{*}$' events. An example of a discriminating variable is shown in Figure 1b [2].

2.3 Event shapes

Event shape variables can be used to characterize hadonic final states in $e^+ e^-$ annihilations. Event shapes assign a number to an event with which particular properties of the hadrons is characterised. For example, event shapes determine whether the distribution of hadrons is pencill-like, planer, sperical etc. The distribution of event shapes is then compared to theoretical calculations. In order for this to be possible, the event shape should be infra-red safe, i.e. insensitive to the emission of soft or collineair gluons.

There are a number of event shape variables that are used. The definition of the most important ones, Thrust and Heavy Jet Mass, are:

Thrust T: This observable is defined by the expression [8]

$$T = \max_{\vec{n}} \left(\frac{\sum_i |p_i \cdot \vec{n}|}{\sum_i |p_i|} \right) \quad . \tag{1}$$

The thrust axis $\vec{n}_T$ is the direction $\vec{n}$ which maximises the expression in parentheses. A plane through the origin and perpendicular to $\vec{n}_T$ divides the event into two hemispheres H_1 and H_2.

Heavy Jet Mass M_H: The hemisphere invariant masses are calculated using the particles in the two hemispheres H_1 and H_2. We define M_H [9] as the heavier mass, divided by $\sqrt{s}$.

In the following, we use the symbol y to denote a generic event shape observable, where larger values of y indicate regions dominated by the radiation of hard gluons and small values of y indicate the region influenced by multiple soft gluon radiation. Note that thrust T forms an exception to this rule, as the value of T reaches one for events consisting of two pencil-like, back-to-back, jets. We therefore occasionally refer to $1 - T$.

Figure 2 show the distributions of the event shape observables $1 - T$, and M_H, corresponding to 189 GeV. With the statistics available at the 189 GeV sample, the event shapes are determined with high precision. However, no deviations from the Monte Carlo model predictions of PYTHIA, ARIADNE and

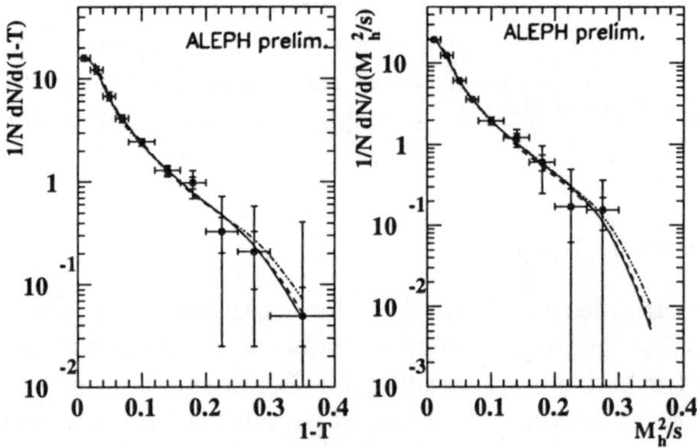

Figure 2: Distributions of the event shapes Thrust and Heavy jet Mass at 189 GeV as measured by the Aleph collaboration. The inner error bars denote the statistical, the outer error bars the systematic uncertainty. The data are compared to the PYTHIA (full line), HERWIG (dashed line) and ARIADNE (dotted line) Monte Carlo models.

HERWIG are observed. Both in the region of large and small values of y the descriptions are adequate. Note that all generators have been tuned at $\sqrt{s} = M_{Z^0}$, and are not re-tuned at these higher energies.

2.4 Determination of α_s

The coupling constant $\alpha_s(Q)$ of QCD depends on the energy scale Q of the process. In e^+e^- annihilation the scale Q is commonly identified with the c.m. energy, i.e. $Q = \sqrt{s}$. This so-called running of the strong coupling constant is a property fundamental to all renormalisable gauge theories. It is a particular feature of QCD that the effects of the running coupling are comparatively large, mainly because the value of the coupling is large. QCD is thus an example of a gauge theory which is well suited for tests of the running behaviour of the couplings in gauge theories in general. Summaries of experimental tests may be found e.g. in [10].

The running of the strong coupling α_s is described by the renormalisation group equation which states that results from a gauge theory should not depend on the energy scale at which the theory is renormalised. From this requirement one obtains a differential equation for the energy dependence of the strong coupling. At one loop level the differential equation and its solution for the

OPAL preliminary

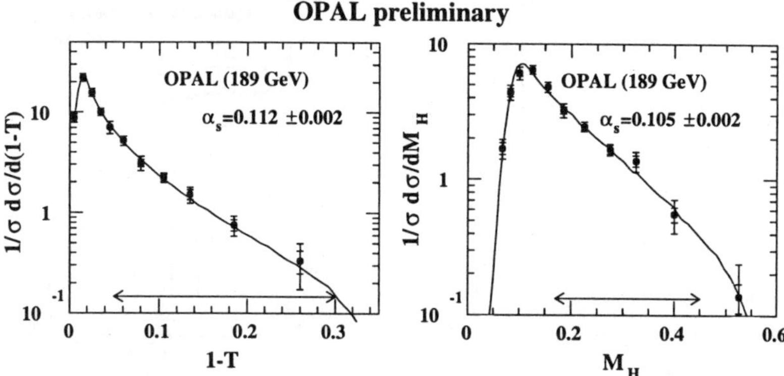

Figure 3: Distributions of the event shape observables thrust $1 - T$ and heavy jet mass M_H are shown together with fits of the $\mathcal{O}(\alpha_s^2)$+NLLA QCD predictions. The fitted regions are indicated by the arrows.

running of α_s read:

$$\mu^2 \frac{\partial \alpha_s(\mu)}{\partial \mu^2} = \beta_0 \alpha_s^2(\mu) \quad \rightarrow \quad \frac{\alpha_s(Q)}{\alpha_s(\mu)} = 1 + 2\beta_0 \alpha_s(Q) \ln(x_\mu) \quad . \tag{2}$$

Here μ is the energy at which the coupling is evaluated, $x_\mu = \mu/Q$ and the first order coefficient is $\beta_0 = (33 - 2n_f)/(12\pi)$ with $n_f = 5$ the number of active quark flavours.

Measurements of the strong coupling strength $\alpha_s(Q)$ are based on fits of the QCD predictions to the corrected distributions for event shapes like $1 - T$ and M_H, but also on the event shapes C, B_T, B_W and y_{23}^D. The theoretical descriptions of these observables are the most complete, allowing the use of combined $\mathcal{O}(\alpha_s^2)$+NLLA QCD calculations. The renormalization scale parameter, $x_\mu \equiv \mu/\sqrt{s}$, is fixed to $x_\mu = 1$, where μ is the energy scale at which the theory is renormalised. The data is corrected to the parton level before the fits to the QCD predictions are made.

All four experiments have performed such QCD fits to the event shapes[1,2]. The details of the fits vary between the experiments, but the resulting value of α_s is consistent between the experiments, as can be seen in Figure 4. The average value of α_s at 189 GeV between the experiments is given by

$$\alpha_s(189 \text{ GeV}) = 0.109 \pm 0.001(\text{stat.}) \pm 0.004(\text{syst.}). \tag{3}$$

When evolved to the scale of M_{Z^0}, using $\mathcal{O}(\alpha_s^3)$ calculations, the values of the weighted average for α_s become $\alpha_s(M_{Z^0}) = 0.121 \pm 0.005$. For comparison, the

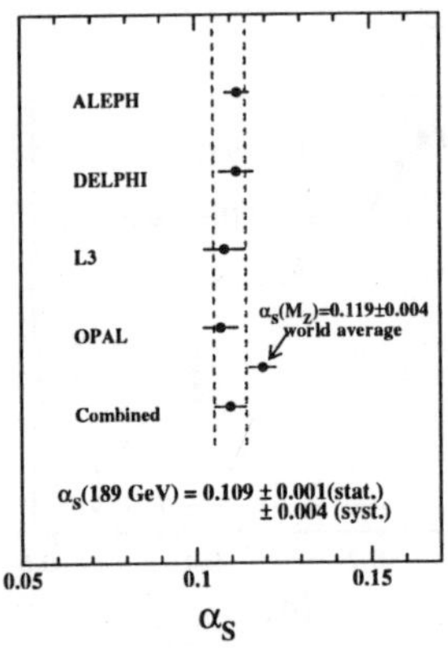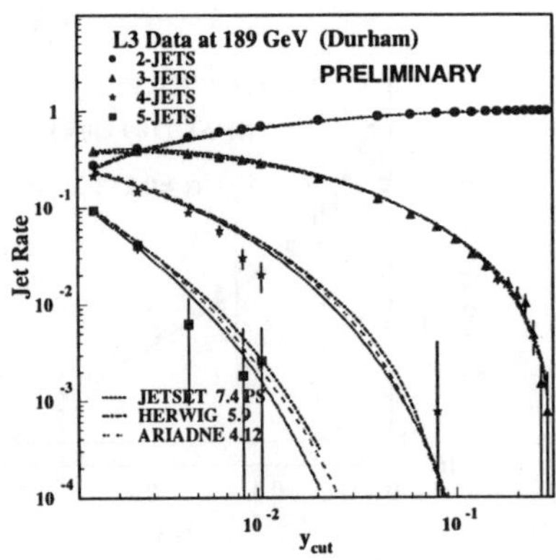

Figure 4: Values of α_s as determined by the four experiments at LEP. On the right hand side, the two-, three-, four- and fivejet rates are shown. The Durham jet clustering is used.

world average value of α_s at $\sqrt{s} = M_{Z^0}$ is equal to $\alpha_s(M_{Z^0}) = 0.119 \pm 0.004$. This indicates clearly that the value of α_s is measured to be lower at LEP-2 energies w.r.t. LEP-1, in excellent agreement to the predictions of QCD.

In Figure 5 the measurement of α_s is shown again, on a logarithmic scale, together with results obtained from the other experiments [10] at lower c.m. energies.

2.5 Jetrates and particle distributions

Jetrates probe intermediate regions between the 'hard' QCD, where radiation of highly energetic gluons dominate, and 'soft' QCD, where multiple gluon emission, each carrying a small fraction of the total momentum, dominates. Jetrates involve clustering of the final state, which is performed iteratively and is terminated at a specific resolution scale y_{cut}. By changing the value of y_{cut}, the final state is resolved into a varying number of jets. For very small values of y_{cut} the iterative process is not continued for long, and many jets are resolved. For large values of y_{cut} the iterative process is stopped when a only a few number of jets are left.

In Figure 4b the fraction of two, three, four and five jets are shown as function of resultion scale y_{cut}. For the whole range of y_{cut} the Monte Carlo described the data well, indicating that both 'hard' and 'soft' QCD are described well.

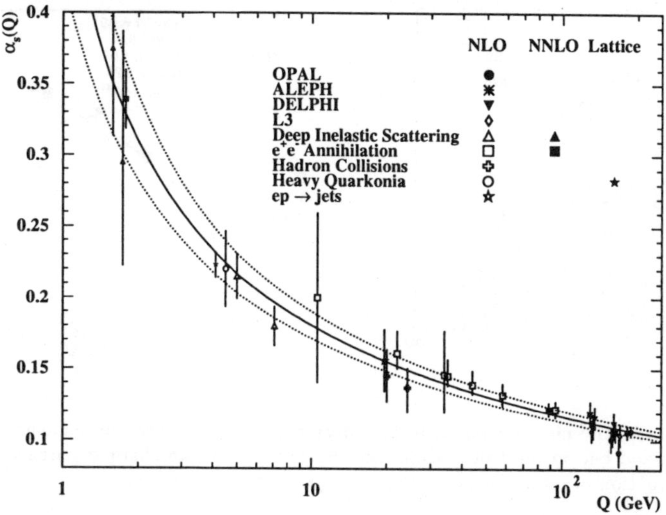

Figure 5: Values of α_s as function of energy. The labels NLO and NNLO refer to the order of calculation used. NLO corresponds to $\mathcal{O}(\alpha_s^2)$ in e^+e^- annihilations, and NNLO to $\mathcal{O}(\alpha_s^3)$. The label Lattice refers to α_s values determined from lattice QCD calculations. The curves show the $\mathcal{O}(\alpha_s^3)$ QCD prediction for $\alpha_s(Q)$ using $\alpha_s(M_{Z^0}) = 0.119 \pm 0.004$; the full line shows the central value while the dotted lines indicate the variation given by the uncertainty.

The dynamics of hadron production in 'soft' QCD can further be probed by using the charged particle multiplicity, which has been found to be very sensitive to the parameters of the QCD models. Figure 6 shows the charged particle distribution, as measured at the c.m. energy $\sqrt{s} = 183$ GeV. Figure 6 also shows the evolution of mean charged particle multiplicity with c.m. energy compared to several QCD models. In this figure we include measurements done by other e^+e^- experiments at similar and lower c.m. energies. The parameters of these models are the same at all energies. We find that the energy dependence of the multiplicity distribution is in agreement with the predictions of parton shower models which include QCD coherence effects. However, parton shower models with no QCD coherence effects like COJETS, or matrix element models as implemented in Jetset, cannot explain the energy dependence.

3　Power Law Corrections

Non-perturbative corrections, or hadronization corrections, to the observables in the process $e^+e^- \to$ hadrons is traditionally performed using QCD-inspired

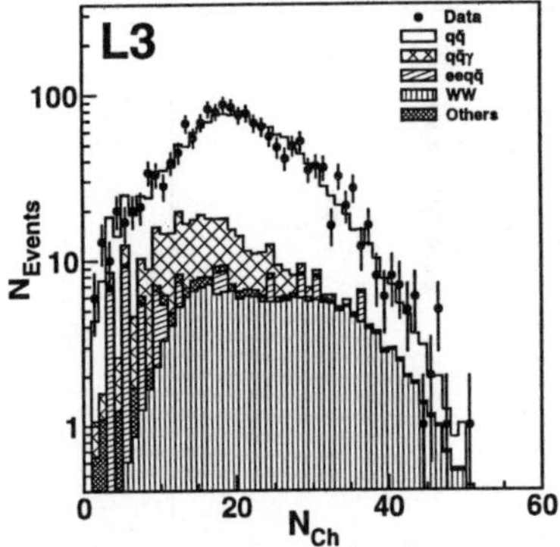

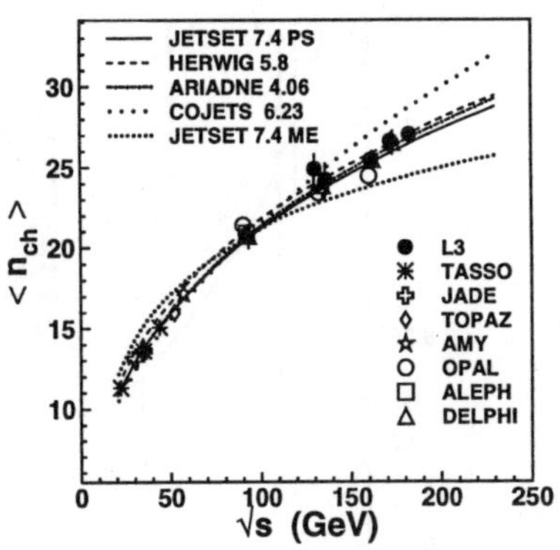

Figure 6: Left, the charged particle distribution at 183 GeV, compared with expectation from signal and background processes. On the right, the mean charged particle multiplicity $\langle n_{\mathrm{ch}} \rangle$, as function of c.m. energy, compared to various QCD models.

Monte Carlo methods, as mentioned above. For event shapes, the non perturbative effects can be sizeable, e.g., at LEP-1 energies corrections of 5-10% are found. Since these variables are extensively used for α_s determinations, non-perturbative effects have to be well understood. They lead to model dependence and thus limitations in the precision of the α_s measurements.

Recently, predictions for a class of event shape observables became available that treat the hadronization corrections analytically. These 'power' corrections [11] have their origin in the perturbative expansion when the overall energy scale Q approaches the Landau pole Λ_{QCD}. The existence of an universal non-singular behaviour of the effective strong coupling at small scales is assumed, which is parametrized by a non-perturbative parameter

$$\alpha_0(\mu_I) = \frac{1}{\mu_I} \int_0^{\mu_I} dk\, \alpha_s(k), \tag{4}$$

with scale $\Lambda_{QCD} \ll \mu_I \ll Q$ (typically $\mu_I = 2$ GeV), which separates the perturbative from the non-perturbative region.

For the mean values of the event shapes y (with y being Thrust, C-parameter, Heavy Jet Mass, Total and Wide Jet Broadening), the 'power-law' prediction are

$$\langle y \rangle = \langle y_{\mathrm{pert.}} \rangle + \langle y_{\mathrm{pow.}} \rangle \qquad \begin{cases} \langle y_{\mathrm{pert.}} \rangle \sim \mathcal{O}(\alpha_s^2) \\ \langle y_{\mathrm{pow.}} \rangle \sim a_f \mathcal{M} \frac{\alpha_0(\mu_I)}{Q} \end{cases} \tag{5}$$

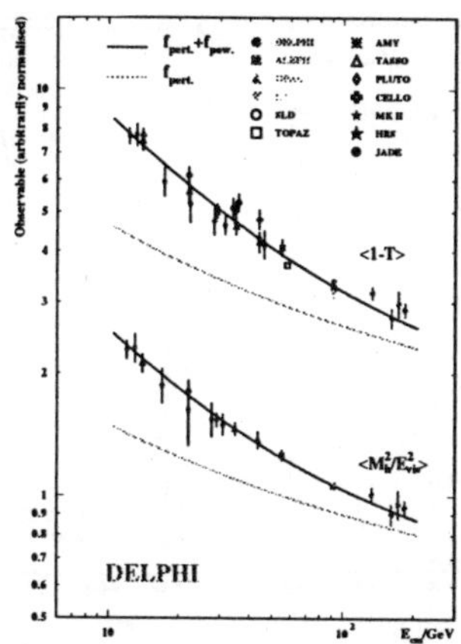

Figure 7: Mean values of $\langle 1 - T \rangle$ and $\langle M_h^2/E_{vis}^2 \rangle$ versus c.m. energy. The full line corresponds to a fit of the data to equations 5. The $\mathcal{O}(\alpha_s^2)$ perturbative predictions, $\langle y\text{pow.}\rangle$, are displayed by the dashed line. The 'power-corrections' correspond to the difference of the full and the dashed lines.

with $\langle y_{\text{pert.}} \rangle$ the usual second order matrix element and $\langle y_{\text{pow.}} \rangle$ the power corrections. The coefficient a_f depends on the event shape, and the Milan factor $\mathcal{M} \sim 1.8$ takes into account two-loop effects. The power correction term has corrections upto $\mathcal{O}(\alpha_s^2)$ which are not displayed. Note that the scale dependence of the power corrections behave as $1/Q$ (hence the name), which can be tested experimentally using LEP-1 and LEP-2 data.

DELPHI [12] have measured mean values for Thrust and Heavy Jet Mass from LEP-1 and LEP-2 data and combined their results with measurements from low energy e^+e^- experiments, in order to extract $\alpha_s(M_{Z^0})$ and α_0 from a fit to equations 5. The fits are shown in Figure 7. Very good fits are obtained with $\alpha_s(M_{Z^0}) \sim 0.118 - 0.120$, and $\alpha_0(2\text{ GeV})$ between 0.40 and 0.55.

For distributions of event shapes observables it has been shown [13] that the non-perturbative corrections lead to a shift of the distribution

$$\frac{1}{\sigma_{tot}}\frac{d\sigma(y)^{corr}}{dy} = \frac{1}{\sigma_{tot}}\frac{d\sigma(y - \Delta y)^{pert}}{dy}, \tag{6}$$

where the shift Δy is given by $\langle y_{\text{pow.}} \rangle$. In Figure 8a fits to the Wide-Jet distributions are made, for various c.m. energies. Good fits are obtained for the power-law fits, as well as for the traditional approach of hadronization corrections from MC models [14]. Figure 8b shows a summary of the results for various event shapes. The value of α_0 is found to be universal (i.e. identical

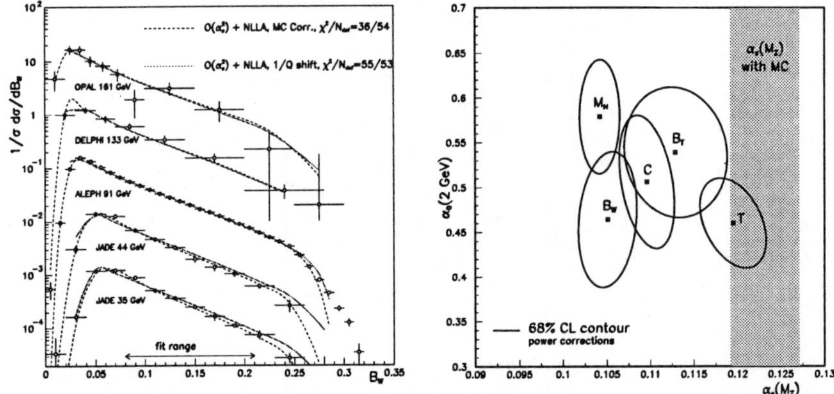

Figure 8: (a) Fits to distributions of the Wide Jet Broadening at various c.m. energies. (b) Summary of the power law fits of event shapes for $\alpha_s(M_{Z^0})$ and $\alpha_0(2\ \text{GeV})$. The average value of $\alpha_s(M_{Z^0})$ using conventional Monte Carlo techniques is indicated by the grey area.

for the various shapes), at the level of 20%. This justifies the description of the power-law corrections for these event shapes. The mean value of $\alpha_s(M_{Z^0})$ (0.108 ± 0.002) is somewhat lower than the one obtained from using hadronization corrections from MC models (0.123 ± 0.004). This difference has still to be understood.

4 Quark versus gluon initiated jets

On the parton level, calculations of gluon emissions from a quark and gluon emission from gluon reveal the difference in colour structure between the two processes. Whereas the matrix element describing the emission of a gluon from a quark involves the SU(3) colour factor $C_F = 4/3$, the emission of a gluon from a gluon involves the colour factor $C_A = 3$. One therefore expects that the properties of jets initiated by gluons are different to the properties of jets initiated by quarks. Specifically, the gluon jets are epected to have a larger multiplicity, and therefore a somewhat softer fragmentation.

Initial studies at OPAL[15] selected symmetric planar three-jet events, with two jets originating from the primary quarks, and one jet originating from hard gluon radiation. The two lower energy jets were required to have an equal energy of 24.4 GeV, and both have an angle of 150° to the highest energy jets (symmetric Y-events). In this configuration, the highest jet has an energy of 42.3 GeV, and due to the nature of gluon radiation originates in

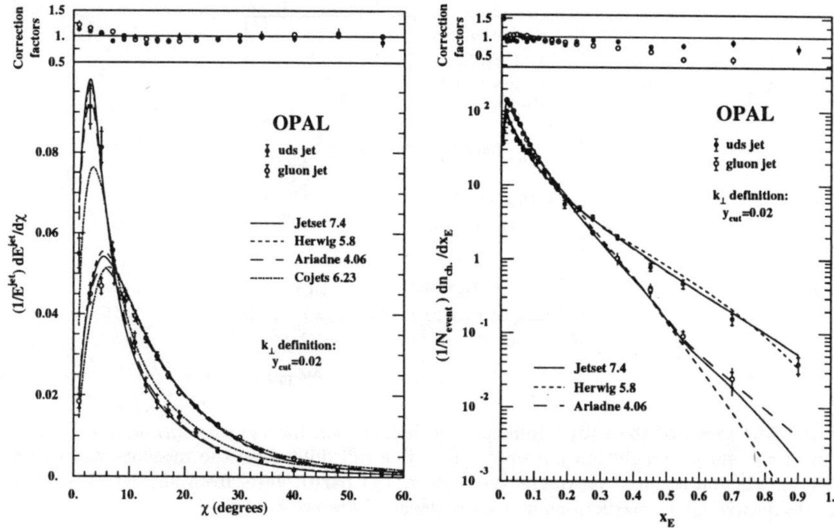

Figure 9: In a) the normalized distribution of jet energy with respect to the jet axis for uds and gluon jets is shown. In b) the particle fragmentation function for the uds and gluon jets is shown.

96% of events from a primary quark. The two lower energy jets, with the same jet-topology, have therefore almost equal probability to have been produced from a primary quark or a primary gluon ('normal mixture'). In a second sample, similar events were selected, but now one of the two lower energy jets was required to be identified as a b-quark-jet (by identification of a secondary vertex). Hence in this second sample, the other low energy jet is 'anti-tagged' as originating from the gluon. By comparing the 'anti-tagged' gluon jets with the 'normal-mixture' jets, one can unfold distributions bin-by-bin to obtain pure gluon and pure quark jets distributions. In Figure 9a a comparison is made between the quark and gluon jets for the particle energy flow around the jet axis, and in Figure 9b the fragmentation function $1/N_{event} dn_{ch}/dx_E$ is shown, with $x_E = E/E^{jet}$. Clearly the gluon jets are broader wrt the quark jets, and the fragmentation of the gluon jets is softer wrt the quark jet[b]. The ratio of charged multiplicity between the gluon and the quark jets is 1.390 ± 0.050 when jets are defined using the Durham clustering algorithm, and 1.135 ± 0.042 when a cone algorithm is used. This ratio is therefore much

---

[b]Similar techniques showed that the shape of the gluon jet is rather similar to the shape of jets originating from b-quarks [15].

174

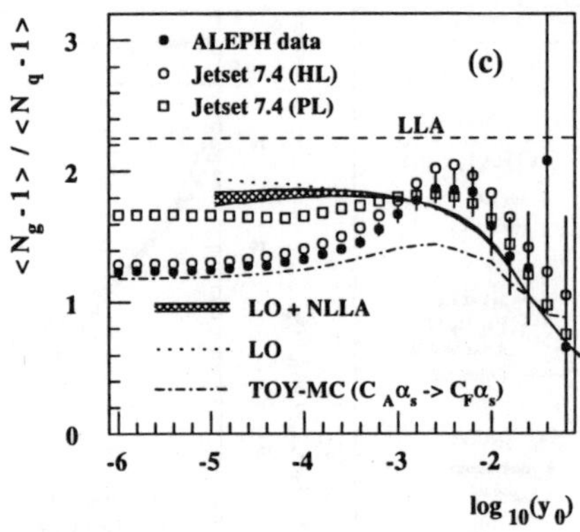

Figure 10: The mean of the subjet multiplicity distribution for the ratio *gluon/quark* as a function of the subjet resolution parameter y_0. The full dots show the measurements and the open circels and squared represent the hadron and parton level from JETSET. The lines show perturbative QCD predictions at various level of precision.

lower than the naive prediction of $C_A/C_F = 2.25$, and depends heavily on the definition of the jets.

In stead of investigating the ratio of charged multiplicity between quark and gluon jets, ALEPH [16] investigated the ratio of subjets between quark and gluon jets. To this end, the identified quark and gluon jets were split-up by performing the iterative clustering on the separate jets, using a resolution y_0, after identifying whether the jet corresponds to a quark or a gluon jet. The ratio $\langle N_g - 1 \rangle / \langle N_q - 1 \rangle$ versus the subjet resolution y_0 is shown in Figure 10. For very small values of y_0, each particle belonging to the jet is resolved and the ratio $\langle N_g - 1 \rangle / \langle N_q - 1 \rangle$ approaches the ratio of the charged multiplicity, much lower than $C_A/C_F = 2.25$ as mentioned in the previous paragraph. However, for larger values of y_0, the perturbative region is approached and the subjet multiplicity ratio grows closer to C_A/C_F. In this region NLLA [17] calculations describe the subjet rate reasonably well.

4.1 Inclusive gluon hemispheres

Alternatively, inclusive gluon jets can be identified in hadronic Z^0 decays as all the particles in a hemisphere opposite to a hemisphere containing two tagged quark jets [18]. To this end, each event is divided into hemispheres using the plane perpendicular to the thrust axis. Jet finding is performed on each of the hemispheres, and the resolution parameter of the jet algorithm is adjusted to

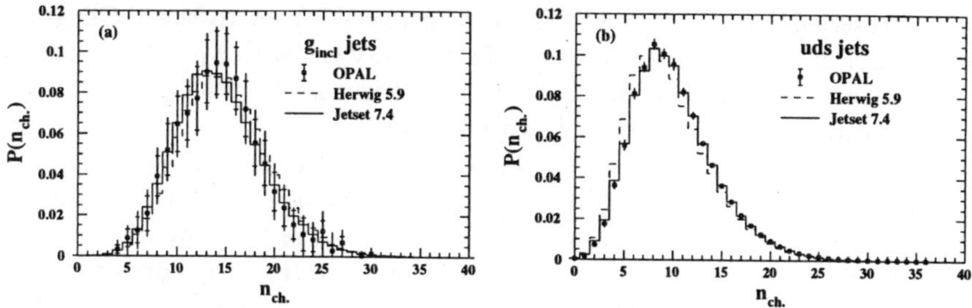

Figure 11: In a) the charged particle distribution for the inclusive gluon hemisphere, in b) the same for the uds-quark hemisphere.

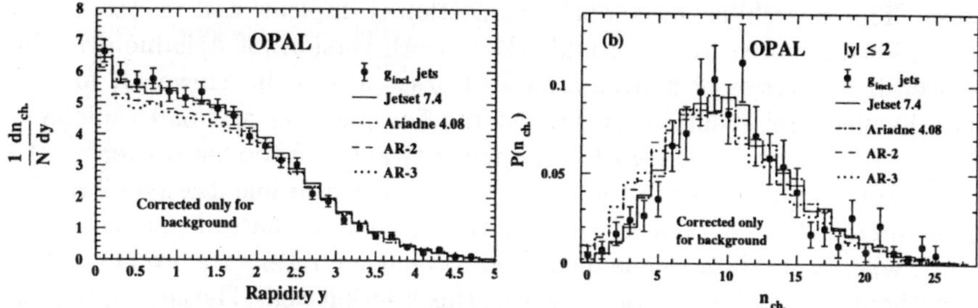

Figure 12: In a) the distribution of charged particle rapidity and in b) the distribution of charged particle multiplicity in the rapidity interval $|y| \leq 2$ for 40.1 GeV $g_{incl.}$ gluon jets. The distributions are corrected for background only.

yield exactly two reconstructed jets in a hemisphere. For each of the two jets a displaced vertex is required to be reconstructed successfully, which indicates that two primary b-quarks have ended up in the same hemisphere, after radiation of a hard gluon. The hemisphere opposite to a one with the two tagged b-quarks corresponds to the 'inclusive gluon' hemisphere. In total, 439 of these events were found, with 82% purity. Note that in the gluon hemisphere no jet-finding is performed. Gluon jets defined in this manner are equivalent to gluon jets produced from a color singlet point source (which are experimentally inaccessible) and thus correspond to the definition employed for most theoretical calculations [19].

For these events, the multiplicity ratio between the quark and gluon hemispheres reaches 1.51 ± 0.04, independent of jet finding. This value corresponds closely to the latest n.n.l.o calculations including energy conservation [20]. Moreover, at restricted phase space, this ratio increases closer to the QCD ratio C_A/C_F. For example, for soft particles ($p < 4$ GeV/c), with p_T value between 0.8 and 3 GeV, the ratio becomes 2.29 ± 0.17, in accordance with $C_A/C_F = 2.25$ and in accordance with theoretical expectations.

It is interesting to compare the gluon hemisphere events in view of colour reconnection. For normally colour connected events, a colour string connects the primary gluon in the 'gluon-hemisphere' with the primary quark, and another colour string connects the gluon to the primary anti-quark (corresponding to the 'large N_c' approximation, with N_c the number of colours, as implemented in most event generators). However, if interference terms of relative order of $1/N_c^2$ are included, colour reconnection can change the colour flow such that the primary gluon in the 'gluon hemisphere' forms a colour singlet by itself (in conjunction with a colour singlet formed by the primary quark and anti-quark). This possebility of colour-reconnection is implemented in Ariadne in two versions, denoted by AR-2 and AR-3. Both versions of Ariadne describe the overall properties of hadronic data at $\sqrt{s} = M_{Z^0}$ well. However, for the 'gluon hemisphere' events, they predict too few particles at small rapidity y (with the rapidity y defined w.r.t. the sphericity axis), or too many events with a low charged multiplicity in the range $|y| < 2$. For example, the two Ariadne models predict that 11-12% of events have less than 5 charged particles with $|y| < 2$, whereas in data this is measured to be $6.4 \pm 2.1\%$. The event generators without colour reconnection describe this fraction well. The corresponding distributions are shown in Figure 12. Hence the 'gluon hemisphere' events are indeed sensitive to colour reconnection properties, and disfavour colour reconnection models as implemented in the AR-2 and AR-2 Monte Carlos.

5 Summary and conclusions

The energy range between LEP-1 and LEP-2 provide excellent tests for various properties of QCD. The distribution of the event shapes is well described by the Monte Carlo generators, which were tuned at $\sqrt{s} = M_{Z^0}$, and the value for α_s at LEP-2 is in excellent agreement with the predictions of QCD. The jetrates and charged particle distributions provide good test of QCD event generators. Models that do not have coherence of parton emission included describe the data less well. Power-law corrections provides a promising alternative to the Monte Carlo based hadronization corrections.

Experimental separation of quark and gluon initiated jets is well possible using the data recorded at LEP-1. Especially utilizing the 'inclusive gluon hemispheres', in which the gluon jet is defined inclusively in the hemisphere opposite to the one containing two tagged quark jets, provide unambiguous determination of differences in quark and gluon jets. Gluon jets are broader and softer w.r.t. quark jets. The ratio of multiplicity in gluon versus quark jet is well described by theoretical calculations. The 'gluon hemisphere' events are sensitive to colour reconnection, but the data disfavours the reconnection models as implemented in Ariadne.

References

1. ALEPH Coll., D. Busculic et al.: "QCD studies with e+e- annihilation data at 189 GeV", Conference Report 99-023 (1999)
   L3 Coll., M.Acciarri et al.: Phys. Lett. B 444 (1998) 569
   Opal Coll, G. Abbiendi et al.: "QCD studies with e^+e^- annihilation data at 172-189 GeV", Physics Note 377(1999)
2. Delphi Coll., P. Abreu et al: "QCD Results from the Measurements of Event Shape Distributions", 99-114 CONF 301 (1999)
3. R. Ellis, W. Stirling and B. Webber: QCD and Collider Physics, Cambridge University Press (1996)
4. T. Sjöstrand: Comput. Phys. Commun. 82 (1994) 74
5. G. Marchesini et al.: Comput. Phys. Commun. 67 (1992) 465
6. L. Lönnblad: Comput. Phys. Commun. 71 (1992) 15
7. P. Mazzanti and R. Odorico: Nucl. Phys. B 394 (1993) 267
8. S. Brandt et al.: Phys. Lett. 12 (1964) 57
9. T. Chandramohan and L. Clavelli: Nucl. Phys. B 184 (1981) 365
10. S. Bethke: Nucl. Phys. B (Proc. Suppl.) 64 (1998) 54
11. Y. Dokshitzer, B. Webber, Phys. Lett. B352 (1995) 451
    Y. Dokshitzer, G. Marchesini, B Webber, Nucl. Phys. B469 (1996) 93.
12. DELPHI Coll., DELPHI 99-19 CONF 219, 1999
13. Y. Dokshitzer, B Webber, Phys. Lett. B404 (1997) 321
14. G. Dissertori, 'Event shapes and power corrections in $e^=e^-$ annihilations', talk presented at DIS99 Workshop, Berlin (1999)
    H. Stenzel, 'Power laws in e^+e^- annihilations', talk presented at the 34th Rencontres de Moriond (1999)
15. OPAL Coll., G. Alexander et al: Zeit. fur Physik C 69 (1996) 543
16. ALEPH Coll., D. Busculic et al.: Phys. Lett. B346 (1995) 389
17. M. Seymour: Phys. Lett. B378 (1996) 279
18. OPAL Coll., G. Abbiendi et al.: CERN-EP/99-028 (1999)
19. J. W. Gary: Phys. Rev. D49 (1994) 4503
20. S. Lupia and W. Ochs: Phys. Lett B 418 (1998) 214
    P. Eden and G. Gustafson: JHEP-09 (1998) 15

JET PHYSICS

JOHN M. HAUPTMAN

Department of Physics and Astronomy
Iowa State University, Ames, IA 50011
hauptman@iastate.edu

The physics of jets and physics with jets is challenging in several respects. Quarks are just as fundamental as leptons, but their experimental identification and reconstruction are very difficult compared to electrons and muons, and even neutrinos. Foremost among the tasks at present and future colliders are the search for new phenomena and stringent tests of QCD, and these tasks require searches in final states containing jets. I discuss some of the triumphs and problems of recent work with jets at high energy colliders, and two outstanding problems: quark-gluon separation and energy scale determination.

1 Introduction

The beginning point for jet physics is the reconstruction of jet four-vectors. This is an old art, whose goals and methods we discuss in general terms. Recent work with jets, largely reconstructed with fixed-cone algorithms, has resulted in many nice measurements both of jets and with jets, including precision jet cross sections, tests of quark compositeness, and tests of next-to-leading order (NLO) QCD in multi-jet events and in the ratios of jet cross sections at different energies. Finally, we discuss two goals for jets physics: a successful scheme for the separation of quark and gluon jets would reap immediate rewards in the search for high mass states, top quark studies, and W/Z reconstruction; and, a scheme for determining the absolute energy scale for jets would allow higher precision cross section measurements in addition to an improved top quark mass measurement.

2 Jet Reconstruction

A quark manifests itself as a bundle, or 'jet', of particles aligned with the original quark momentum vector. However, there are deep theoretical and experimental issues which render this identification ambiguous in several respects. 'Hard radiation' of energetic gluons in the initial state may produce multiple jets not aligned with the original quark. The emission of 'soft radiation' from the quark of interest generates extraneous particles, and both hard and soft radiation from the quarks in the beam generate 'underlying events', all of which confuse reconstruction algorithms.

A good jet reconstuction algorithm should possess several desirable properties, as recently articulated by Pierce and Webber [1] and by Soper [2]:

- insensitivity to low-energy gluon radiation ('infrared safety');

- insensitivity to particles emitted along the same direction ('collinear safety');

- minimum number of 'spurious jets' formed, for example, by the incorrect association of particles from beam jets or underlying events;

- insensitivity to 'hadronization corrections' between physics quantities at the physics level (partons) and the measurement level (hadrons);

- invariance to boosts along the beam axis;

- suitable for soft gluon summations in the theory;

- allows factorization of initial state singularities;

- be simple and elegant.

The experimental measurement of a jet, almost always with fine-grained calorimeters, both solves some of the above theoretical problems and creates further problems. A calorimeter with finite resolution in both energy and angle automatically solves the 'infrared' and 'collinear' safety problems since energies of individual particles are summed in finite solid angular detectors. These same finite resolutions exacerbate the problem of formation of 'spurious jets' wherein particles from several sources are not distinguishable. 'Hadronization corrections' remain due to both theoretical uncertainties in the QCD calculations of mass corrections, as for example in the W mass [3], and to a much larger degree in the experimental capability to understand the interactions of particles in the materials of calorimeters to establish an absolute energy scale for jets. Difficulties of this latter sort derive from the different responses of calorimeters to electromagnetic and hadronic particles ('e/h' differs from unity [4]) and from the basic difficulty of any reconstruction algorithm to include all the particles of the jet, and no other particles.

These are difficult criteria to attain, and at the Tevatron the CDF and D0 collaborations plan to maintain a degree of commonality in jet reconstruction. [5]

It seems to me that the goals of jet reconstruction are two: (i) the jet four-vectors reconstructed must be robust with respect to the 'soft stuff', including instrumental noise, radioactive decays in the calorimeter mass, underlying events and soft bremsstrahlung; and, (*ii*) the jet four-vectors must be Lorentz invariant both with respect to boosts along the beam axis and for boosts into the rest frame of the jet. One must be able to test the structure of a jet in its rest frame where the particles of the jet are nearly isotropic. It seems to me that only a k_T algorithm, with decisions based on the invariant masses of particles and subjets, can accomplish these goals.

2.1 Coordinate systems

The geometry of collider detectors over the past 30 years has been cylindrical with very few exceptions, *viz.*, only a couple of spherical detectors appearing at e^+e^- colliders. This implies cylindrical coordinates for any particle or jet with energy E, polar angle θ and azimuthal angle ϕ. However, the QCD cross section varies rapidly as $1/\theta^4$ at small θ. The kinematic variable in which QCD particle distributions are nearly flat is rapidity, y,

$$y = \frac{1}{2} \ln \frac{(E + p_L)}{(E - p_L)}, \tag{1}$$

where E is the energy of the particle and p_L is its longitudinal momentum. Rapidity was introduced as a useful variable in inclusive reactions such that a Lorentz boost along the longitudinal axis simply adds a constant to the rapidity. Rapidity differences are therefore invariant to a Lorentz boost.

E and p_L are not easily known in a collider environment, and therefore 'pseudorapidity', η, which is equal to y as $m \to 0$, or as $\beta \to 0$, is used exclusively. In this ultra-relativistic limit, $\cos \theta = E_L/E$, so that pseudorapidity η is

$$\eta \approx \frac{1}{2} \ln \frac{1 + \cos \theta}{1 - \cos \theta} = \frac{1}{2} \ln \frac{\cos^2 \theta/2}{\sin^2 \theta/2}, \tag{2}$$

or,

$$\eta = -\ln \tan \theta/2. \tag{3}$$

2.2 Problems with Cone Algorithms

The simplest cone algorithm begins by isolating a region of (η, ϕ) space in the calorimeter about the supposed jet coordinates, (η_j, ϕ_j), of radius $\mathcal{R}$, where

$$\mathcal{R} = \sqrt{(\eta - \eta_j)^2 + (\phi - \phi_j)^2}. \tag{4}$$

$\mathcal{R}$ is usually chosen in the interval 0.5 to 1.0, depending upon the physics problem. The jet energy is estimated as the summed energy of all calorimeter cells within the cone, and η_j, ϕ_j of the jet are the energy-weighted means of the calorimeter η, ϕ cells within the cone. This simple

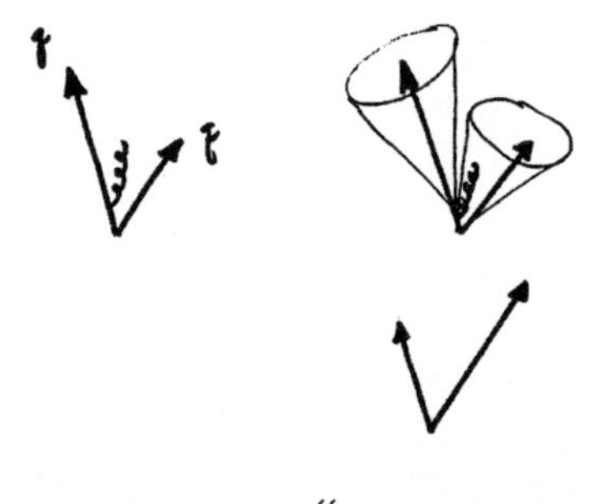

Figure 1: The parton level has two energetic quarks, the first more energetic than the second; the first quark radiates a gluon between the two quarks. The cone algorithm mistakenly includes the calorimeter towers from the gluon into the second jet, resulting in two jets, but the second jet is now higher energy than the first jet, for a wrong answer.

procedure may be iterated since η_j, ϕ_j are a function of the initial cone. More sophisticated and better defined cone algorithms [6,7] based upon well-defined theoretical work [8] have been developed and used for a decade.

The first problem with cones is that two separate jets may overlap; special rules are required to resolve this overlap and to apportion calorimeter energies to one jet or the other. There can be discontinuous effects at the numerical boundary, specified by $\mathcal{R}$, of the jets. The cone algorithm begins to fail the 'simplicity' criterion.

Secondly, cone algorithms can be sensitive to the 'soft stuff' between jets, as illustrated in Fig.1, in which two quark jets are easily separated by $\approx 2\mathcal{R}$, and the higher energy jet has radiated a gluon in between the two jets. The algorithm puts its cones around the calorimeter cells, but includes the radiated gluon into the cone of the second jet. The final solution is two jets, but the first jet is lower energy than the second jet, for a wrong answer.

In this way, the four-vector solutions for jets reconstructed with cones are sensitive to the details of lower energy gluon radiation and the energy deposition in the calorimeter cells between the jets.

2.3 Problems with k_T Algorithms

More discreet algorithms began with the JADE algorithm in e^+e^- collisions where the jets are almost always well separated in coordinate space. There are several k_T algorithms, and the familiar ones are JADE,[9] Durham,[10] Soper-Ellis,[11] Cambridge,[12] Cambridge with angular-ordering,[1] Cambridge with 'solf-freezing',[13] and the venerable LUND clustering algorithm.[14]

The main thrust of these algorithms is similar: begin with a set of particles, calorimeter cells, or proto-jets, *i.e.*, those basic elements which when combined will form the jet four-vector. These elements are combined in successive approximation such that at each stage there exists a list of elements $(E_{T\ell}, \eta_\ell, \phi_\ell)$, and a list of jets (E_{Tj}, η_j, ϕ_j), where the index j labels the supposed jets and E_T is the energy transverse to the beam axis. There are physical criteria, involving kinematic proximity or invariant masses of each new element with respect to other elements or to those jets already in the list of jets, to determine if an element is to be merged with another element, added to the list of jets, or itself labelled a jet. In the end, all elements are grouped into jets.

For illustration, the Soper-Ellis algorithm proceeds as follows: for each element ℓ, define

$$d_\ell = E_{T,\ell}^2, \qquad (5)$$

and for each pair of elements (ℓ, m) define

$$d_{\ell,m} = \min[E_{T,\ell}^2, E_{T,m}^2][(\eta_\ell - \eta_m)^2 + (\phi_\ell - \phi_m)^2]/R^2, \quad (6)$$

where R is a parameter, say $R \approx 1$, to keep the angular factor of the E_T^2 terms of order 1, or smaller. The algorithm then proceeds:

1. find the smallest of the d_ℓ and $d_{\ell,m}$;

2. if the smallest is a d_ℓ, remove ℓ from the list of elements and add it to the list of jets;

3. if the smallest is a $d_{\ell,m}$, combine elements ℓ and m, and then compute the energy and energy-weighted means for η and ϕ,

$$E_T = E_{T,\ell} + E_{T,m} \qquad (7)$$

$$\eta = (\eta_\ell E_{T,\ell} + \eta_m E_{T,m})/(E_{T,\ell} + E_{T,m}) \qquad (8)$$

$$\phi = (\phi_\ell E_{T,\ell} + \phi_m E_{T,m})/(E_{T,\ell} + E_{T,m}) \qquad (9)$$

4. go to 1

The basic criterion in this algorithm for merging elements into larger elements, which eventually with further iterations necessarily become jets, is the proximity in (η, ϕ) space weighted by the factor

$$\min[E_{T,\ell}^2, E_{T,m}^2]. \qquad (10)$$

Several other metrics are listed in Table 1. The JADE and Cambridge algorithms use invariant mass explicitly.

Table 1: k_T Algorithm Metrics

Algorithm	Metric
JADE	$d_{ij} = 2E_i E_j(1 - \cos\theta_{ij})$
Durham	$d_{ij} = \min[E_i, E_j](1 - \cos\theta_{ij})$
Soper and Ellis	$d_{ij} = \min[E_{Ti}^2, E_{Tj}^2]\Delta R_{ij}^2/R^2$
Cambridge	$y_{ij} = 2E_i E_j/E_{\text{vis}}^2(1 - \cos\theta_{ij})$
Pierce and Webber	$d_{ij} = \min[E_{Ti} E_{Tj}]R_{ij}$

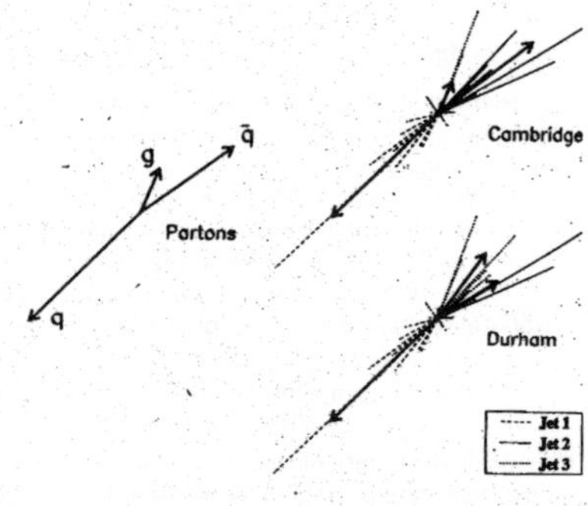

Figure 2: The parton level has two energetic quarks, with the antiquark radiating a gluon. The Durham algorithm results in two nearly equial energy jets to absorb the the energy of the antiqaurk; the Cambridge algorithm more closely reconstructs the separate antiquark and gluon jets.

The original JADE algorithm uses the invariant mass of two massless elements, a sensible choice. One can generate new algorithms with more clever choices of metrics. However, these k_T algorithms suffer from a subtle instability [13] in which less energetic jets attract additional elements that could form spurious jets. Removing the softer of two resolved jets, to prevent it from accreting extra elements, is called 'soft freezing', and is discussed in the above paper, and illustrated in Fig.2. Noise fluctuations and debris from underlying events can both be accreted into larger jets, or if large enough, labelled as jets themselves. In this respect, k_T algorithms are also sensitive to the 'soft stuff'.

Each of these discreet k_T algorithms has a set of rules and criteria both for the merging of elements and the removal of elements into the jet list. These rules begin to make k_T algorithms fail the 'simplicity' criterion, also.

2.4   Other Methods

The cone and other discreet algorithms will be used in further running at the Tevatron for reasons of consistency with earlier data, for security because experimenters understand them in great detail, including their corresponding energy scales, and for their relative robustness. However, other methods ought to be explored at the Tevatron before the advent of LHC data to assess their strengths and weaknesses.

k_T algorithms, or similar algorithms derived from them, should be used in jet reconstruction and decision making. Using invariant masses for decisions is smart, but problems remain. All k_T algorithms use local decisions, that is, merging decisions and jet labelling decisions are made element-by-element, without consideration of the whole jet. Methods which use the entire reconstructed putative jet as a whole ought to be explored. For example, given the ensemble of elements (particles or proto-jets), one can boost to the jet center-of-mass frame where one should find a nearly uniform distribution, and clumps in this distribution would be subjets. There are 'maximum entropy' techniques in pattern recognition and image reconstruction in noisy data which may be useful.

3   Nice Measurements with jets

A wide range of excellent measurements with jets have been made: cross section measurements in E_T, in the two-jet mass and in the two-jet angular distribution have been made 1.8 TeV at the Tevatron; cross sections in E_T at both 0.63 and 1.8 TeV show moderate agreement with next-to-leading order (NLO) QCD calculations; and, cross section measurements at HERA [15] show excellent agreement with NLO CQD. Measurements of the strong coupling constant, α_s, distributions in 6-jet events, and quark substructure tests are shown.

The leading order (LO) diagram is just Rutherford scattering; the NLO diagram allows the radiation of one gluon from each line, Fig. 3. Some of the terminology of jet cross section measurements is depicted in Fig. 4.

The underlying event is generated by the interactions of the 'spectator quarks' of the incident hadrons; the scattered quarks or gluons materialize in the strong coupling regime as jets of hadrons in the laboratory. Precision measurements coupled with precision calculations (for wise choices of the factorization and renormalization scales, μ_f and μ) can lead to more precise parton distribution functions (PDFs) of the quarks and gluons in the proton.

A good example of the precision and reach of a jet cross section measurement is shown in Fig. 5. The basic

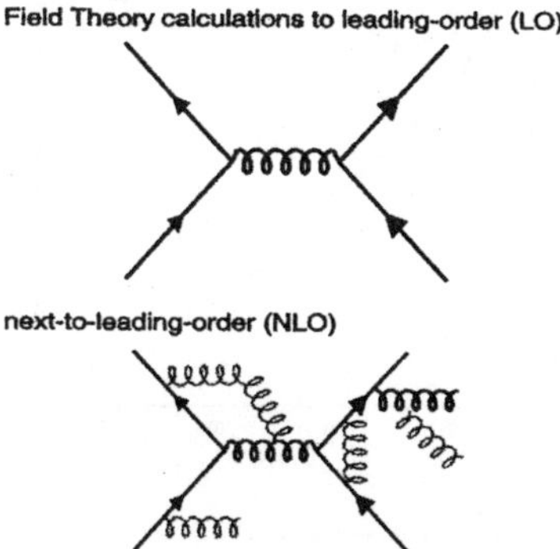

Figure 3: The leading-order (LO) QCD process is essentially elastic Rutherford scattering; the next-to-leading order (NLO) processes involve one additional gluon.

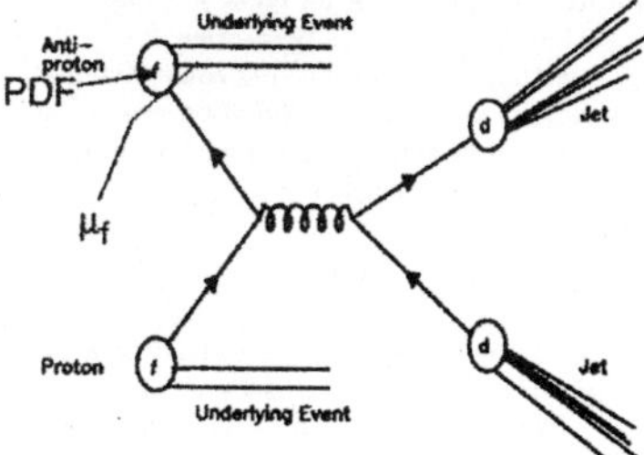

Figure 4: A glossary of common terms: the 'parton distribution function' (PDF) specifies the momentum distribution of the partons (both quarks and gluons) within the proton; the 'factorization scale' separates long-distance from short-distance effects in the QCD calculation; the 'renormalization scale' specifies the energy, or distance, scale of the hard parton scattering; the underlying event results from the break-up of the spectator quarks in the protons; and, the outgoing partons from the collision 'fragment' non-perturbatively into the jets measured in the laboratory.

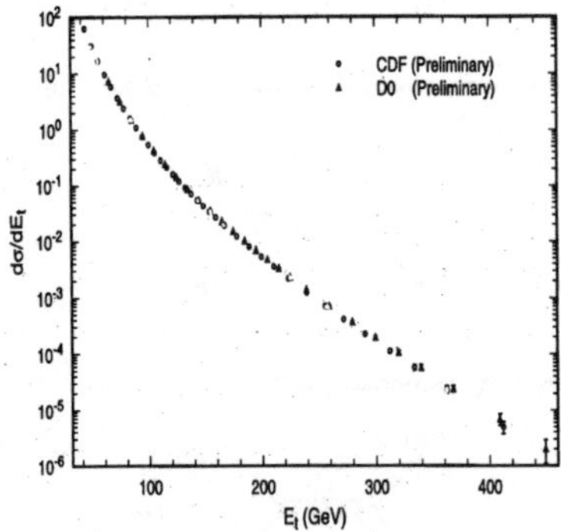

Figure 5: Measured jet cross sections by the D0 and CDF collaborations over eight orders of magnitude in 1.8 TeV collisions at the Tevatron.

jet cross section,

$$\frac{d\sigma}{d\eta dE_T} \qquad (11)$$

near the central region of the CDF and D0 detectors, $\eta_{jet} \leq 0.7$, is shown over eight orders of magnitude up to an E_T of 0.45 TeV. A comparison of the D0 cross section for $\eta_{jet} \leq 0.5$ with the JETRAD[16] event generator, using the CTEQ3M [19] PDFs with a renormalization scale of $\mu = E_T^{\max}/2$ displays excellent agreement up to $E_T = 0.5$ TeV, as shown in Fig. 6

A direct comparison of jet cross sections at two energies, 0.63 and 1.8 TeV, has been done[17] as a stringent test QCD with the expectation that some systematic uncertainties will cancel in the ratio of cross sections. The best comparison is made by plotting the dimensionless cross section *versus* a scaled E_T:

$$\frac{E_T^3}{2\pi} \frac{d^2\sigma}{dE_T d\eta} \quad versus \quad x_T = \frac{E_T}{\sqrt{s}/2}. \qquad (12)$$

This measurement has been extensively compared to QCD calculations, and one of many such plots[17] is shown in Fig. 7. The agreement with QCD calculations is not good, even allowing for wide variations in the renormalization scales at different energies and for several different PDFs.

3.1 Measurement of the running of α_s

Excellent and high-precision measurements of the strong coupling constant, α_s, have been made at the LEP Collider.[20] The CDF Collaboration has made a very nice

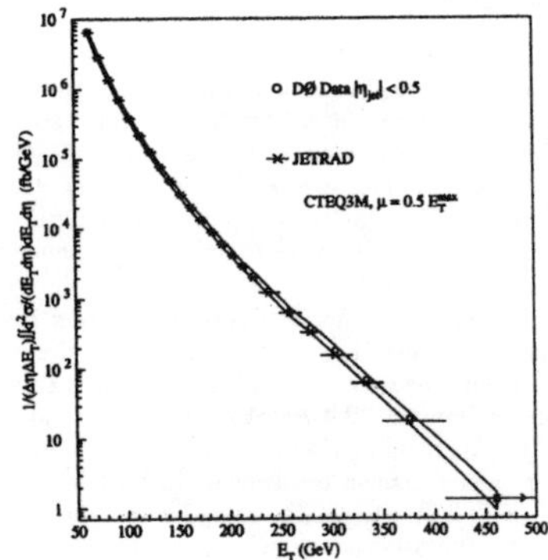

Figure 6: The D0 measured jet cross section in E_T compared to the JETRAD NLO calculation using PDF structure functions CTEQ3M at a renormalization scale of $\mu = E_T^{\max}/2$, where $E_T^{\max}$ is the highest E_T jet in the event.

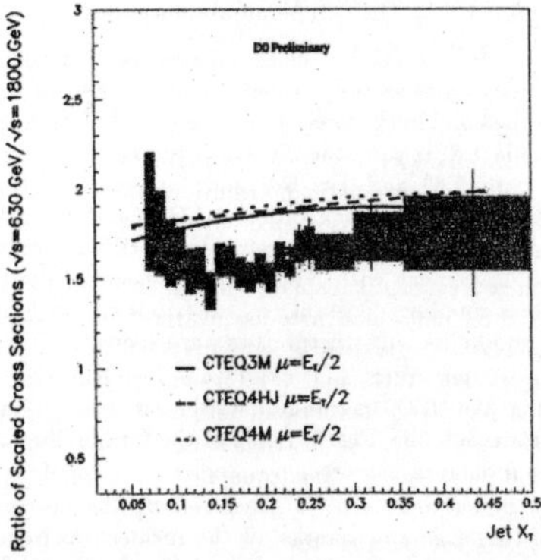

Figure 7: Ratio of the invariant jet cross section at $\sqrt{s} = 0.63$ TeV to the cross section at $\sqrt{s} = 1.8$ TeV as a function of scaled E_T. Three structure functions (PDFs) are shown, and the NLO JETRAD calculation was performed at a renormalization scale of $\mu = E_T^{\max}/2$.

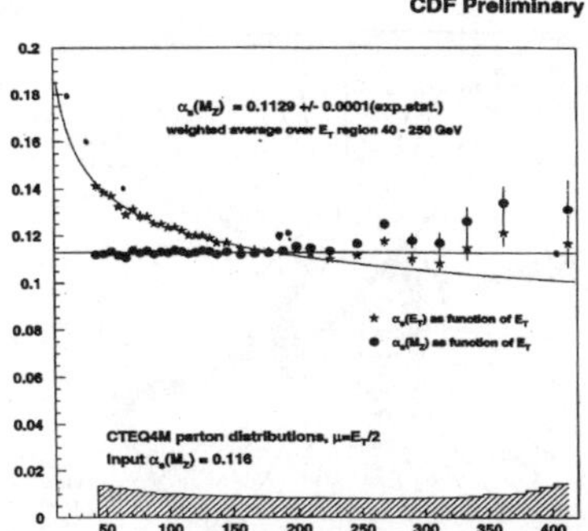

Figure 8: A preliminary estimate of the strong coupling constant α_s as a function of jet energy, including a check estimate of $\alpha_s(M_Z)$ as a function of jet E_T. These data are not sensitive to as much to the absolute value of α_s as to the energy dependence, or the running of α_s.

preliminary measurement, displayed in Fig. 8, of the energy dependence of α_s over nearly a factor of 10 in energy, from 50 to 400 GeV in E_T. This measurement, although never intended by CDF to be an absolute measurement, displays the running of α_s in QCD.

3.2 Six-jet events in CDF

A *tour de force* with multiple jets is the CDF measurement [21] of many kinematic distribution functions among the reconstructed six jets of an event. A picture of four of these six-jet events is shown in Fig. 9. The character of these measurements is displayed in Fig. 10, showing the ratio of three particular three-jet invariant masses to the six-jet invariant mass, with comparisons to two physics generators, HERWIG and NJETS, and to phase space. Clearly HERWIG does very well both on the global shape and the details of these distributions, NJETS does less well, and phase space is inadequate.

3.3 Quark substructure tests

The search for quark substructure at the highest Tevatron energies has been performed somewhat independently in three variables: summed jet transverse energies above 0.5 TeV, high jet-jet masses and jet-jet angular distributions. The signature for quark substructure is a deviation from the QCD expectation for point-like quarks, and this deviation must scale dimensionally as an energy squared divided by the new interaction energy

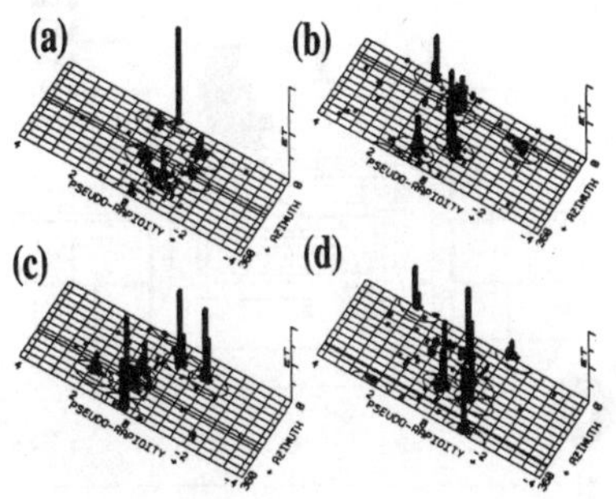

Figure 9: Four six-jet events collected and reconstructed by the CDF Collaboration. Although not clear from this figure, the six jets in these events are clearly well-reconstructed. An analysis of mass and angular distributions has been done with comparisons to theoretical calculations.

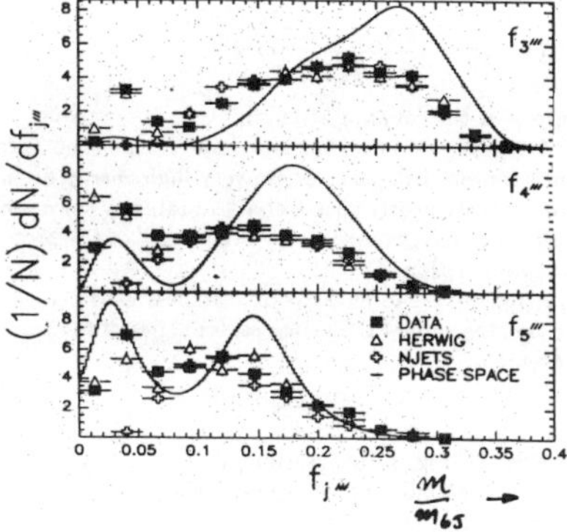

Figure 10: The distribution of the ratio of one particular 3-jet submass relative to the 6-jet mass, with comparisons to two physics generators, HERWIG and NJETS, and to phase space. HERWIG does very well both on the global shape and in the details of these distributions, NJETS does less well, and phase space is inadequate.

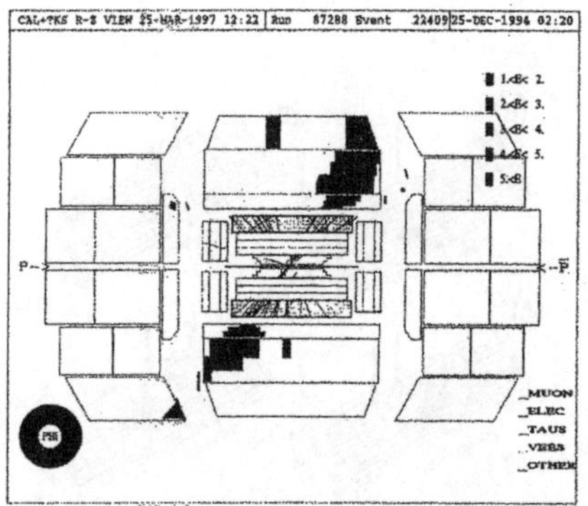

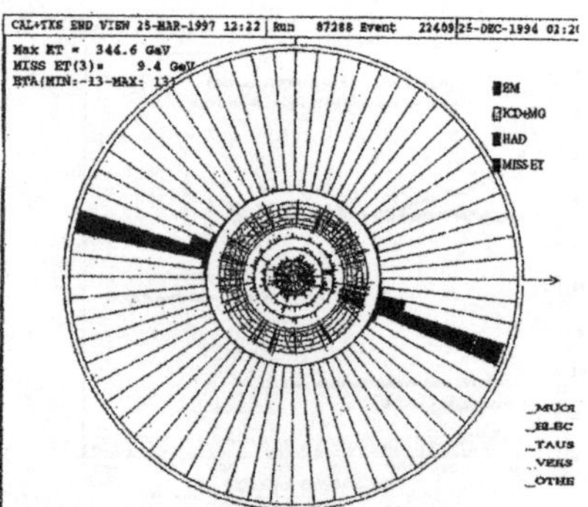

Figure 11: One D0 event at very high transverse energy, nearly at the kinematic reach of the Tevatron. These very high energy events are necessarily nearly back-to-back in η and ϕ since each quark is sampled from the very high-x end of the proton structure function.

scale squared, Λ^2,

$$\frac{\text{DATA} - \text{QCD}}{\text{QCD}} \propto \tilde{g}^2 \frac{E_T^2}{\Lambda^2}, \tag{13}$$

where $\tilde{g}^2$ is the coupling strength.

A typical event near the kinematic reach of the Tevatron is shown in Fig. 11. These very high energy events are necessarily nearly back-to-back in rapidity since each quark is sampled from the very high-x end of the proton structure function.

A recent test of quark compositeness used the variable H_T, the sum of the transverse energies of the jets of an event:

$$H_T = \sum_{\text{jets}} E_T^{\text{jet}}. \tag{14}$$

The event in Fig. 11 has a value of $H_T \approx 947$ GeV. H_T is a robust variable which is insensitive to the splitting and merging of jets, does not depend on the number of jets, and suffers only small bias by the exclusion of low E_T jets, $E_T < 20$ GeV, from the sum. The direct event distribution from D0 is shown in Fig. 12 spanning three orders of magnitude from 0.5 to 1.0 TeV in H_T and evidently nearly exponential. A direct comparison of these high H_T events with a NLO QCD calculation using JE-TRAD is shown in Fig. 13. There is no hint in these data

at the highest energies of any deviation from QCD, using the latest PDFs and NLO calculations.

The present lower limit on a new energy scale Λ for quark compositeness from all methods is now

$$\Lambda \approx 2 - 3 \; TeV, \tag{15}$$

placing a spatial limit of the size of a light quark at about 1/10,000 of the proton radius. We should teach our high school students that not only is an atom almost all empty space, but that a proton is also empty space, and for the same reason.

4 Important work with jets

4.1 Quark-gluon separation

Gluons are a distraction. Essentially all important high-mass physics searches (top quark, SUSY, technicolor, higgs, etc.) look for quark jets in the final state. Recent successful searches by D0 and CDF for 'top to all jets',

$$t\bar{t} \to W^+W^- b\bar{b} \to 6 \; quark \; jets, \tag{16}$$

dealt with enormous gluon jet backgrounds, and the critical experimental challenge was to overcome this background. The problem of quark-gluon separation is important and difficult. A D0 measurement using a k_T

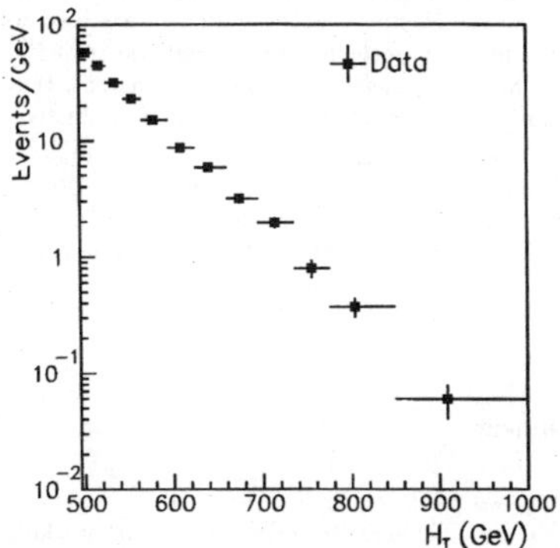

Figure 12: The measured distribution of the variable H_T in the D0 experiment. The event distribution is nearly exponential over three orders of magnitude from $H_T = 0.5$ TeV to $H_T = 1.0$ TeV.

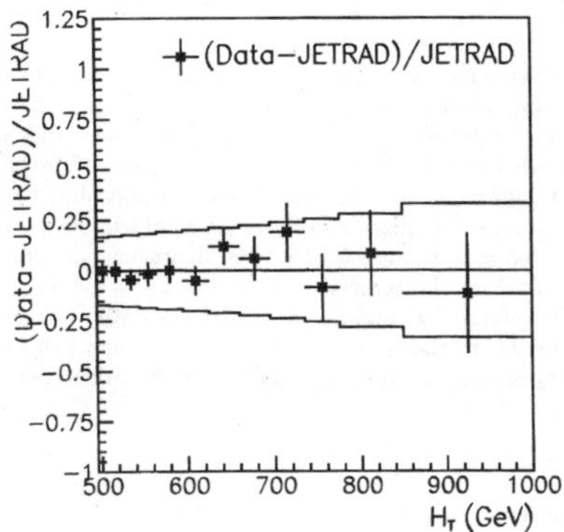

Figure 13: Direct comparison of high H_T events with a NLO QCD calculation using JETRAD. There is no hint in these data at the highest energies of any deviation from QCD, using the latest PDFs and NLO calculations.

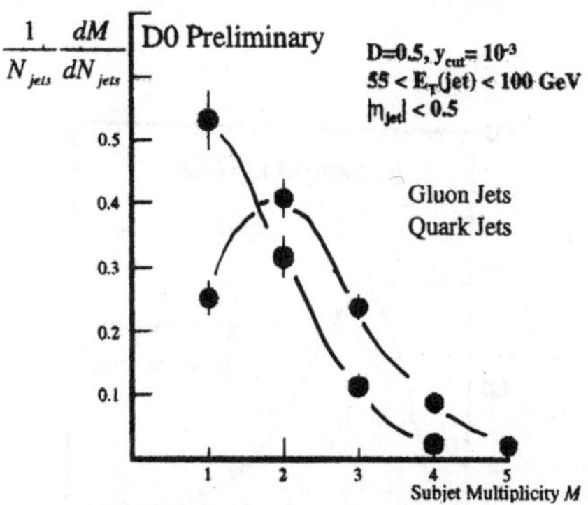

Figure 14: The distributions of subjet multiplicity as reconstucted from a k_T algorithm for quark and gluon jets. The separation of these jets in D0 is done statistically . Differences in rate of about a factor of two are evident at both high and low subjet multiplicity.

algorithm counting the number of sub-jets within a jet [22] achieves roughly a factor of two separation, shown in Fig. 14. The OPAL Collaboration counts the number of charged particles in a jet,[23] and achieves a comparable or better quark-gluon separation, as shown in Fig. 15.

In the end, quark-gluon separation may only be employed as a weight for individual jets, or used in neural network algorithms, to enhance a signal relative to background. This is good; however, there is potential for very large gains if a k_T-like algorithm can someday achieve rejections of 5-10 for gluon jets relative to quark jets while maintaining high quark jet acceptance.

5   Jet energy calibration

One of the most arduous problems with jets is absolute measurement of jet energies. The problem is difficult at low energies due to spatially ill-defined jets and by uncertainties in calorimeter response to low energy particles. At high energies there is no easy calibration standard, the highest mass objects known which decay to jets being the W and Z near 100 GeV. Therefore, the energy scale uncertainty diverges and reaches about $\pm 2\%$ absolute at 500 GeV, as displayed in Fig. 16.

This is impressive; however, jet cross sections fall as a high power ($\approx 6-8$) of E_T and the fractional uncertainty in the cross section is that same power times the uncertainty in the energy scale. This limits the knowledge of the top quark mass, the quark compositeness energy scale, and others. Recent work in D0 is illustrative of

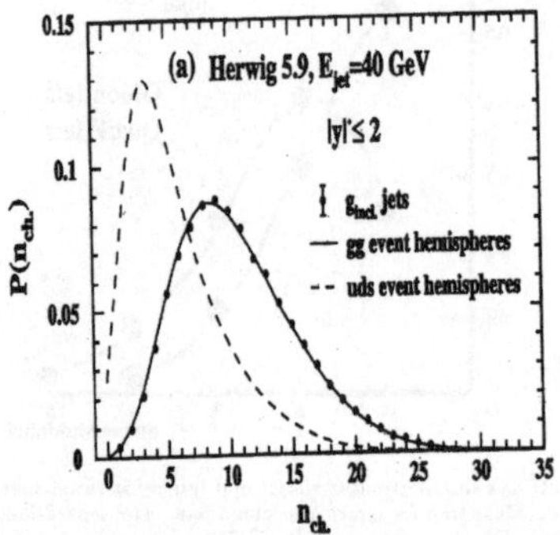

Figure 15: The OPAL measurement of charged particle multiplicity in cleanly tagged quark and gluon jets at LEP. Discrimination factors as large as 10 are evident at high and low multiplicities.

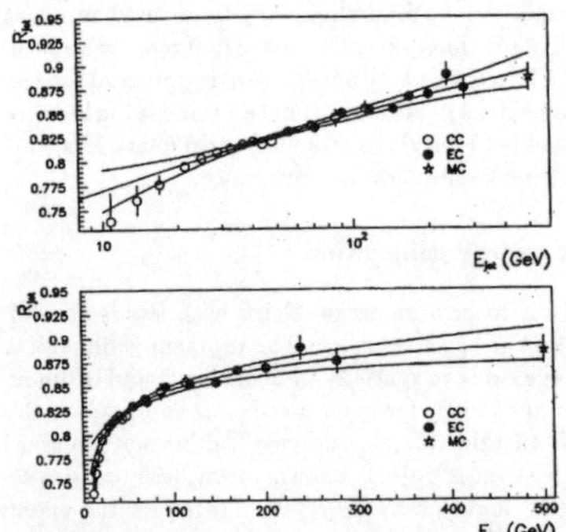

Figure 16: The uncertaintly in the jet energy scale determined in the D0 experiment on both a logarithmic energy scale (upper plot) and a linear scale (lower plot). The large uncetainty at low E_T is largely instrumental, and at high E_T the scale is difficult to determine due to lack of a known energy or mass standard, and to dwindling statistical samples.

the level of precision presently achievable.[17] It is only an accident that the optimum in energy scale uncertainty in E_T is at half the W mass. Many data sets are used to determine the energy scale and, over a wide range of E_T, balancing the electromagnetic energy of a γ against that of a jet in $p\bar{p} \to \gamma\ jet$ events has proven to be productive.

A new idea for jet calibration is to use $Z \to b\bar{b} \to jj$ events.[18] In Run 2 at the Tevatron both CDF and D0 will be able to tag b-jets, and with event samples of several hundred Z's up to $E_T \approx 100$ GeV, both experiments will be able to reduce their systematic uncertainties at high jet energy.

6   Comments

Permit me two brief and final comments. It was mentioned that the CDF and D0 collaborations are working jointly on jet reconstruction algorithms. This is good – experiments will be able to cross-check each other. However, there is a danger that Tevatron jet physics will lose a degree of healthy diversity. It is my recollection that in the early years LEP data had *too few* differences among the four experiments. Maybe this was because the experiments used similar algorithms, codes and procedures, and possibly because physicists talk to each other! Experiments are *supposed* to differ from each other by 2σ every twenty, or so, measurements. So, 'CD0F' physics may not be a good idea if taken too far.

Absolute measurements are difficult but critical to good tests of QCD. Good energy resolution is also important in the measurement of the top quark mass, which limits the indirect inference on the Higgs mass, and in the search for new narrow states of SUSY or technicolor. Experimentally, the energy and the integrity of jets, resolved by the all-important reconstruction algorithms, are crucial. But, on the theoretical side, especially for direct QCD comparisions, higher order QCD calculations ought to be done, so that in the near future the somewhat arbitrary renormalization scale used in these calculations will play a smaller role.

Acknowledgments

I would like to thank the physics staff at the University of Michigan for bringing together such a good small conference on many diverse topics in physics, and Tina Wells and her staff for its smooth running. I also thank John Krane for help and advice. This work was supported by the US Department of Energy under grant DE-FG02-92ER40730.

References

1. A.T. Pierce and B.R. Webber, "Comparisons of New Jet Clustering Algorithms for Hadron-Hadron Collisions," hep-ph/9807523, 29 July 1998, and Cavendish-HEP-98/10, 8 December 1998.
2. D.E. Soper, Meeting of the 'Jet Algorithm Working Group,' 3-4 June 1999.
3. S. Keller, "Theoretical Uncertainties Associated with the Extraction of M_W at Hadron Colliders," XXXIIIrd Rencontres de Moriond on "QCD and High Energy Hadronic Inteactions", and hep-ph/9804287.
4. C. Caso *et al*, *Review of Particle Physics*, Particle Data Group, *Eur. Phys. Jour.* C **3**, 159-160 (1998).
5. Jerry Blazey, "Jet Physics", Run II Workshop on QCD and Weak Boson Physics, Fermilab, 3-4 June 1999.
6. B. Flaugher and K. Meier, Proc. 1990 Summer Study on High energy Physics, Snowmass, Colorado, (World Scientific, 1992) p. 125.
7. J.E. Huth, *et al*, *ibid.*, p. 134.
8. G. Sterman and S. Weinberg, PRD **39**, 1436 (1977).
9. JADE Collaboration, W. Bartel *et al*, PL **B123**, 460 (1993); and, *Z. Phys.* **C33** 23 (1986).
10. Yu.L. Dokshitzer, Workshop on Jet Studies at LEP and HERA, Durham, December 1990, *J. Phys* **G17** 1537 (1991).
11. S.D. Ellis and D.E. Soper, "Successive combination jet algorithm for hadron collisions," PRD **48**, 3160 (1993).
12. Yu.L. Dokshitzer, G. Leder, S. Moretti and B. Webber, JHEP **08** 001 (1997).
13. S. Bentvelsen and I. Meyer, "The Cambridge jet algorithm: features and applications," EPJC **C4**, 623 (1998).
14. T. Sjöstrand, Comp. Phys. Commun. **28** 1436 (1977).
15. C. Glasman, for the H1 and ZEUS Collaborations, "Jet Physics at HERA," hep-ex/9905063, 31 May 1999; and, S. Maxfield, B. Pötter and L. Sinclair, "Jet cross sections at HERA – Current Issues," hep-ex/9904030 v2, 30 April 1999, GLAS-PPE/1999-03.
16. W. Giele, E. Glover and D. Kosovar, NPB **304**, 633 (1993).
17. J. Krane, "The ratio of inclusive jet cross sections at $\sqrt{s} = 630$ GeV and $\sqrt{s} = 1800$ GeV," PhD thesis, 1998, available at D0.
18. Suggestion by Bill Murray at this conference.
19. H.L. Lai *et al*, PRD **55**, 1280 (1997).
20. C. Caso *et al*, Particle Data Group, *ibid.*, p. 81.
21. F. Abe, *et al*, "Properties of six-jet events with large six-jet mass at the Fermilab Proton-Antiproton Collider," Fermilab-PUB-97/093-E, April 16, 1997.
22. Rob Snihur, "Subjet multiplicity in quark and gluon jets," D0 collaboration, hep/ex-9907059, July 1999.
23. OPAL Collaboration, "Experimental properties of gluon and quark jets from a point source," hep/ex-9903027, 15 Mar 1999.

Gamma-Ray Bursts: An Overview of Recent Observational Progress

T. A. Mckay, C. Akerlof, B. Kehoe, A. Pawl
University of Michigan Department of Physics, 500 East University
Ann Arbor, MI 48109, USA
E-mail: tamckay@umich.edu

R. Balsano, J. Bloch, D. Casperson, S. Fletcher, G. Gisler,
J. Hills, J. Szymanski, J. Wren
Los Alamos National Laboratory
Los Alamos, NM, 87545, USA

S. Marshall
Lawrence Livermore National Laboratory
Livermore, CA 94550, USA

B. Lee
Fermi National Accelerator Laboratory
Batavia, IL 60510, USA

Gamma-ray bursts have, since their discovery in 1969, been the archetypal astro-physical mystery. Despite the detection of thousands of events, our knowledge of the origin and nature of GRBs remained minimal for nearly 30 years. Progress in understanding gamma-ray bursts has undergone explosive growth since the observation in 1997 of the first optical afterglow of a burst. The discovery of afterglows was followed in 1999 by the first simultaneous optical detection of a GRB. These discoveries constitute the beginning of a new field, the multiwavelength study of GRBs. We review here some highlights of what we have learned over the last two years, and look ahead towards an observational program likely to settle most of the remaining GRB questions.

1 Introduction and Background

What is a gamma-ray burst (GRB)? The definition is, for now, purely phe-nomenological. A GRB is characterized by a short, intense burst of gamma-rays. The distribution of timescales for bursts suggests two classes; those which are short (with a characteristic time of 10s) and those which are are very short (with characteristic times of <1s). We refer to these classes as long and short bursts in what follows. Their very brief appearance is a primary characteristic of GRBs. While photon energies contained in a burst may cover a very broad range, the νF_ν spectrum typically peaks at a few MeV. Most of the energy of a typical burst emerges as gamma-rays. While all GRBs, by definition, share this simple phenomenology, there is no guarantee that all GRBs emerge from the same kind of source object.

What follows is a review of selected recent advances in our observational understanding of GRBs. For a recent review of GRB models, see for example Piran [1].

1.1   A Brief History

The study of gamma-ray bursts, born of the cold war, has had a most unusual path of development. GRBs were serendipitously discovered by a team of Los Alamos scientists monitoring the nuclear test ban treaty [2]. For 20 years after this GRBs were so little understood that no one knew whether they were important or not. A series of satellite experiments gradually accumulated a large, but very inhomogeneous set of GRB observations. By 1990, we had observed many bursts, but still could say nothing confident about their origin or energy scale.

This embarrassing situation began to change with the launch of NASA's Compton Gamma-Ray Observatory (CGRO) in 1991 [3]. The Burst and Transient Source Experiment (BATSE) onboard CGRO has, since then, assembled the first large and homogeneously observed set of GRBs; more than 2400 to date. Interest in GRBs increased significantly with announcements [4] from the BATSE team of the extreme isotropy of GRB arrival directions, combined with the spatial inhomogeneity implied by their intensity distribution. These results were important because they strongly suggested a cosmological origin for GRBs. Such distant sources require enormous energy releases to appear as bright flashes at the Earth. This discovery elevated GRBs from the arcane to the important.

By 1995, thousands of GRBs had been observed by gamma-ray detectors. Light curves had been pondered by many groups, and detailed statistical analyses of burst arrival directions were the subject of lengthy debate. Despite this wealth of information there remained no confident identification of GRBs with a class of source objects [5]. New information, specifically multiwavelength information, was clearly required.

1.2   GRBs and Multiwavelength Observations

It has been known since the first detection of GRBs that the key to understanding them would be detection in other wave bands. Why are these observations so elusive? First, there is a substantial mismatch between the ability of BATSE to localize bursts and the typical field of view of other instruments. An average burst is localized by BATSE to within about a 5° circle. This is much larger than the $\sim\!10'$ field of view of a typical optical telescope. Second, there is the unpredictable and very brief appearance of GRBs. Bursts occur about once a day on the sky. Any single observatory site has the opportunity to observe a burst at night only about once every two weeks. Despite considerable effort (see for example Vrba[6]), no counterpart to a GRB was observed at any other wavelength for more than 25 years.

Multiwavelength studies of astrophysical objects reveal crucial details of physics. But at least as important, they allow us to bring a powerful suite of instruments, from Keck Telescope spectroscopy to VLBI imaging to bear on a problem. The synergistic use of the instrumentation assembled by the astronomical community over the last few decades has been critical for recent progress in the GRB field. It is an example of interdisciplinary, and interagency, cooperation worth advertising.

The last five years have led to advances against each of the impediments to multiwavelength GRB observations. The remainder of this work is devoted to explaining these advances and outlining what we have learned from them.

2 GCN, Beppo-SAX, and Afterglows

2.1 Advances in Satellite Triggers

A first step towards achieving multiwavelength GRB observations was the fortuitous appearance in 1993 of the GCN system (then called BACODINE[7]). When the failure of an onboard tape system required a realtime downlink for CGRO data, it became possible to provide really rapid burst notification. Since then GCN has provided, via internet socket connections, $<\!5\mathrm{s}$ notification for most BATSE bursts, and it is expected to play a similar role for future missions[8]. GCN essentially solved the problem of rapid burst announcement.

Interestingly, GCN has evolved to play another crucial role; as the archive of record for burst observations. Followup of a burst happens rapidly, and globally, and keeping track of what everyone else is doing is crucial. The GCN circular system[9] has become the tool of choice for this purpose.

A second major advance towards multiwavelength observations was the launch in April 1996 of the joint Italian/Dutch x-ray satellite Beppo-SAX[10]. Beppo-SAX is named for Giuseppe Occialinni, in yet another reminder of the connection between particle physics and astrophysics. It combines two features important for GRB studies; a coded-aperture mask Wide-Field Camera system, capable of localizing bursts to $<\!5'$, and a set of narrow field x-ray telescopes, which can localize an x-ray afterglow to $<\!2'$. Note the enormous and critical advance which this represents. Going from $5°$ to $5'$ completely alters the approach to afterglow studies.

There are two caveats however. Beppo-SAX is not capable of rapid notification. Data is downlinked once an orbit, and additional time is required to repoint the telescope for narrow-field observations. This leads to a delay between a burst and notification of an accurate position of about four hours. Beppo-SAX can tell us where a burst occured, but only long after it is over. In addition, Beppo-SAX can trigger only on long bursts (with durations $\geq$5s), so it cannot tell us if the short bursts are a different class. Nevertheless, this ability to localize bursts on a scale comparable to the FOV of large telescopes has made possible the first big revolution in GRB studies: the detection of optical afterglows.

2.2   X-ray and Optical Afterglows

Beppo-SAX localizations, combined with agressive ground based optical follow-up, led to the first discovery of an X-ray[11] optical afterglow in February 1997[12]. Three months later, another burst gave rise not only to both optical[13] and radio [14] afterglows, but for the first time allowed measurement of a burst redshift[15].

What have we learned from afterglows? So far, X-ray afterglows have been detected for about two dozen bursts[16]. Of these about half have had optical counterparts detected. This number remains a bit soft, as followup has hardly been uniform. It is apparent that afterglows last for weeks. Generally, though not always, they fade in a smooth power law with index between -1 and -2. This is a very rapid decline, suggesting that some of the events lacking optical afterglows may merely fade more rapidly.

Interestingly, afterglow intensity seems uncorrelated with any reasonable measure of gamma-ray intensity. This is an important point, as it emphasizes that the afterglow and the burst itself are *different* phenomena. Afterglows are a somewhat generic result of releasing a large amount of energy in a small volume. The relativistic shocks which must follow such an injection of energy give rise to the afterglow. As a result, details of the afterglow may depend more on the local environment than the burst mechanism itself.

A cartoon of afterglow phenomenology, useful for understanding the observational challenges, is shown in Figure 1. They fade very rapidly, which places a tremendous premium on rapid response observations. In the first four hours, spectra for these objects can be obtained on 2-4m telescopes. After this much larger instruments are required. Extrapolations backward in time are, in most cases, purely speculative. In the case of 990123 (the only burst with simultaneous optical data), the extrapolation is an underestimate of the optical brightness.

The most important thing we have learned from afterglows is the distance

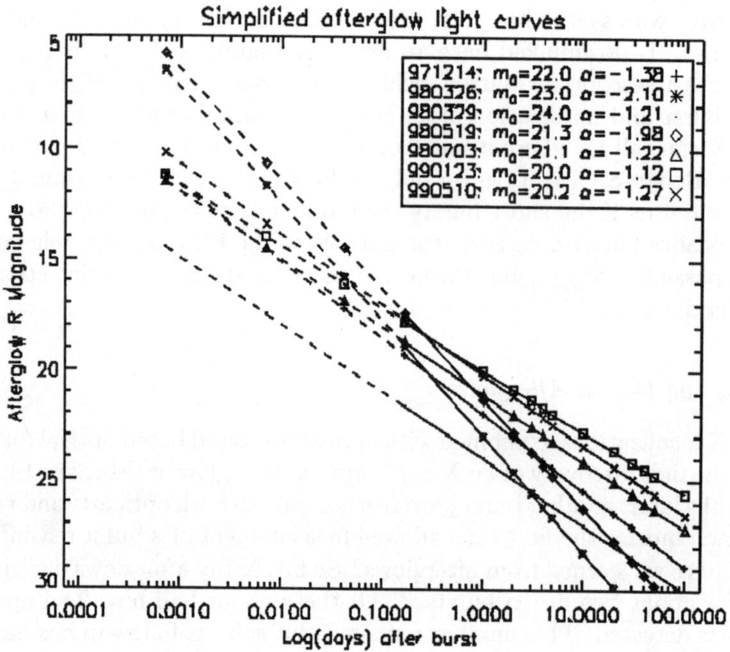

Figure 1: Cartoon of afterglow phenomenology. These curves are generated by fitting the R magnitude and decay slope at 1 day after each burst and naively extending the light curve forward and backward in time. Dashed lines for times less than 0.1d after the burst reflect the great uncertainty of early emission.

scale. There are 9 robustly measured afterglow distances. Eight of these have z>0.4 and seven have z>0.7. If we assume isotropic emission, the energy releases implied by these distances are truly stunning; typically 10^{51} ergs and as large as a 2×10^{54} ergs. For comparison, the rest energy of the sun is 1.8×10^{54} ergs. It is worth reiterating that most (> 90%) of this energy emerges as gamma-rays.

A key question which afterglows can address is the isotropy of GRB emission. Any beaming of gamma-ray emission could imply significantly lower total energy requirements. Beaming models [17] make two generic predictions about afterglow behavior. First, if GRBs are beamed into a small angle, the intrinsic event rate must be subtantially higher than the observed event rate. In addition to constraining source models, this also suggests a possible class of orphan afterglows; optical transients unassociated with observed GRBs. The search

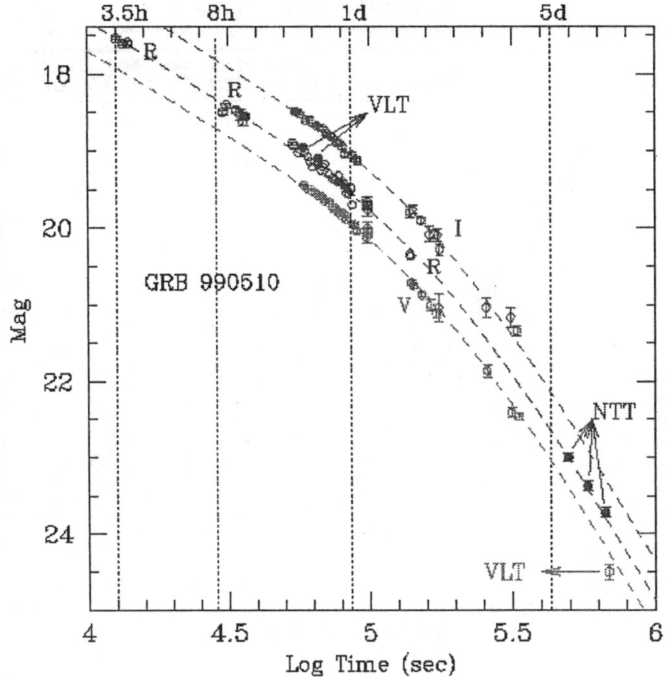

Figure 2: Example light curve for GRB990510 from Marconi et al. GCN #329.

for these will be quite complex [18], but is being undertaken by several groups.

In addition, beaming models generally predict that the fading afterglow will change character when the jet of material cools sufficiently to allow it to expand sideways. This will give rise to a rather sudden steepening of the light curve. There have been hints of this in several afterglows [19], and a very clear detection in GRB990510 [20,21] (see Figure 2 [22]). This is a particularly tantalizing result, as an early transition to such a rapid fade might explain the absence of an observed optical afterglow in some cases. While this does not prove conclusively that GRBs are beamed, it makes it clear that high quality afterglow data for a large number of events will be useful in settling these issues.

Accurate afterglow positions have allowed us to study in detail the burst location [23]. It is clear that most bursts are associated with detectable host galaxies. They are not, however, all centrally located, which probably rules out any association with AGN.

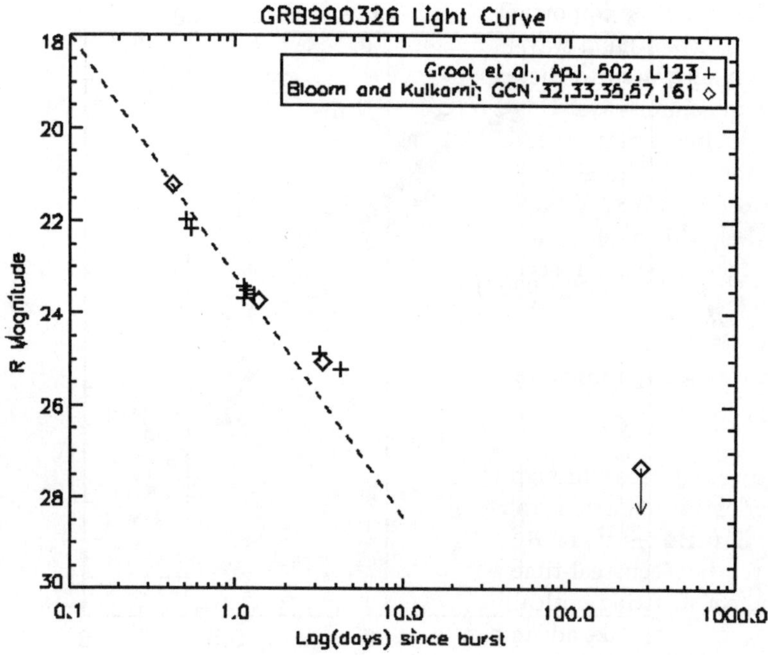

Figure 3: Light curve for GRB990326. Early flattening in the light curve was interpreted as the appearance of the host galaxy. Later observations proved this was incorrect. The suggestion has been made that this brightning represents the appearance of an underlying supernova.

Several points suggest a possible connection between GRB and supernovae. Most important is GRB980425; a burst which occured in apparent coincidence with SN 1998bw in a galaxy at z=0.0085. While the identification of the burst with this object now seems robust [24], it is clearly anomalous both as a GRB (it is very faint) and as a supernova [25] (it is extremely luminous, especially in radio). So perhaps this is another class of object. More interesting, if there is a supernova connection, the light from the underlying event may become dominant at late times. There is a hint of such behavior [26] in GRB990326, one of the fastest fading GRB yet recorded (see Figure 3). Again, there is nothing completely conclusive here, but it does argue for more extensive late time observations.

While afterglows have settled a number of key GRB questions, and completely revitalized the field, they cannot tell us everything we want to know about GRBs. Most important, they reveal the central engine of the burst in, at best, a second hand way. We noted above that the afterglow brightness is essentially unrelated to the gamma-ray brightness. Yet most of the energy is emitted in gamma-rays. Unveiling the details of this most important period of GRBs requires prompt studies of their evolution.

In addition, afterglow detection, for now, requires a Beppo-SAX localization. Since Beppo-SAX cannot trigger on short bursts, we have no information on the behavior of short bursts. Finally, earlier localization of bursts would greatly facilitate spectroscopy. For all these reasons, prompt optical observations of GRBs have been sought for many years, and by many people[6].

3   Simultaneous Optical Observations

3.1   The Robotic Optical Transient Search Experiment

A third recent breakthrough in GRB studies was the detection of contemporaneous optical emission from GRB990123 by the Robotic Optical Transient Search Experiment[27], or ROTSE. This instrument was designed to solve the mismatch between real-time BATSE localization errors and optical fields of view by constructing optical instruments which match typical BATSE errors. In this way, it can take advantage of rapid GCN notification of bursts, and later use Beppo-SAX or IPN localizations to focus analysis and confirm discovered transients.

The ROTSE I telescope, pictured in Figure 4, began operations in March 1998. It images a $16° \times 16°$ field of view to $m_v = 15$. ROTSE I has made prompt observations of more than 20 burst locations over the past year, and recently made a clear detection of optical emission from a GRB.

3.2   GRB990123

On January 23rd 1999, both BATSE[28] and Beppo-SAX[29] detected a very strong GRB (in the top 1% of the BATSE fluence distribution). ROTSE I imaging of this location began 20s after the burst trigger. About four hours after the burst Beppo-SAX announced a $5'$ localization for this event, which led almost immediately to the discovery of a bright optical afterglow[30]. The following morning analysis of automatically obtained ROTSE I data revealed a bright and rapidly varying prompt optical transient. The localization of this burst is illustrated in Figure 5. The initial GCN localization for this burst was $9°$ from the final position, primarily because BATSE triggered on a very weak precursor to this bright burst. More than a day after the burst, a redshift of $z = 1.61$ was measured from a number of absorption lines[31].

Figure 4: ROTSE I telescope in operation. Four Canon 200mm F1.8 lenses instrumented with 2048^2 CCDs are co-mounted on a rapid slewing mount.

The relation between the observed optical and gamma-ray emission during the burst is shown in the inset to Figure 6. The peak optical brightness of the burst is M_{abs}=-36.4, making this burst, briefly, the brightest optical object ever observed. It is about one thousand times brighter at its peak than the brightest quasars. The optical burst is violent, rising by a factor of 15 in 20s. There is not sufficient time resolution in the optical data to make a detailed comparison to the gamma-ray light curve, but it is at least possible to say that the two do not peak at the same time.

After reaching a peak brightness of 9th magnitude, the optical burst fades precipitously, with a decline seen in ROTSE I data to gradually slow until it merges smoothly into the afterglow. Figure 6 shows this transition, and emphasizes that, for this burst at least, the optical burst is substantially brighter than might be expected by extrapolating back the late time afterglow.

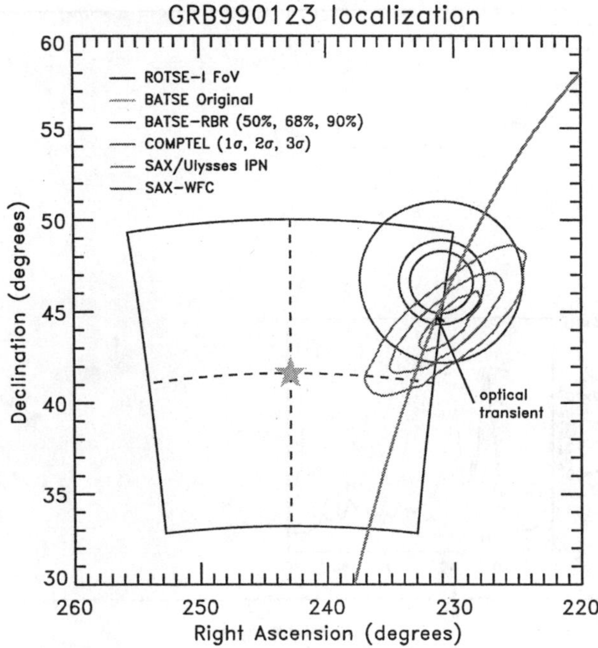

Figure 5: The localization of GRB990123 is shown as obtained by a variety of means. The SAX 5' localization is much smaller than the head of the arrow.

3.3 Observations of additional events

The optical burst was so bright that it nearly saturated the ROTSE I system. ROTSE I responds regularly, about once every 15 days, to GCN triggers. Since March 1998, it has recorded 27 GRB triggers, 18 of which are prompt responses. Analysis of 8 of these bursts which have restricted localizations (from IPN arcs or Beppo-SAX positions) has been completed [32]. No additional optical bursts have been discovered. The limits as a function of time for these bursts are shown in Figure 7.

In five of these bursts there is sufficient sensitivity to observe an optical burst as bright as that from GRB990123, yet none is seen. Since GRB990123 was so bright in gamma-rays, we naively expect that its optical emission would also be bright. This might explain the missing optical bursts. As a first test we simply rescale our optical magnitude limits by

$$-2.5 log_{10}(Fluence/Fluence_{GRB990123}) \qquad (1)$$

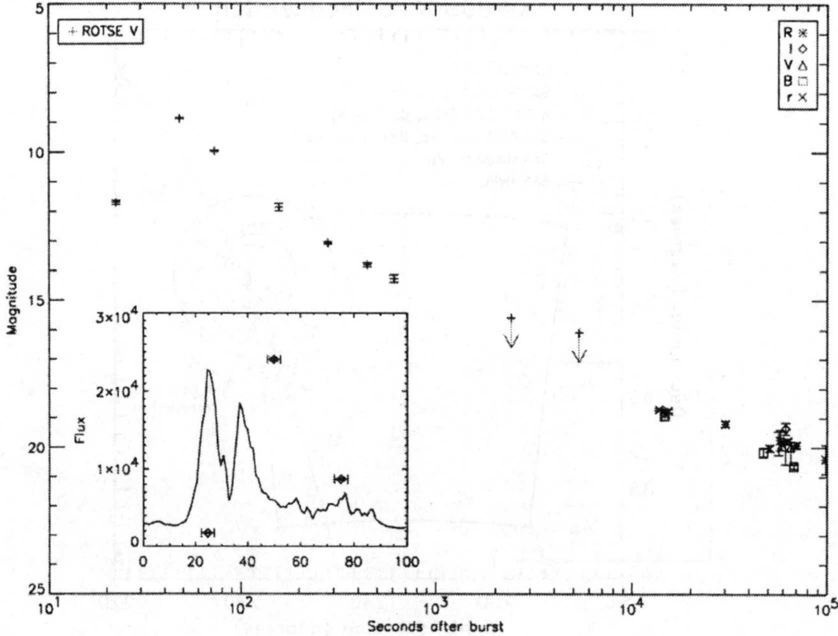

Figure 6: Inset: Gamma-ray and optical light curves of GRB990123 compared. The horizontal error bars on ROTSE points indicate the periods sampled. The main figure compares the ROTSE prompt observations (all points before 10^4s) to the later afterglow observations (drawn from the GCN archive).

The results are shown in Figure 8. ROTSE I clearly has sufficient sensitivity to observe several of these scaled bursts. Their absence forces the conclusion that there is no strong correlation between the brightness of the optical burst and the gamma-ray burst.

4 Conclusions and a Look to the Future

What have we learned from this? Most important, optical bursts exist. They had been predicted [33,34] within the canonical relativistic fireball model for GRBs. The origin of the optical burst in GRB990123 has been generally attributed to a reverse shock [35,36,37] though it remains possible that they are formed in the same internal shocks which generate the gamma-rays. The optical emission is not directly related to the gamma-ray emission, so there is

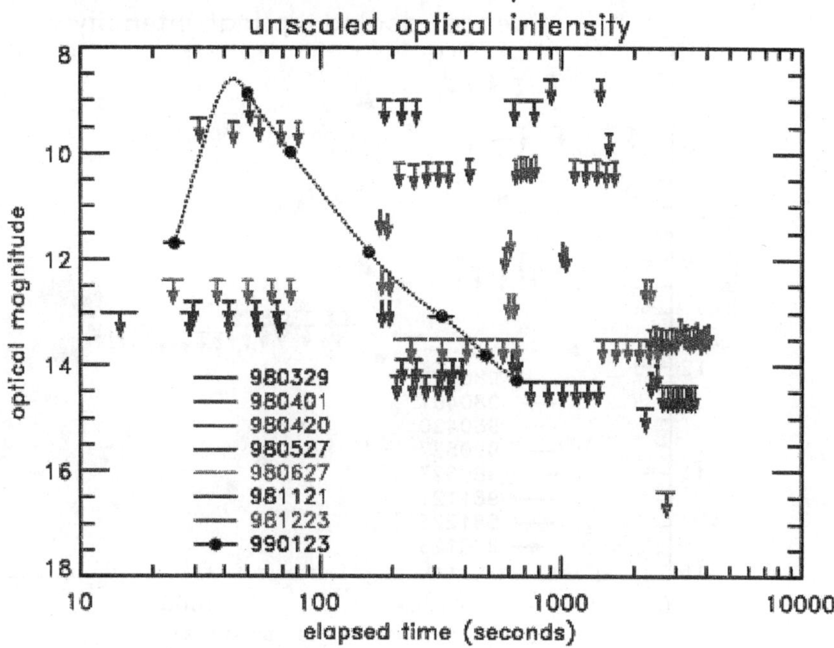

Figure 7: ROTSE optical limits for seven bursts compared to the observed optical burst from GRB990123 (dashed line).

rich early spectral evolution to be studied in bursts. The appearance of optical bursts promises the ability to localize bursts to sub-arcsecond accuracy; holding out the prospect of detailed studies of the transition from burst to afterglow, and perhaps providing bright, high redshift probes with intrinsically featureless spectra.

The field continues to expand, with a future driven by the prospect of good (5′) prompt localizations for many bursts. The HETE II mission [38] is expected to launch in January 2000. It will provide perhaps 40 prompt localizations a year. Less certain are additional missions such as Ballerina and SWIFT [39]. This last, if approved, would launch in about 2004 and distribute a 5′ localization once a day for several years. In this era, the job of following up GRBs will be enormous. The community will need to be both well organized and supported at a substantial level if we are to take full advantage of the opportunity which missions like SWIFT will provide.

Rapid response telescopes are also moving ahead. ROTSE II [40], Super-

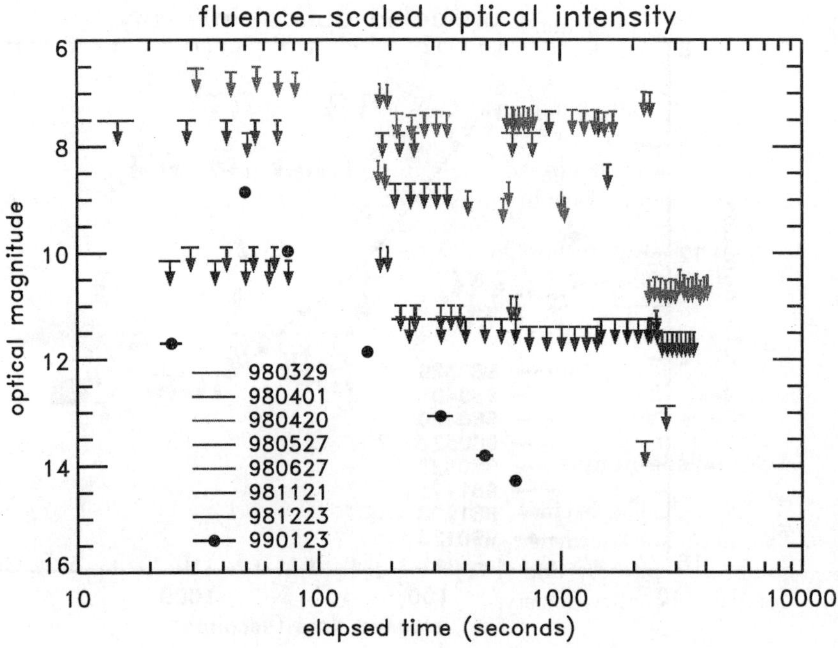

Figure 8: ROTSE optical limits for seven bursts, rescaled by the ratio of their gamma-ray fluence to that of GRB990123 compared to the observed optical burst from GRB990123.

LOTIS[41] and others promise to move rapid response sensitivity to 18th magnitude this year. In addition, the ROTSE project is addressing the relatively low event rate at a single site by constructing a global network of 8-10 0.45m telescopes, in part as preparation for future GRB missions.

4.1 Key Questions

In conclusion: what are the key questions now, and how do we expect to address them in the coming years?

- Is burst emission beamed? The answer to this question could alter the GRB energy scale by a factor of hundreds. It will be answered both by searches for orphan afterglows and by more detailed studies of afterglow evolution.

- Are bursts associated with supernovae? This will require additional late-time followup, and is perhaps best settled with additional late HST imaging.

- What is the burst mechanism? Our best information is most likely to come from more, and more detailed, early time observations. To achieve this we must upgrade rapid response equipment until it is sensitive to bursts of average brightness. Rapid localization is crucial here.

- What are the short bursts, and more generally, is there more than one class of bursts?

The GRB field has been, for the last 2 years, dominated by surprises. Given the sparse state of our current knowledge, we ought to expect this to continue.

Acknowledgments

We thank Jochen Greiner for his excellent web summary of localized GRBs[16]. The authors acknowledge support from NASA grant NAG5-5101, from NSF grants AST 9970818 and AST 9703282, from the Research Corporation, and from the Planetary Society.

References

1. T. Piran, Phys. Rep. **314**, 575 (1999).
2. R.W. Klebasedal, I.B.Strong, and R.A. Olsen, *ApJ* **182**, L85 (1973).
3. http://cossc.gsfc.nasa.gov/cossc/cossc.html
4. C.A. Meegan, *et al.* in *Proc. of the 2nd Huntsville GRB Workshop*, ed. G. Fishman, J. Brainerd, and K. Hurley (AIP, New York, 1994).
5. G.J. Fishman and C.A. Meegan, ARA&A **33**, 415 (1995).
6. F.J. Vrba, in *Proc. of the 3rd Huntsville GRB Workshop*, ed. C. Kouveliotou, M. Briggs, and G. Fishman (AIP, New York, 1996).
7. S.D. Barthelmy, *et al.*, in *Proc. of the 2nd Huntsville GRB Workshop*, ed. G. Fishman, J. Brainerd, and K. Hurley (AIP, New York, 1994).
8. S.D. Barthelmy, *et al.*, in *1999 AAS HEAD Meeting* (#31, #17.06).
9. http://gcn.gsfc.nasa.gov/gcn/gcn_main.html
10. http://www.tesre.bo.cnr.it/Research/SAX
11. E. Costa, *et al.* Nature **387**, 261 (1997).
12. J. Van Paradijs, *et al.*, Nature **386**, 686 (1997).
13. S.G. Djordovski, *et al.*, Nature **387**, 876 (1997).
14. D.A. Frail, *et al.*, Nature **389**, 261 (1997).
15. M.R. Metzger, *et al.*, Nature **387**, 878 (1997).

16. http://www.aip.de/People/JGreiner/grbgen.html
17. J. Rhoads, *ApJL* **487**, L1 (1997).
18. E. Woods and A. Loeb, *ApJ* **508**, 760 (1998).
19. S. Kulkarni, *et al.*, Nature **398**, 389 (1999).
20. K. Stanek, *et al.*, GCN **318**, (1999).
21. J. Bloom, *et al.*, GCN **323**, (1999).
22. G. Marconi, *et al.*, GCN **329**, (1999).
23. J. Bloom, *et al.*, *ApJL* **518**, L1 (1999).
24. L. Piro, *et al.*, GCN **155**, (1998).
25. T. Galama, *et al.*, Nature **395**, 670 (1998).
26. J. Bloom, *et al.*, astro-ph/9905301 , (1999).
27. C. Akerlof, *et al.*, Nature **398**, 400 (1999).
28. BATSE trigger 7343.
29. L. Piro, *et al.*, GCN **199**, (1999).
30. S. Odewahn, *et al.*, GCN **201**, (1999).
31. D.D. Kelson, *et al.*, IAUC **7096**, (1999).
32. C. Akerlof, *et al.*, In preparation for *ApJL* , (1999).
33. P. Meszaros and M. Rees, *ApJ* **476**, 232 (1997).
34. R. Sari and T. Piran, *ApJ* **520**, 641 (1999).
35. R. Sari and T. Piran, *ApJL* **517**, 109 (1999).
36. E. Fenimore, E. Ramirez-Ruiz, and B. Wu, *ApJL* **518**, 73 (1999).
37. P. Meszaros and M. Rees, MNRAS **306**, L39 (1999).
38. http://space.mit.edu/HETE
39. http://swift.gsfc.nasa.gov
40. http://www.umich.edu/~rotse
41. http://hubcap.clemson.edu/~ggwilli/LOTIS

DARK MATTER, DARK ENERGY, AND FUNDAMENTAL PHYSICS

Michael S. TURNER

Astronomy & Astrophysics Center, Enrico Fermi Institute
The University of Chicago
5640 So. Ellis Avenue, Chicago, IL 60637-1433, USA

NASA/Fermilab Astrophysics Center, Box 500
Fermi National Accelerator Laboratory
Batavia, IL 60510-0500, USA

E-mail: mturner@oddjob.uchicago.edu

More than sixty years ago Zwicky made the case that the great clusters of galaxies are held together by the gravitational force of unseen (dark) matter. Today, the case is stronger and more precise: Dark, nonbaryonic matter accounts for 30%±7% of the critical mass density, with baryons (most of which are dark) contributing only 4.5% ± 0.5% of the critical density. The large-scale structure that exists in the Universe indicates that the bulk of the nonbaryonic dark matter must be cold (slowly moving particles). The SuperKamiokande detection of neutrino oscillations shows that particle dark matter exists, crossing an important threshold. Over the past few years a case has developed for a dark-energy problem. This dark component contributes about 80% ± 20% of the critical density and is characterized by very negative pressure ($p_X < -0.6\rho_X$). Consistent with this picture of dark energy and dark matter are measurements of CMB anisotropy that indicate that total contribution of matter and energy is within 10% of the critical density. Fundamental physics beyond the standard model is implicated in both the dark matter and dark energy puzzles: new fundamental particles (e.g., axion or neutralino) and new forms of relativistic energy (e.g., vacuum energy or a light scalar field). A flood of observations will shed light on the dark side of the Universe over the next two decades; as it does it will advance our understanding of the Universe and the laws of physics that govern it.

1   In the Beginning ...

The simplest universe would contain just matter. Then, according to Einstein, its geometry and destiny would be linked: a high-density universe ($\Omega_0 > 1$) is positively curved and eventually recollapses; a low-density universe is negatively curved and expands forever; and the critical universe ($\Omega_0 = 1$) is spatially flat and expands forever, albeit at an ever decreasing rate.

As described by Sandage, such a universe is today characterized by two numbers: the expansion rate $H_0 \equiv \dot{R}(t_0)/R(t_0)$ and the deceleration parameter $q_0 \equiv -\ddot{R}(t_0)/H_0^2 R(t_0)$ where $R(t)$ is the cosmic scale factor and t_0 denotes the age of the Universe at the present epoch. Through Einstein's equations the deceleration parameter and density parameter are related: $q_0 = \Omega_0/2$. There is a consensus that we are finally closing in on the expansion rate: $H_0 = 65 \pm 5\,\mathrm{km\,sec^{-1}\,Mpc^{-1}}$ (or $h = 0.65 \pm 0.05$).[1] Type Ia supernovae seem to have provided the first reliable measurement of the deceleration parameter[2] – and a surprise: the Universe is accelerating not decelerating. So much for a simple Universe.

We have known for thirty years our Universe is not as simple as two numbers; it is much more interesting! In 1964 Penzias and Wilson discovered the cosmic microwave background radiation (CMB). Today the CMB is a minor component, $\Omega_{\mathrm{CMB}} = 2.48h^{-2} \times 10^{-5}$ which modifies the relationship between the density parameter and deceleration parameter only slightly. However, the CMB changes the early history of the Universe in a profound way: Earlier than about $40,000\,\mathrm{yrs}$ the dynamics of the Universe are controlled by the energy density of the CMB (and a thermal bath of other relativistic particles) and not matter, with the temperature being the most important parameter for describing the events taking place.

Not only do we live in a very interesting Universe, but also fundamental physics is crucial to understanding its past, present and future. Figure 1, which summarizes the present make up of the Universe, makes the point well:[3] in units of the critical density CMB photons and relic relativistic neutrinos contribute about 0.01%; bright stars contribute about 0.5%; massive neutrinos contribute more than 0.3% (SuperK), but less than about 15% (structure formation); baryons (total) contribute $4.5 \pm 0.5\%$; matter of all forms contributes $35 \pm 7\%$; and dark energy contributes $80 \pm 20\%$. By matter I mean particles with negligible pressure (i.e., nonrelativistic, or in terms of a temperature, $T \ll mc^2$); by dark energy I mean stuff with pressure whose magnitude is comparable to its energy density but negative.

While cosmology is much more than two numbers, the second of Sandage's two numbers is still very interesting and at the heart of much of what is most exciting today. Allowing for a Universe with more than just matter in it, the deceleration parameter becomes:

$$q_0 = \frac{\Omega_0}{2} + \frac{3}{2} \sum_i \Omega_i w_i \tag{1}$$

where $\Omega_0 \equiv \sum_i \rho_i/\rho_{\mathrm{CRIT}}$, Ω_i is the fraction of critical density contributed by component i and $p_i \equiv w_i \rho_i$ characterizes the pressure of component i (e.g., matter, $w_i = 0$, radiation, $w_i = \frac{1}{3}$ and vacuum energy, $w_i = -1$), and $\rho_{\mathrm{CRIT}} = 3H_0^2/8\pi G = 1.88h^2 \times 10^{-29}\,\mathrm{g\,cm^{-3}}$. Note, the energy density in component i evolves as $R^{-3(1+w)}$: R^{-3} for matter, R^{-4} for radiation, and constant for vacuum energy.

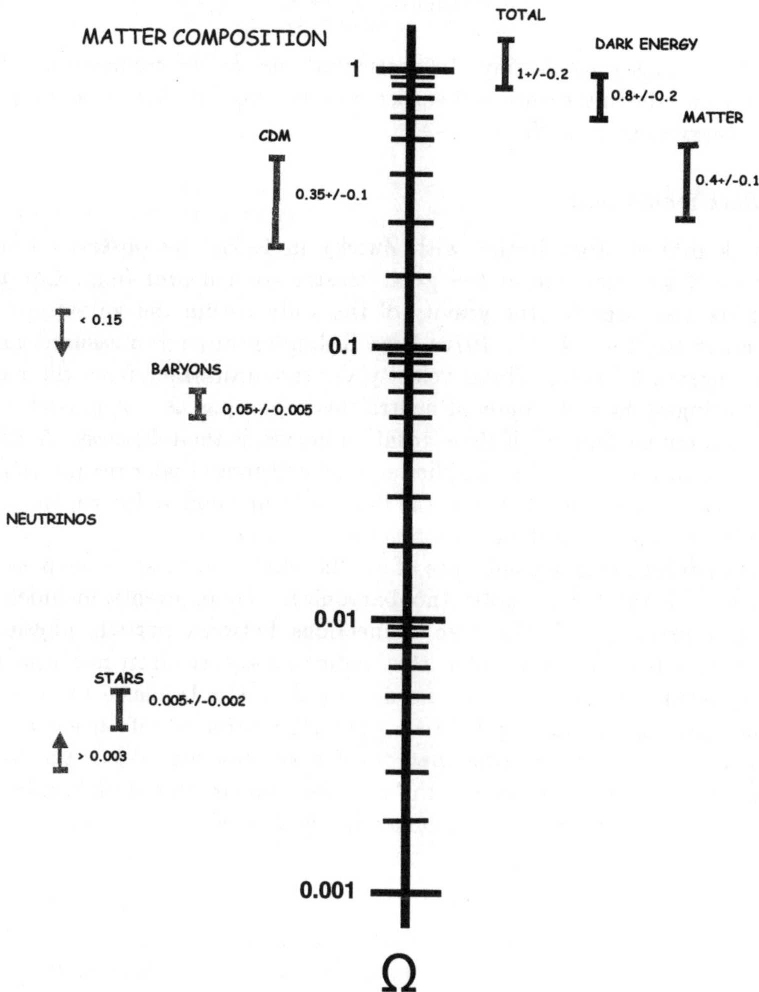

Figure 1: Summary of matter/energy in the Universe. The right side refers to an overall accounting of matter and energy; the left refers to the composition of the matter component. The contribution of relativistic particles, CBR photons and neutrinos, $\Omega_{\rm rel} h^2 = 4.170 \times 10^{-5}$, is not shown. The upper limit to mass density contributed by neutrinos is based upon the failure of the hot dark matter model of structure formation, and the lower limit follows from the evidence for neutrino oscillations. H_0 is taken to be $65\,{\rm km\,s^{-1}\,Mpc^{-1}}$.

The density parameter Ω_0 determines the geometry of the Universe:

$$R_{\mathrm{CURV}} = \frac{H_0^{-1}}{|\Omega_0 - 1|}, \tag{2}$$

but not necessarily its destiny. In particular, the simple connection between geometry and destiny mentioned earlier does not hold if there is a component to the energy density with $w_i < -\frac{1}{3}$.[4]

1.1 Dark matter past

The dark matter story begins with Zwicky in 1935. He observed that the velocities of galaxies within the great clusters of galaxies (e.g., Coma and Virgo) are too large for the gravity of the stars within the galaxies to hold the clusters together. In the 1970s Vera Rubin and others[5] measured galactic rotation curves (circular orbital velocity vs. radial distance from the galactic center) using stars and clouds of neutral hydrogen gas as test particles. The most conspicuous feature of these rotation curves is their flatness. According to Newtonian mechanics this implies an enclosed mass that rises linearly with galactocentric distance. However, the light falls off rapidly. Hence, the matter that holds ordinary spiral galaxies together must be "dark."

In the early 1980s, a confluence of events spurred interest in the possibility that the dark matter is exotic (nonbaryonic). Those events included: the growing appreciation of the deep connections between particle physics and cosmology, a Russian experiment that indicated the electron neutrino had a mass of around $30\,\mathrm{eV}$ (the mass needed to close the Universe for $h \sim 0.6$), and the growing case for gap between the dark matter density needed to hold the Universe together and what baryons can account for. While the Russian experiment proved to be wrong, the case for nonbaryonic dark matter grew and the inner space/outer space connection flourished.

1.2 Dark matter present

The case for nonbaryonic dark is now very solid and follows from the inequality, $\Omega_M = 0.35 \pm 0.07 \gg \Omega_B = 0.0045 \pm 0.005$. Briefly, here is where we stand. Big-bang nucleosynthesis provides the best accounting of the baryons. A precise determination of the primeval abundance of deuterium has allowed the baryon density to be very accurately pegged: $\Omega_B = (0.019 \pm 0.001)h^{-2} \simeq 0.045 \pm 0.005$.[6] From this follows the best determination of the total matter density.

The ratio of baryons to total mass in clusters has been determined from a sample of more than 40 clusters using x-ray and Sunyaev-Zel'dovich measurements: $f = (0.075 \pm 0.002)h^{-3/2}$.[7] (The fact that only about 15% of the matter known to be in clusters can be accounted for as baryons is already strong evidence for nonbaryonic dark matter.) Making the assumption that clusters provide a fair sample of matter, a very reasonable assumption given their large size, one can equate f to Ω_B/Ω_M and use the BBN value for Ω_B to infer: $\Omega_M = 0.35 \pm 0.07$.

There is plenty of supporting evidence for this value of the mean matter density[3] It comes from studying the evolution of the abundance of clusters (with redshift), measurements of the power spectrum of large-scale structure, relating measured peculiar velocities to the observed distribution of matter, and observations of the outflow of material from voids. Further, every viable model for explaining the evolution of the observed structure in the Universe from density inhomogeneities of the size detected by COBE and other CMB anisotropy experiments requires nonbaryonic dark matter.

We have a very strong case that the bulk of the nonbaryonic dark matter is cold dark matter (slowly moving particles). This is based upon the many successes of the cold dark matter scenario for the formation of structure in the Universe, as well as the many failures of the hot dark matter scenario. We also have two very compelling – and highly testable – particle candidates: the axion and the neutralino[8] A very light axion (mass $\sim 10^{-6}\,\mathrm{eV} - 10^{-4}\,\mathrm{eV}$) is motivated by the use of Peccei-Quinn symmetry to solve the strong CP problem. A neutralino of mass $50\,\mathrm{GeV}$ to $500\,\mathrm{GeV}$ is motivated by low-energy supersymmetry.

On the experimental side, we now have the first evidence for the existence of particle dark matter. The SuperKamiokande Collaboration has presented a very strong case for neutrino oscillations based upon the direction dependent deficit of atmospheric muon neutrinos, which implies at least one of the neutrinos has a mass greater than about $0.1\,\mathrm{eV}$[9] This translates into a neutrino contribution to the critical density of greater than about 0.3% (about what stars contribute). *The issue is no longer the existence of particle dark matter, but the quantity of particle dark matter. An important threshold has been crossed.*

There are now experiments operating with sufficient sensitivity to directly detect particle dark matter in the halo of our own galaxy for the two most promising CDM candidates: axions and neutralinos[8] The axion dark matter experiment at Livermore National Laboratory is slowly scanning the favored mass range; the DAMA experiment in Gran Sasso and the CDMS experiment in the Stanford Underground Facility (soon to be relocated in the Soudan Mine in Northern Minnesota) are now probing a part of neutralino parameter space that is favored by theory.

1.3 Baryonic dark matter and MACHOs

There are actually two dark-matter problems; the second being the discrepancy between the mass density contributed by bright stars (about 0.5% of the critical density) and the BBN-determined mass density of about 4.5% of the critical density. As this discussion will illustrate, a baryon inventory is much easier to do at 1 sec, when the baryons exist as a smooth soup of hadronic matter, than today, when they are dispersed in stars, stellar remnants, hot gas, cold gas, and so on.

At redshifts of around 3 to 4, most of the baryons were still in gas in the intergalactic medium (IGM). This is what numerical simulations of CDM say and what observations of the IGM at high redshift reveal. At this time, structure was just beginning to form and can be observed by studying the absorption of matter between us and distant quasars. The baryon accounting based upon these observations does indeed account for essentially all the baryons, though assumptions must be made and the uncertainties are not as small as at BBN.[10]

In clusters of galaxies today the accounting is complete: most of the baryons are in the hot, intracluster gas that glows in x-rays. The gas outweighs stars by about 10 to 1. However, only about 5% of galaxies are in the great clusters of galaxies, so this leaves the accounting very incomplete. Globally, only about 1/3 of the BBN baryon density can be accounted for, in the form of stars, cold gas, and warm gas within galaxies. The other 2/3 is *presumed* to be in hot intergalactic gas and/or warm gas associated with galaxies. One of the challenges for astrophysics is to complete the baryon accounting today by detecting this gas. Efforts will involve both x-ray and UV instruments looking for absorption or emission lines associated with the gas.

A dark horse possibility for the dark baryons is dark stars (low-mass objects that never lit their nuclear fuels or the end points of stellar evolution such as white dwarfs, neutron stars and black holes that have exhausted their nuclear fuels). Such objects in the halo of our own galaxy can be detected by microlensing. Microlensing of stars in the bulge of galaxy and in the Large and Small Magallenic Clouds by dark, foreground objects has been detected by the EROS, MACHO, DUO and OGLE groups. This is one of the exciting developments of the decade: These rare (one in a million or so stars is being lensed at any time) brightenings have provided a new probe of the dark side of the Universe. Already binary lenses, a black-hole candidate, planets and important information about the structure of the galaxy (strong evidence for a bar at the center) have been revealed, and one very intriguing mystery remains.

While the handful of events toward the SMC can be explained as "self lensing," foreground objects in the SMC lensing SMC stars, the more than twenty occurrences of microlensing of LMC stars are not so easily understood. Because the LMC is (thought to be) more compact, self lensing is less important. If one interprets the LMC lenses as a halo population of dark objects, they would account for about 50% of own halo. The mass inferred from the timescale of the brightenings (about $0.5\,M_\odot$) and the stringent limits to the number of main-sequence stars of this mass points to white dwarfs. (Recent HST observations give evidence for a handful of nearby, fast-moving white dwarfs, consistent with a halo population of white dwarfs.)

Beyond that, nothing else makes sense for this interpretation. Since white dwarf formation is very inefficient there should be 6 to 10 times as much gas left over as there are white dwarfs. This of course would exceed the total mass budget of the halo by a wide margin. The implied star formation rate exceeds the measured star formation rate in the Universe by more than an order of magnitude. And where are their siblings who are still on the main sequence?

Since microlensing only determines a line integral of the density of lenses toward the LMC, which is heavily weighed by the nearest 10 kpc or so, it gives little information about where the lenses are. Its limitations for probing the halo are significant: It cannot probe the halo at distances greater than the distance to the LMC (50 kpc), and as a practical matter it can only directly probe the innermost 15 kpc or so of the halo. Recall, the mass of the halo increases with radius and the halo extends at least as far as 200 kpc.

Alternative explanations for the LMC lenses have been suggested.[11] An unexpected component of the galaxy (e.g., a warped and flaring disk, a very thick disk component, a heavier than expected spheroid, or a piece of cannibalized satellite-galaxy between us and the LMC) which is comprised of conventional objects (white dwarfs or lower-main sequence stars); LMC self lensing (the LMC is being torn apart by the Milky Way and may be more extended than thought); or a halo comprised of $0.5\,M_\odot$ primordial black holes formed around the time of the quark/hadron transition (which also acts as the cold dark matter). For all but the last, very speculative explanation, the mass in lenses required is less than 10% of the halo.

Because the cold dark matter framework is so successful and a baryonic halo raises so many problems (in addition to those above, how to form large-scale structure), I am putting my money on a CDM halo. More data from microlensing is crucial to resolving this puzzle. The issue might also be settled by a dazzling discovery: direct detection of halo neutralinos or axions or the discovery of supersymmetry at the Tevatron or LHC.

2 Dark Energy

The discovery of accelerated expansion in 1998 by the two supernova teams (Supernova Cosmology Project and the High-z Supernova Team) was the most well anticipated surprise of the century. It may also be one of the most important discoveries of the century. Instantly, it made even the most skeptical astronomers take inflation very seriously. As for the hard-core, true-believers like myself, it suffices to say that there was a lot of dancing in the streets.

2.1 Anticipation

In 1981 when Alan Guth put forth inflation most astronomers responded by saying it was an interesting idea, but that its prediction of a flat universe was at variance with cosmological fact. At that time astronomers argued that the astronomical evidence pointed toward $\Omega_M \sim 0.05 - 0.10$ (even the existence of a gap between Ω_B and Ω_M was debatable). Inflationists took some comfort in the fact that the evidence was far from conclusive; it was largely based upon the mass-to-light ratios of galaxies and clusters of galaxies, and it did not sample sufficiently large volumes to reliably determine the mean density of matter. As techniques improved, Ω_M rose. Especially encouraging (to inflationists) were the determinations of Ω_M based upon peculiar velocity data (large-scale flows). They not only probed larger volumes and the mass more directly, but also by the early 1990s indicated that Ω_M might well be as large as unity.

Even so, beginning in the mid 1980s, the Omega problem ($\Omega_M \ll 1$) received much attention from theorists who emphasized that the inflationary prediction was a flat universe ($\Omega_0 = 1$), and not $\Omega_M = 1$ (though certainly the simplest possibility). A smooth, exotic component was suggested to close the gap between Ω_M and 1 (smooth, so that it would not show up in the inventory of clustered mass). Possibilities discussed included a cosmological constant (vacuum energy), relativistic particles produced by the recent decay of a massive particle relic and a network of frustrated topological defects.[12]

By 1995 it seemed more and more unlikely that $\Omega_M = 1$; especially damning was the determination of Ω_M based upon the cluster baryon fraction discussed earlier. On the other hand, the CDM scenario was very successful, especially if $\Omega_M h \sim 1/4$ (the shape of the power spectrum of density inhomogeneity today depends upon this product because it determines the epoch when the Universe becomes matter dominated). Add to that the tension between the age of the Universe and the Hubble constant, which is exacerbated for large values of Ω_M. ΛCDM, the version of CDM with a cosmological constant ($\Omega_M \sim 0.4$ and $\Omega_\Lambda \sim 0.6$), was clearly the best fit CDM model (see Figure 2). And it has a smoking gun signature: accelerated expansion ($q_0 = \frac{1}{2} - \frac{3}{2}\Omega_\Lambda$).

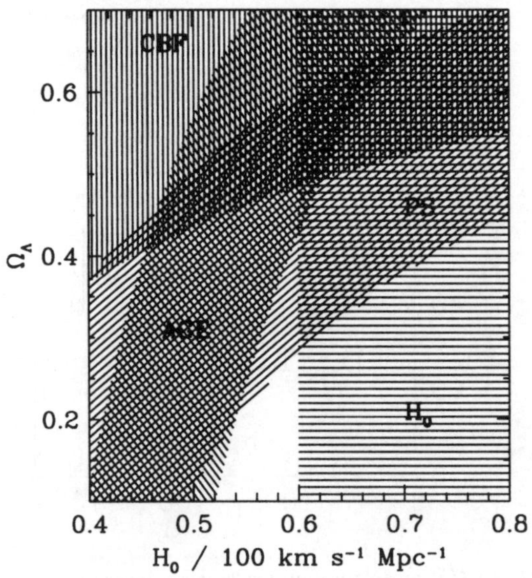

Figure 2: Constraints used to determine the best-fit CDM model: PS = large-scale structure + CBR anisotropy; AGE = age of the Universe; CBF = cluster-baryon fraction; and H_0= Hubble constant measurements. The best-fit model, indicated by the darkest region, has $H_0 \simeq 60 - 65\,\mathrm{km\,s^{-1}\,Mpc^{-1}}$ and $\Omega_\Lambda \simeq 0.55 - 0.65$.

At the June 1996 Critical Dialogues in Cosmology meeting at Princeton, in the CDM beauty contest the only mark against ΛCDM was the early result from the Supernova Cosmology Project indicating that $\Omega_\Lambda < 0.5\,(95\%)$.[14]

After the Princeton meeting the case grew stronger as CMB anisotropy results began to define the first acoustic peak at around $l = 200$, as predicted in a flat Universe (the position of the first peak scales $l = 200/\sqrt{\Omega_0}$). Today, the data imply $\Omega_0 = 1\pm0.1$[13] (see Figure 3). With results from the Boomerang Long-duration Balloon experiment expected in January, the DASI experiment at the South Pole next summer, and the launch of the MAP satellite in the Fall of 2000, we can expect a truly definitive determination of Ω_0 soon.

The smoking-gun confirmation came in early 1998 with the results from the two supernova groups indicating that the Universe is speeding up, not slowing down. Everything now fit together: inflation and the flat universe, the CMB determination that $\Omega_0 \sim 1$ and the cluster measurement of $\Omega_M \sim 0.4$, and the successes of CDM, and ΛCDM in particular (see Figures 2-4). In the minds of theorists like me, the only surprise was that it took the cosmological constant to make everything work. Everything was pointing in that direction,

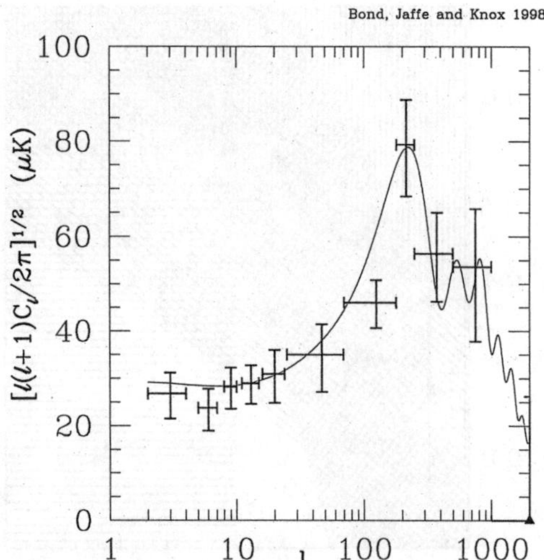

Figure 3: Summary of CMB anisotropy measurements, binned to reduce error bars. The theoretical curve is for the ΛCDM model with $H_0 = 65\,\mathrm{km\,s^{-1}\,Mpc^{-1}}$ and $\Omega_M = 0.4$ (Figure courtesy of L. Knox).

and were it not to the checkered history of the cosmological constant, there would have been no surprise at all.

2.2 The dark-energy problem

At the moment, a crucial element in the case for accelerated expansion and dark energy is the "independent confirmation" based upon the otherwise discrepant numbers $\Omega_0 \sim 1$ and $\Omega_M \sim 0.4$. Balancing the books requires a component that is smooth and contributes about 60% of the critical density. In order that it not interfere with the growth of structure, its energy density must evolve more slowly than matter so that there is a long matter-dominated era during which the observed structure today can grow from the density inhomogeneities measured by COBE and other CMB anisotropy experiments. Since $\rho_X \propto R^{-3(1+w_X)}$, this places an upper limit to w_X:[15] $w_X < -\frac{1}{2}$, and in turn, an upper limit to q_0: $q_0 < \frac{1}{2} - \frac{3}{4}\Omega_X < 0$ for $\Omega_X > \frac{2}{3}$ and a flat Universe.

Because of the checkered history of the cosmological constant – cosmologists are quick to invoke it to solve problems that later disappear and particle physicists have failed to compute it to an accuracy of better than a factor of 10^{55} – there is an understandable reluctance to accept it without some skep-

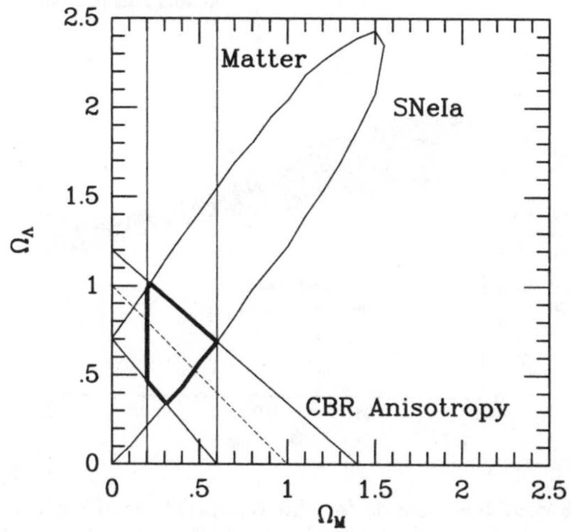

Figure 4: Two-σ constraints to Ω_M and Ω_Λ from CBR anisotropy, SNe Ia, and measurements of clustered matter. Lines of constant Ω_0 are diagonal, with a flat Universe indicated by the broken line. The concordance region is shown in bold: $\Omega_M \sim 1/3$, $\Omega_\Lambda \sim 2/3$, and $\Omega_0 \sim 1$.

ticism. To wit, other possibilities have been suggested: For example, a rolling scalar field (essentially a mini-episode of inflation),[16] or a frustrated network of very light topological defects (strings of walls).[17]

My preference is to characterize it as simply and most generally as possible, by its equation of state: $p_X = w_X \rho_X$, where w_X is -1 for vacuum energy, $-\frac{N}{3}$ for frustrated topological defects of dimension N, and time-varying and between -1 and 1 for a rolling scalar field. The goal then is to determine w_X and test for its time variation.[18]

In determining the nature of dark energy, I believe that telescopes and not accelerators will play the leading role – even if there is a particle associated with it, it is likely to be extremely difficult to produce at an accelerator because of its gravitational or weaker interactions with ordinary matter. Specifically, I believe that type Ia supernovae will prove to be the most powerful probe. The reason is two fold: first, the dark energy has only recently come to be important; the ratio $\rho_M/\rho_X = (\Omega_M/\Omega_X)(1 + z)^{-3w_X}$ grows rapidly with redshift. Secondly, dark energy does not clump (or at least not significantly), so its presence can only be felt through its effects on the large-scale dynamics of the Universe. Type Ia supernovae have the potential of reconstructing the recent history of the evolution of the scale factor of the Universe and from it, to shed light on the nature of the dark energy. Figures 5 and 6 show the simulated

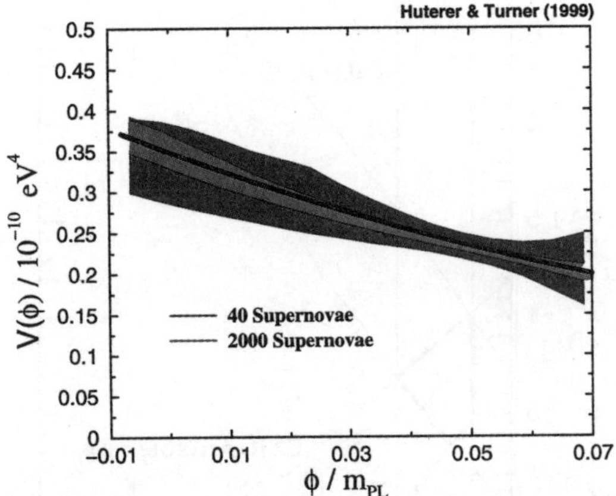

Figure 5: The 95% confidence bands for the simulated reconstruction of the scalar-field potential for a quintessence model; the solid curve is the original potential.

reconstruction of two dark-energy models: a quintessence model (scalar field rolling down a potential) by means of supernovae and a variable equation of state.

Once one is convinced with high confidence that there is dark energy out there (the next round of CMB anisotropy results will be crucial) and that type Ia supernovae are standardizable candles (more study of nearby supernovae), the next step is a dedicated assault, probably a satellite based telescope (which I like to call DaRk-Energy eXplorer or D-REX) to collect 1000s of supernovae between redshift 0 and 1. By carefully culling the sample and doing good follow up and accurate photometry one will able to address determine Ω_X, w_X and address the time variation of w_X.[19]

3   Looking Forward

The two dominant ideas in cosmology over the past 15 years have been inflation and cold dark matter. They have provided the field with a grand guiding paradigm which has spurred the observers and experimenters to put in place a remarkable program that keep the field of cosmology lively for at least two decades.

Over the past few years this paradigm has begun to be tested in a significant way, with many more important tests to come. The first tests have been encouraging. The first acoustic peak in the CMB power spectrum indicates a flat Universe and is consistent with the scale-invariant inflationary power spec-

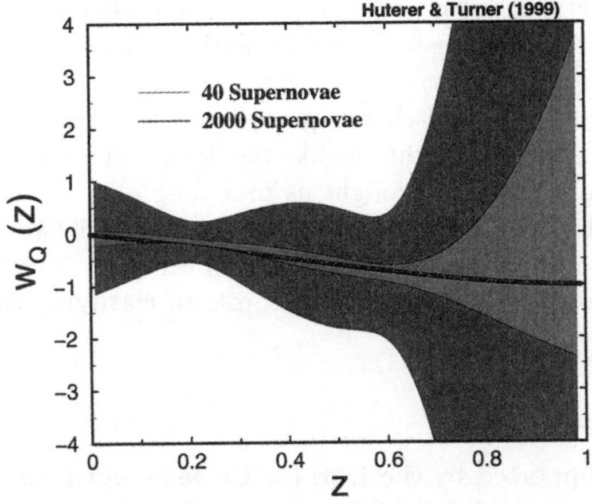

Figure 6: The 95% confidence bands for the simulated reconstruction of the equation of state, $w(z)$; the solid curve is the input equation of state. Note, because dark energy becomes less important (relative to matter) at high redshift, it becomes more difficult to probe its properties.

trum which predicts a series of acoustic peaks. The discovery of accelerated expansion provided the evidence for the component that balanced the books: our flat universe = 40% dark matter + 60% dark energy. This is only the beginning of this great adventure.

Central to cosmology and the connection between cosmology and fundamental physics are the two dark problems: dark matter and dark energy. The dark matter problem is more than sixty years old and quick mature. We have divided the dark matter problem into two distinct problems, dark baryons and nonbaryonic dark matter, and narrowed the possibilities for each. The baryons are most likely in the form of diffuse, hot gas. The nonbaryonic dark matter is most likely slowly moving, essentially noninteracting particles (cold dark matter), with relic elementary particles from the earliest moments being the leading candidate. Foremost among them are the axion and neutralino.

We could still be in for some surprises: the CDM particles could be something more exotic (primordial black holes or superheavy particles produced in the reheating process at the end of inflation). Likewise, the simple and thus far very successful assumption that the only interactions of the CDM particles that are relevant today are gravitational, could be wrong. There are some hints otherwise: The halo profiles predicted for noninteracting CDM are too cuspy at the center.[20] The resolution could be astrophysical or it could involve fundamental physics. Perhaps, it is indicating that the CDM particles have significant interactions today (scattering or annihilations) that round off the central cusps. It is intriguing to note that neither the axion nor the neutralino has such interactions.

By comparison, the dark-energy problem is in its infancy. The evidence for it, while solid, is not air tight. Unlike the dark-matter problem where sixty years of detective work have brought us to a couple of very specific suspects, the possibilities for the dark energy are wide open. But two things are clear: as with the dark-matter problem, the solution certainly involves fundamental physics, and telescopes will play a major role in clarifying the nature of the dark energy.

Acknowledgments

This work was supported by the DoE (at Chicago and Fermilab) and by the NASA (through grant NAG 5-7092 at Fermilab).

References

1. See e.g., J.R. Mould et al, astro-ph/9909260.
2. A.G. Riess, et al., *Astron. J.* **116**, 1009 (1998); S. Perlmutter et al, *Astrophys. J.*, in press (1999) (astro-ph/9812133).
3. See e.g., M.S. Turner, astro-ph/9811454.
4. L. Krauss and M.S. Turner, *Gen. Rel. Grav.*, in press (astro-ph/9904020).
5. *Dark Matter in the Universe* (IAU Symposium #117), eds. J. Knapp and J. Kormendy (Reidel, Dordrecht, 1987).
6. S. Burles et al., *Phys. Rev. Lett.* **82**, 4176 (1999).
7. J. Mohr et al, *Astrophys. J.*, in press (1999) (astro-ph/9901281).
8. B. Sadoulet, *Rev. Mod. Phys.* **71**, S197 (1999).
9. Y. Fukuda et al, *Phys. Rev. Lett.* **81**, 1562 (1998).
10. See e.g., D. Weinberg et al, *Astrophys. J.* **490**, 546 (1997).
11. E. Gates and G. Gyuk, astro-ph/9911149.
12. E.W. Kolb and M.S. Turner, *The Early Universe* (Addison-Wesley, Redwood City, CA, 1990).
13. S. Dodelson and L. Knox, astro-ph/9909454.
14. *Critical Dialogues in Cosmology*, ed. N. Turok (World Scientific, Singapore, 1997); L. Krauss and M.S. Turner, *Gen. Rel. Grav.* **27**, 1137 (1995).
15. M.S. Turner, astro-ph/9904049.
16. M. Bronstein, *Phys. Zeit. Sowjet Union* **3**, 73 (1933); M. Ozer and M.O. Taha, *Nucl. Phys.* **B 287** 776 (1987); K. Freese et al., *ibid* **287** 797

(1987); L.F. Bloomfield-Torres and I. Waga, *Mon. Not. R. astron. Soc.* **279**, 712 (1996); J. Frieman et al, *Phys. Rev. Lett.* **75**, 2077 (1995); K. Coble et al, *Phys. Rev.* **D55**, 1851 (1996); R. Caldwell et al, *Phys. Rev. Lett.* **80**, 1582 (1998); B. Ratra and P.J.E. Peebles, *Phys. Rev.* **D 37**, 3406 (1988).

17. A. Vilenkin, *Phys. Rev. Lett.* **53**, 1016 (1984); D. Spergel and U.-L. Pen, *Astrophys. J.* **491**, L67 (1997).

18. M.S. Turner and M. White, *Phys. Rev.* **D 56**, R4439 (1997); S. Perlmutter, M.S. Turner and M. White, *Phys. Rev. Lett.* **83**, 670 (1999).

19. D. Huterer and M.S. Turner, *Phys. Rev.* **D 60**, 081301 (1999); and in preparation.

20. B. Moore, *Nature* **370**, 629 (1994).

ATMOSPHERIC NEUTRINOS

K. KANEYUKI

Department of Physics, Tokyo Institute of Technology,
2-12-1, Ookayama, Meguro, Tokyo, 152-8551, Japan
E-mail: kaneyuki@hp.phys.titech.ac.jp

Fully contained (F.C.) and partially contained (P.C.) atmospheric neutrino data of a 52kton·yr exposure of the Super-Kamiokande detector exhibit a zenith angle dependent deficit of muon neutrinos. The zenith angle distribution of upward through going and stopping muon data are also suggest the neutrino oscillation. The data are in good agreement with a two-flavor $\nu_\mu - \nu_\tau$ oscillation hypothesis. The recent Macro (upward muon) and Soundan-II (F.C.) results are consistent with the Super-Kamiokande results.

1   Atmospheric neutrinos

Atmospheric neutrinos detected by large underground detectors provide a good probe to test the neutrino properties. Originally, atmospheric neutrinos were studied as the background of nucleon decay experiments. These neutrinos are produced by the decay chain of mesons, which are generated by the collisions of the primary protons and nucleus at the top of the atmosphere. Neutrinos travel downward for the distance of the depth of atmosphere ($\sim 10km$) and upward for the diameter of the earth ($\sim 10^4 km$). The observed zenith angle is directly related to the neutrino path length. Therefore, they can be utilized to test the neutrino oscillation hypothesis from $\Delta m^2 = 10^0$ to $10^{-4} eV^2$. Kamiokande[1] and IMB[2] reported the deficit of the μ-like data. Recently, Super-Kamiokande[34] and Soudan-II [5] confirmed this anomaly. The zenith angle distributions of the Super-Kamiokande data strongly suggest the neutrino oscillation hypothesis.

2   Super-Kamiokande

The Super-Kamiokande is a 50 kton Water Cherenkov detector at the Kamioka observatory deep underground (2700m.w.e.) in Japan. The detector consists of a cylindrical tank to store 50 kton water, viewed by 11,146 of 50cm diameter PMTs attached at wall of the inner detector and 1885 of 20cm diameter PMTs in the outer detector. The fiducial volume of the atmospheric neutrino and the proton-decay analysis is defined as to be 2m inside from the inner detector wall, 22.5 kton.

Sub-GeV	Data	M.C.
1 ring	3678	4371.6
e-like	1826	1754.0
μ-like	1852	2617.6
Multi-GeV		
F.C.		
1 ring	790	901.3
e-like	439	414.3
μ-like	351	487.0
P.C.	450	656.4

Table 1. Summary of F.C.+P.C. data sample. Total detector exposure is 52kton·yr. Expected number of atmospheric neutrino events are based on 40yr M.C. in Honda's flux calculation.

Using a 52kton·yr exposure of the Super-Kamiokande detector, total of 6982 F.C. events and 470 P.C. events were selected in the fiducial volume. Total live time was 848 days. Each single ring F.C. event was identified as e-like or μ-like according to its cherenkov ring pattern and its opening angle, and used for the atmospheric neutrino and proton decay analysis. Table 1 shows the summary of each data sample. The absolute values predicted by Monte Carlo (M.C.) has large uncertainty, coming from the large uncertainty in the atmospheric neutrino flux calculation and neutrino interaction cross sections. Therefore, the double ratio, $R = (\mu/e)_{data}/(\mu/e)_{M.C.}$ is used to cancel the absolute uncertainty. Super-Kamiokande obtained the R for Sub-GeV ($E_{vis} < 1.3 GeV$) as

$$\frac{(\mu/e)_{data}}{(\mu/e)_{M.C.}} = 0.680^{+0.023}_{-0.022}(stat.) \pm 0.053(syst.),$$

and for Multi-GeV ($E_{vis} > 1.3 GeV$) + P.C. as

$$\frac{(\mu/e)_{data}}{(\mu/e)_{M.C.}} = 0.678^{+0.042}_{-0.039}(stat.) \pm 0.080(syst.).$$

Super-Kamiokande confirmed the small R values for Sub-GeV and Multi-GeV data samples with much more statistics than the KAMIOKANDE and other experiments.

The absolute normalization of the expectation has a large uncertainty. However, the shape of the zenith angle distribution has much smaller uncertainty for high energy neutrinos. Especially, most of the systematic uncertainties and biases are canceled if the up/down asymmetry is considered. The

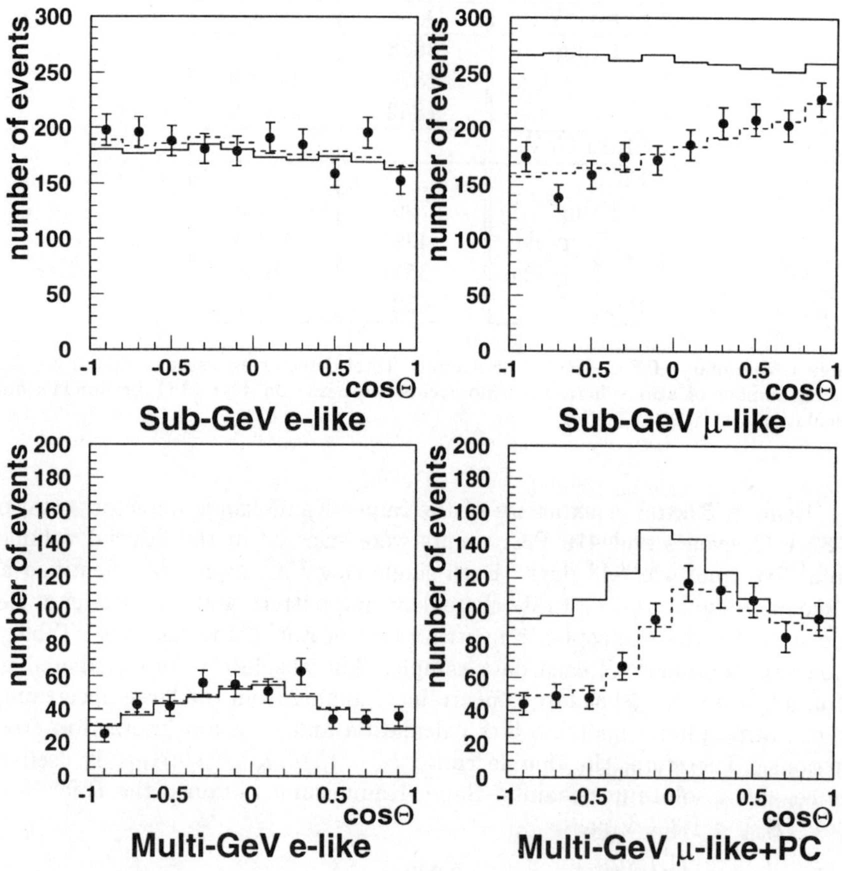

Figure 1. Zenith angle distribution of Sub-GeV e-like (upper left), Sub-GeV μ-like (upper right), Multi-GeV e-like (lower left) and Multi-GeV μ-like+P.C. (lower right). Solid histograms show the atmospheric neutrino Monte Carlo expectation and dashed histograms show the best fit of the $\nu_\mu - \nu_\tau$ oscillation with $(\Delta m^2, \sin^2 2\theta) = (3.5 \times 10^{-3} \text{eV}^2, 1.0)$

azimuthal distribution of the atmospheric neutrino agrees with the Monte Carlo expectation well[6]. Figure 1 shows the zenith angle distributions of Sub-GeV e-like, μ-like, Multi-GeV e-like and μ-like+P.C. events obtained by

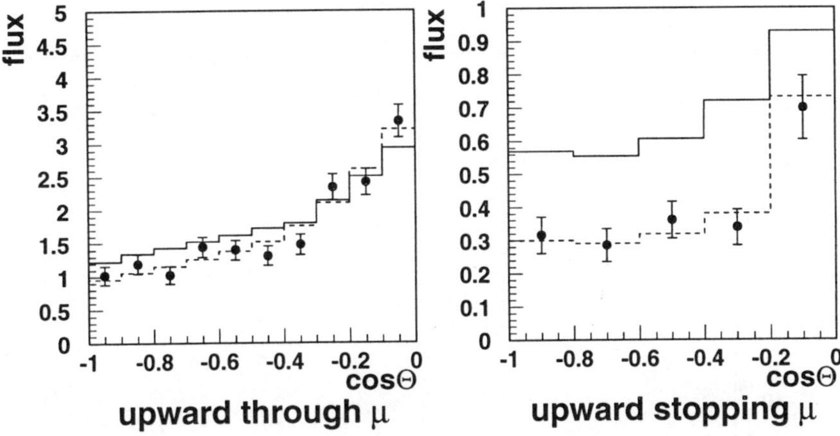

Figure 2. Zenith angle distribution of upward through going muon (left) and of upward stopping muon (right). Solid histograms show the atmospheric neutrino Monte Carlo expectation and dashed histograms show the best fit of the $\nu_\mu - \nu_\tau$ oscillation with $(\Delta m^2, \sin^2 2\theta) = (3.5 \times 10^{-3} \mathrm{eV}^2, 1.0)$

Super-Kamiokande. Solid histograms show the atmospheric neutrino Monte Carlo expectation. These distributions clearly show that the upward-going μ-like events are smaller than the expectation while the downward-going μ-like events are consistent with the expectation. The distributions of e-like events agree with the expectation well. The up/down asymmetry defined as $A = (\mathrm{up} - \mathrm{down})/(\mathrm{up} + \mathrm{down})$ for the Multi-GeV+P.C. data is $A = -0.32 \pm 0.04(stat.) \pm 0.01(syst.)$. No systematic biases can explain this large up/down asymmetry. These distributions strongly suggest that about half of muon neutrinos have disappeared while traveling a distance of $10^3 \sim 10^4 \mathrm{km}$.

The upward through going and stopping muons are also generated by the atmospheric neutrinos. The down going muons produced by the neutrinos can't be distinguished from the cosmic ray muon background. Thus, only the upward going muon can be utilized for the neutrino oscillation analysis. The neutrino path length for the horizontally direction is about $10^3 \mathrm{km}$ and the vertically direction is about $10^4 \mathrm{km}$. The energy spectra of the parent neutrinos of upward stopping and through going muon says the mean neutrino energy

222

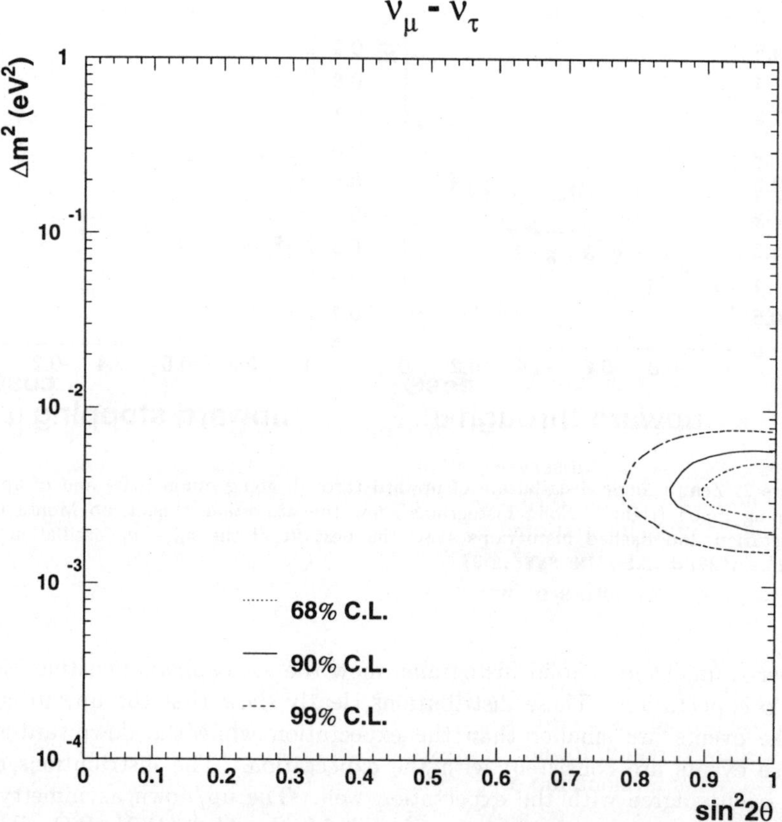

Figure 3. Allowed region with 68% C.L.(dotted), 90%C.L.(solid) and 99%C.L.(dashed) for F.C.+P.C.+upward through going and stopping muon. The best fit is obtained at $(\Delta m^2, \sin^2 2\theta) = (3.5 \times 10^{-3} \text{eV}^2, 1.0)$ with χ^2_{min} =67.5/82d.o.f.

are about 10GeV and 100GeV, respectively. Figure 2 (a) shows the zenith angle distribution of the upward through going muons in Super-Kamiokande. A total of 1028 events were observed after the 923 days live time. The observed flux is $1.70 \pm 0.05 \pm 0.02$ ($\times 10^{-13} \text{cm}^{-2}\text{s}^{-1}\text{sr}^{-1}$), and expectation based on the Bartol and GRV94DIS is 1.97 ± 0.44. The shape of the distribution is not well represented by the theoretical prediction without neutrino oscillation. The χ^2 for no oscillation is 24.7 with 10 d.o.f.

Figure 2 (b) shows the zenith angle distribution of the upward stopping muons. A total of 245 upward stopping muon events were observed after the 902 days live time. The observed flux is $0.40 \pm 0.03 \pm 0.02$, and expectation is 0.73 ± 0.16. The upward stopping muon data shows the large discrepancy compared with the neutrino Monte Carlo prediction.

Super-Kamiokande observed the small R in the Sub-GeV and Multi-GeV+P.C. data samples, and a large up/down asymmetry in the Multi-GeV+P.C. sample. Observed zenith angle distributions of F.C., P.C., upward through going and stopping muons strongly suggest a $\nu_\mu - \nu_\tau$ oscillation, because zenith angle distributions of e-like data are consistent with the expectation. Thus, a $\nu_\mu - \nu_\tau$ oscillation analysis were performed by using the following χ^2 for F.C., P.C., upward through going and stopping muons,

$$\chi^2(\sin^2 2\theta, \Delta m^2) = \sum \frac{(N_{data} - N_{fit})^2}{\sigma^2} + \sum \frac{\alpha_i^2}{\sigma_{\alpha_i}^2}, \tag{1}$$

where N_{data} is the observed number of events or flux in each zenith angle and momentum bin, N_{fit} is the fitted expected number of events or flux, calculated by using parameters, α_i, for each oscillation parameters ($\sin^2 2\theta$, Δm^2), and σ_{α_i} is the systematic uncertainty. The χ^2 value was minimized by varying parameters α_i within errors for each oscillation parameter. The minimum χ^2 value was obtained at ($\sin^2 2\theta$, Δm^2) = $(1.0, 3.5 \times 10^{-3} \text{eV}^2)$ with 67.5/82d.o.f. in the physical region. The dashed histograms in Fig.1,2 show the fitted expected zenith angle distribution at the χ^2_{min} point, and it reproduces the data well. Figure 3 shows the 68%, 90% and 99% confidence level allowed region for $\sin^2 2\theta$ and Δm^2 obtained by this oscillation analysis. If we assume no neutrino oscillation, χ^2 value is 214/82 d.o.f.

3   Macro

Macro (Monopole Astrophysics and Cosmic Ray Observatory) is the large tracking detector in Gran Sasso underground laboratory. The detector consists of six super-modules with overall dimensions of $12\text{m} \times 77\text{m} \times 10\text{m}$. Each super-module consists of planes of streamer tube for tracking, liquid scintillation counter for T.O.F. measurement and crushed rock absorber.

From the 607 upward going muon events which have more than 1GeV, Macro obtained the following ratio

$$\frac{data}{M.C.} = 0.74 \pm 0.031(stat.) \pm 0.044(syst.) \pm 0.12(theoritical).$$

They obtained χ^2_{min} is 12.5/8 d.o.f. in the physical region at $(\Delta m^2, \sin^2 2\theta)$ $= (2.5 \times 10^{-3} \text{eV}^2, 1.0)$. The probability for no oscillation hypothesis is 0.36%.

They obtained the consistent allowed region with Super-Kamiokande from this oscillation analysis [8].

Macro also analyzed the low energy atmospheric neutrino events, internal up-going events, up-going stopping events and internal down-going events. The statistics of these sample are not so large. However, they Macro obtained the deficit of the data compare to the prediction for each data sample. The $\nu_\mu - \nu_\tau$ oscillation expectation with $(\Delta m^2, \sin^2 2\theta) = (2.5 \times 10^{-3} \text{eV}^2, 1.0)$ also reproduce the zenith angle distribution of low energy data [9].

4 Soudan-II

Soudan-II detector is a 960 ton fine-grained gas tracking calorimeter. The detector is 8 m wide, 5.5 m high and 16 m long, and it consists of 224 modules. A total of 15 thousand of the drift tube are used to measure the position and the dE/dx. The detector is surrounded by a 1700 m^2 active shield to veto the cosmic ray muons.

From the 4.6 kilo ton year detector exposure, Soudan-II obtained the double ratio using single track events and show events as,

$$\frac{(track/shower)_{data}}{(track/shower)_{M.C.}} = 0.68 \pm 0.11(syst.) \pm 0.06(stat.).$$

Soudan-II used the high resolution sample for the neutrino oscillation analysis. The zenith angle distribution of ν_e data is consistent with the no oscillations, while that of ν_μ shows a deficit of the measured number of events with respect of the prediction. However, no dramatic up/down asymmetry was observed on ν_μ data. From this data, Soudan-II obtained the consistent allowed region with Super-Kamiokande [10].

5 $\nu_\mu - \nu_\tau$ v.s. $\nu_\mu - \nu_s$

The ν_τ charged current interaction in the detector is strongly suppressed by the large mass of τ in the atmospheric neutrino energy region. Thus, the $\nu_\mu - \nu_\tau$ oscillation represent the deficit of ν_μ events. If the fourth and sterile neutrino (ν_s) exists, the $\nu_\mu - \nu_s$ oscillation may also reproduce the data well. The $\nu_\mu - \nu_\tau$ and $\nu_\mu - \nu_s$ oscillation can be distinguished by (1) ν_τ charged current interaction in the detector, (2) ν_τ neutral current interaction in the detector and (3) matter effects in the earth. The effect (1) is suppressed by the large τ mass and is difficult to distinguish from the atmospheric neutrino background. Thus, we have to wait long baseline experiments. The effect (2) may checked by the neutral current single π production. However, the

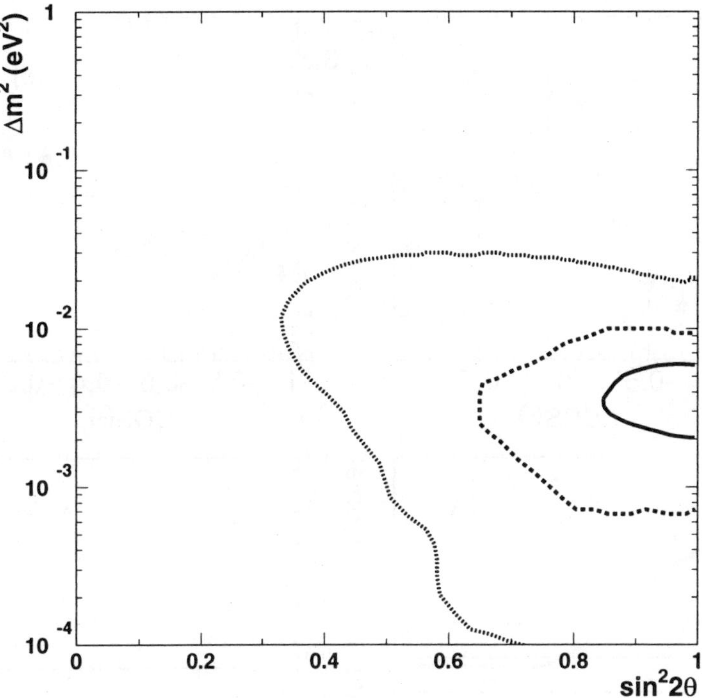

Figure 4. Allowed region with 90% C.L. for Super-Kamiokande F.C.+P.C.+upward through going and stopping muon (solid), for Macro upward through going muon (dashed) and for Soudan-II F.C. high resolution data (dotted).

uncertainty of the neutral current cross section is large ($\sim 20\%$). This uncertainty will be reduced by a large number of π^0 data obtained by the K2K front detector. The effect (3) may be utilized by using the difference of the zenith angle distribution of the high energy atmospheric neutrino events.

The $\nu_\mu - \nu_s$ oscillation is suppressed by the MSW effect in the earth. Thus, the zenith angle distribution of P.C. and upward through going muon show the difference for each hypothesis[11]. Figure 5 (a) shows the zenith angle distribution of P.C. data with Evis > 5.0GeV compared with $\nu_\mu - \nu_\tau$ (dashed) and $\nu_\mu - \nu_s$ (dotted) oscillation expectation with $\Delta m^2 = 10^{-2.5}$ and

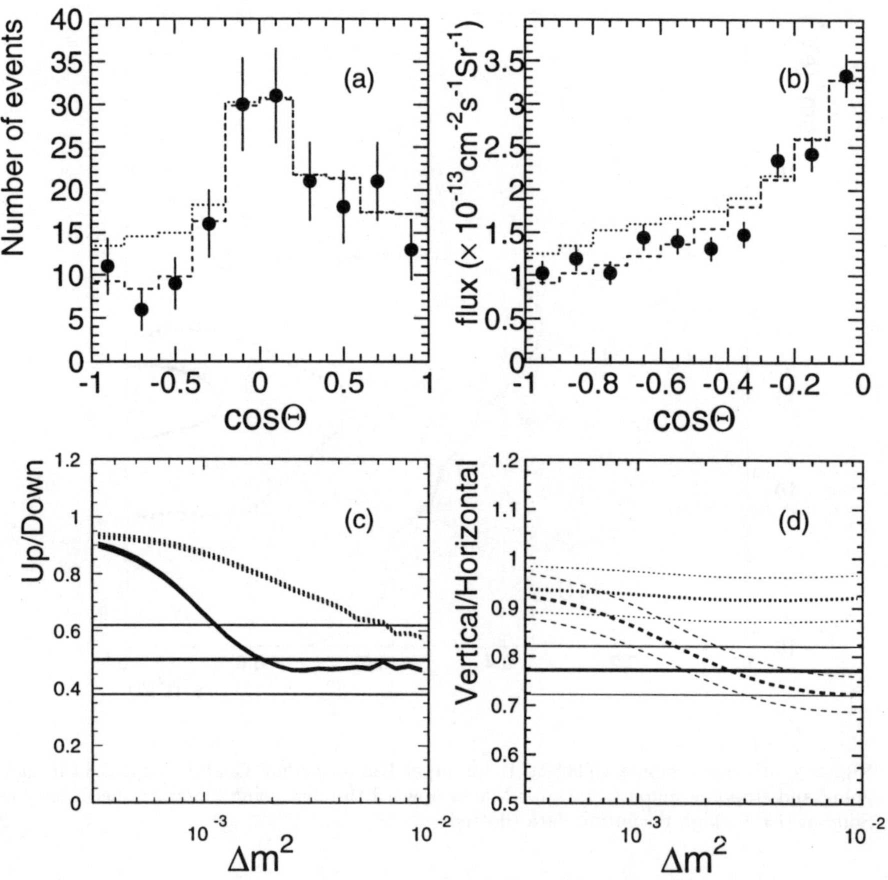

Figure 5. (a) Zenith angle distribution of P.C. data (Evis > 5.0GeV). The dashed histogram shows the expectation for $\nu_\mu - \nu_\tau$ oscillation and the dotted histogram shows the same for $\nu_\mu - \nu_s$ oscillation. (b) Same distributions for the upward through going muon. (c) The expected up/down ratio of P.C. for $\nu_\mu - \nu_\tau$ and $\nu_\mu - \nu_s$ oscillation as a function of Δm^2 at $\sin^2 2\theta = 1.0$. The horizontal line shows the data with the statistical error. (d) The expected upward/horizontal ratio of upward through going muon.

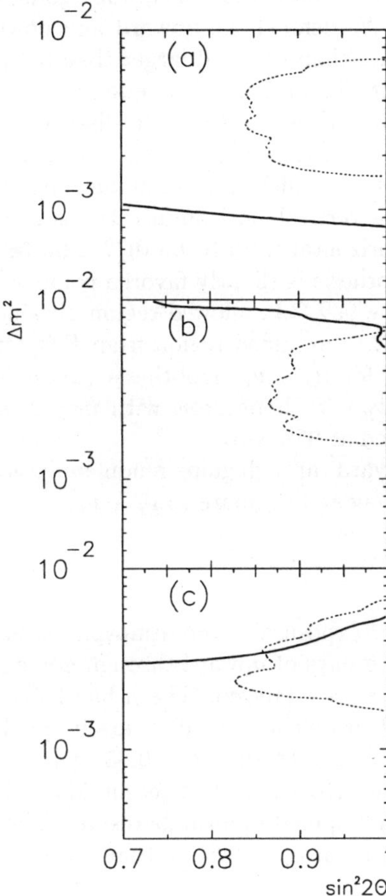

Figure 6. (a) The 90%C.L. allowed region obtained by using F.C. data compared with 95% C.L. excluded region from F.C.+upward through going muon for transfer-matrix S transforms the state function $f(x)$ and $\nu_\mu - \nu_\tau$ hypothesis. (b) Same distribution for $\nu_\mu - \nu_s$ hypothesis with $m_{\nu_\mu} < m_{\nu_s}$. (c) Same distribution for $\nu_\mu - \nu_s$ hypothesis with $m_{\nu_\mu} > m_{\nu_s}$.

$\sin^2 2\theta = 1.0$. The $\nu_\mu - \nu_s$ oscillation of upward going events are suppressed by the MSW effect. We defined the upward and downward to be cosine of the zenith angle smaller than -0.4 and larger than 0.4. Figure 5 (c) show the up/down ratio as a function of Δm^2. The data gives up/down $= 0.50 \pm 0.12 \pm 0.01$. Figure 5 (b) shows the zenith angle distribution of upward through going data compared with the expectation. We defined the horizontal and vertical to be cosine of the zenith angle of muon is larger or smaller than -0.4. Figure 5 (d) show the vertical/horizontal ratio as a function of Δm^2. The data gives vertical/horizontal ratio to be $0.77 \pm 0.07 \pm 0.04$. For both data samples, $\nu_\mu - \nu_\tau$ hypothesis is slightly favored compare to $\nu_\mu - \nu_s$.

Figure 6 shows the 90%C.L. allowed region obtained by using F.C. data compared with 95% C.L. excluded region from F.C.+upward through going muon combined chi^2 for $\nu_\mu - \nu_\tau$ hypothesis (a), $\nu_\mu - \nu_s$ hypothesis with $m_{\nu_\mu} < m_{\nu_s}$ (b) and $\nu_\mu - \nu_s$ hypothesis with $m_{\nu_\mu} > m_{\nu_s}$ (c). The $\nu_\mu - \nu_s$ hypothesis is disfavored at 95% C.L.

Recent Macro upward through going muon analysis also show that $\nu_\mu - \nu_\tau$ hypothesis is slightly favored compare to $\nu_\mu - \nu_s$.

6 Conclusion

The total of 52kton·yr exposure of the atmospheric neutrino F.C. and P.C. data and 923 (902) live days of upward through going (stopping) muon data in Super-Kamiokande were analyzed. The μ-like F.C. and P.C. data exhibit a zenith angle dependent deficit. The data are in good agreement with two-flavor $\nu_\mu - \nu_\tau$ oscillations with $\sin^2 2\theta > 0.85$ and $2 \times 10^{-3} < \Delta m^2 < 6 \times 10^{-3} \mathrm{eV}^2$ at 90% confidence level. The recent Macro and Soudan-II results are consistent with the Super-Kamiokande results. The high energy P.C. and upward through going muon samples show that $\nu_\mu - \nu_s$ hypothesis is disfavored at 95% C.L.

References

1. K.S.Hirata *et al. Phys. Lett.* B **205**, 416 (1988).
2. D.Casper *et al. Phys. Rev. Lett.* **66**, 2561 (1991).
3. Y.Fukuda *et al. Phys. Lett.* B **433**, 9 (1998).
4. Y.Fukuda *et al. Phys. Lett.* B **436**, 33 (1998).
5. W.W.M.Allison *et al. Phys. Lett.* B **391**, 491 (1997).
6. Y.Fukuda *et al. Phys. Rev. Lett.* **82**, 5194 (1999).
7. Y.Fukuda *et al. Phys. Rev. Lett.* **81**, 1572 (1998).
8. M.Ambrosio *et al. Phys. Lett.* B **434**, 451 (1998).

9. F.Surdo *et al.* Proceedings of 26th ICRC (1999).
10. A.Mann *et al.* To be appeared in the proceedings of XIX International Symposium on Lepton and Photon Interactions at High Energies, (1999).
11. R.Foot *et al.* *Phys. Rev.* D **58**, 013006 (1998).

REACTOR AND ACCELERATOR BASED NEUTRINO OSCILLATION EXPERIMENTS

R. MASCHUW

Institut für Strahlen- und Kernphysik, Universität Bonn, Nussallee 14–16,
53115 Bonn, Germany
E-mail: maschuw@iskp.uni-bonn.de

This paper describes the current status and recent results of both reactor and accelerator based neutrino oscillation experiments. The reactor experiments CHOOZ and Palo Verde are sensitive to mixing parameters relevant to the atmospheric neutrino problem whereas the future KamLAND experiment may even test the large mixing angle MSW-solution of the solar neutrino problem. In the cosmological relevant region $\Delta m^2 \geq 1\,eV^2$ the high energy neutrino experiments CHORUS and NOMAD so far have not found a signal for $\nu_\mu \to \nu_\tau$-oscillations. The controversial results of LSND and KARMEN on $\bar{\nu}_\mu \to \bar{\nu}_e$-oscillations require new experiments testing the region $10^{-1}\,eV^2 \leq \Delta m^2 \leq 1\,eV^2$. The potential of terrestrical Long BaseLine neutrino experiments is discussed in view of the recent SuperKamiokande atmospheric neutrino results.

1 Introduction

Atmospheric and solar neutrino experiments are linked to the phenomenon of neutrino oscillations as these would explain their results deviating subtantially from our standard model expectations. Accelerator and reactor neutrino oscillation experiments on the other hand, being intrinsicly dedicated to the question of neutrino mass and mixing can be tuned to cover specific areas of the oscillation parameter space. A nondegenerate neutrino mass spectrum $m_i \neq m_j$ and the existence of mass mixing $\nu_l^{fl} = \sum U_{li} \times \nu_i^m$ (i = 1,2,3; l $= e, \mu, \tau$) are the necessary prerequisits for neutrino flavour oscillations to occur. Assuming a mass hierachy in the neutrino sector $m_1 \ll m_2 \ll m_3$ and restricting to a two-generation mixing scheme the oscillation probability for a given neutrino flavour to transmute into some other flavour is given by $P_{osc} = \sin^2 2\Theta * \sin^2 1.27\,\Delta m^2[eV^2] * \{L[km]/E[GeV]\}$. The oscillation parameters Θ and Δm^2 are the mixing angle between the neutrino mass eigenstates and the difference of the squared masses of these eigenstates, $\Delta m^2 = \left| m_2^2 - m_1^2 \right|$, respectively. L denotes the distance from the neutrino source and E represents the neutrino energy. Not knowing if and where nature might have choosen Δm^2 and $\sin^2 2\Theta$ to be, the experimental parameter L/E has to be varied and occasionally to meet $L/E \approx \Delta m^2$ to get an oscillation probability significantly different from zero. The appearance of a new

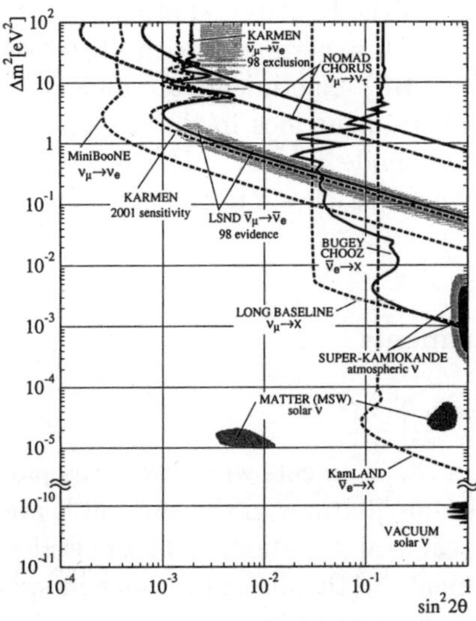

Figure 1. Status of neutrino oscillation experiments in a two flavour mixing scheme. Hatched areas: evidence claims; solid lines: exclusion limits; dashed lines: future sensitivities.

neutrino flavour and the measurement of its L/E-dependence would be the most convincing signal for neutrino oscillations to exist. Simultaneously the corresponding disappearance probability {1-P(L/E)} would cause a depletion of the original neutrino flux and alter its spectral shape depending on Δm^2 in a disappearance experiment.

The general status of the experimental search for neutrino oscillations is reflected in Fig. 1. If no oscillation is observed then statistical and systematic uncertainties determine an upper limit for the oscillation probability which is transformed into an exclusion or sensitivity curve in the two dimensional plot of Δm^2 vs. $\sin^2 2\Theta$. This has been the case for all accelerator or reactor based oscillation experiments performed so far (solid exclusion curves in Fig. 1) except one, the LSND experiment, which in the channel $\bar{\nu}_\mu \to \bar{\nu}_e$ has found an excess of $\bar{\nu}_e$ above background. Atmospheric and solar neutrinos both show a deficit of neutrinos compared to what is expected. If these findings are interpreted in terms of evidence for neutrino oscillations, these could have been caused by oscillation parameters within the hatched areas of Fig. 1. The dashed lines reflect the future prospects of neutrino oscillation experiments with terrestrial neutrino sources indicating their sensitivities to the oscillation parameters. The LSND/KARMEN controversy on the oscillation $\bar{\nu}_\mu \to \bar{\nu}_e$ needs to be resolved by new experiments with higher significance. Accelerator based Long BaseLine (LBL) experiments are underway to test the oscillation hypothesis of atmospheric neutrinos in the disappearance channel $\nu_\mu \to x$ as well as the appearance mode $\nu_\mu \to \nu_\tau$. The reactor based long baseline experiment KamLAND even reaches into the parameter space of solar neutrino oscillations.

2 Reactor Experiments

2.1 CHOOZ and Palo Verde

Reactor based neutrino experiments which were the pioneering experiments in the search for neutrino oscillations in the early 80's are using $\bar{\nu}_e$ from the fission products in nuclear power stations. Their energies of up to about 8 MeV restrict these experiments to the dissappearance mode $\bar{\nu}_e \to x$ as oscillated $\bar{\nu}_\mu$ or $\bar{\nu}_\tau$ of this energy cannot produce muons or taus in charged current interaction processes. The technique of those experiments is therefore to place a detector of suitable size at some distance or even different distances from the reactor and measure precisely the $\bar{\nu}_e$-energy spectrum. A neutrino flux deficit and a characteristic spectral distortion would indicate an oscillation with specific mixing parameters $\sin^2 2\Theta$ and Δm^2. The detection reaction for $\bar{\nu}_e$ is the classical inverse beta decay $^1\mathrm{H}(\bar{\nu}_e, e^+)n$ where both the e^+ and the n are detected in coincidence in a liquid scintillator calorimeter. Due to the low $\bar{\nu}_e$-energies reactor oscillation experiments are sensitive to relatively low Δm^2-values however, the disappearance mode restricts the sensitivity on $\sin^2 2\Theta$ hardly to be below 10^{-1}. To enhance the Δm^2-sensitivity current experiments use larger and larger detector masses at increasing distances from the source.

The currently best experiment, CHOOZ [1], named after its location in France, has applied a 5 metric tons Gd-loaded scintillator in an acrylic vessel viewed by 192 eight-inch phototubes through an unloaded scintillator buffer volume immersed in an optically separated active cosmic muon veto shield. This calorimeter is housed in a pit with 300 m.w.e. overburden about 1 km away from a power station with two reactors of $8.5\,\mathrm{GW}_{th}$ total thermal power. Having taken data since 1997 CHOOZ has not found any significant deviation from what was expected, neither in the total flux nor in the spectral shape. The corresponding 90 % CL exclusion curve for the dissappearance oscillation $\bar{\nu}_e \to x$ is indicated in Fig. 2. It covers completely the $\nu_\mu \to \nu_e$-interpretation of the former Kamiokande atmospheric neutrino result and thus exclude this oscillation channel by the argument of CP conservation.

A competing similar experiment, Palo Verde [2] in the US has started data taking in 1998. Contrary to CHOOZ it is a segmented detector of 66 acrylic modules containing a total of 16 metric tons of Gd-loaded scintillator. It is surounded by a water shielding buffer and enclosed in a muon veto. Each module is viewed by five-inch phototubes from either end looking for coincidences between the e^+- and the $Gd(n,\gamma)$-capture signal. The detector is located in a shallow laboratory (32 m.w.e) at 750 m from one and 850 m from two other reactors of the Palo Verde power station. From the first 70 days of data taking again no flux depletion has been observed. The corresponding 90 % CL exclusion curve is also shown in Fig. 2 and might approach the CHOOZ sensitivity after another year of data taking.

2.2 *The KamLAND detector*

The most advanced low energy neutrino project is probably the KamLAND [3] experiment currently being installed at the Kamiokande site. There the old KI/III water Cherenkov detector is replaced by a huge liquid scintillation detector i.e. a plastic balloon of 13 m diameter filled with $1200\,m^3$ isoparaffin based liquid scintillator. Surrounded by a pure isoparaffin buffer volume it is viewed by 1300 17-inch phototubes mounted on a 18 m diameter sphere. Because of its target mass this detector can measure electron antineutrinos from nuclear reactors at distances as far as 200 km. The five most relevant nuclear power stations at distances betwen 170 km and 210 km will provide about 750 $\bar{\nu}_e$-events per year. The experiment is therefore ideally suited to look for $\bar{\nu}_e \rightarrow x$-oscillations with an unprecedented sensitivity to very low Δm^2-values of $\Delta m^2 \geq 7 \times 10^{-6}\,eV^2$. This ultimate reactor neutrino oscillation experiment will thus be able to prove the large mixing angle MSW solution of the solar neutrino problem as indicated in Fig. 1. At a later stage KamLAND may also investigate neutrinos from the earth crust, supernova neutrinos or even solar neutrinos thus serving as a multipurpose detector contributing to the field of geophysics, astrophysics and particle physics.

3 Accelerator Based Neutrino Oscillation Experiments

3.1 *General Scheme*

Accelerator based neutrino oscillation experiments in general use muon neutrinos ν_μ or $\bar{\nu}_\mu$ from the decay of pions and kaons being produced when the particle beam of a proton facility hits some production target. At rather low proton energies up to about 1 GeV the pions might even be stopped in the production target providing a decay at rest (DAR) neutrino source of ν_μ, ν_e and $\bar{\nu}_\mu$ with well defined energy spectra from the decay chain $\pi^+ \to \mu^+ + \nu_\mu$ followed by $\mu^+ \to e^+ + \nu_e + \bar{\nu}_\mu$. At high energy facilities pions and kaons are normally horn-focussed into a decay tunnel where behind a massive muon and hadron absorber they provide a beam of ν_μ or $\bar{\nu}_\mu$ with only a small contamination of electron type neutrinos. Appropriate neutrino detectors are located hundreds of meters or even kilometers away from the target looking for either the dissappearance $\nu_\mu \to x$ of the original neutrino flavour or the appearance of a new flavour $\nu_\mu \to \nu_e$ or $\nu_\mu \to \nu_\tau$. Detection reactions are typically quasi elastic or deep inelastic CC and NC scattering processes or at lower energies, distinct neutrino nuclear reactions with well defined final nuclear states. In the past the search for neutrino oscillations at accelerators has often been just a byproduct of the investigation of the properties of electro-weak interaction and the structure of matter at the large neutrino facilities of CERN, Fermilab and others. Recent experiments, however, are dedicated neutrino oscillation experiments taylored to explore certain regions of the mixing parameter space of specific interest.

3.2 CHORUS and NOMAD — the cosmological Δm^2-region

The mass region $m \geq 1\,\mathrm{eV}$ is often attributed as the cosmologically relevant region in view of a hot or mixed dark matter scenario. This had been the main motivation for CHORUS [4] and NOMAD [5] at CERN to probe the region of $\Delta m^2 \geq 1\,\mathrm{eV}^2$ in a search for $\nu_\mu \to \nu_\tau$, assuming the heaviest neutrino to be the major component of ν_τ. Both experiments are using the SPS wide band neutrino beam with energies ranging from 10 to 40 GeV and very small $\bar{\nu}_\mu$- and ν_e-contamination (5 % and 1 % resp.). Evidence for a $\nu_\mu \to \nu_\tau$-appearance would be the detection of ν_τ via the CC-reaction $\nu_\tau N \to \tau^- X$ with subsequent leptonic or semileptonic τ-decay.

The heart of the CHORUS detector is an emulsion target of 770 kg total mass where the topology of the ν_τ-reaction can directly be observed. Tracking devices behind the emulsion sheets are guiding back to the reaction vertex in the emulsion. There, powerful scanning sytems with $< 1\ \mu$m resolution will identify the reaction topology looking for a typical decay kink in the τ-track after a decay length of a few tenth of a millimeter. The stack of emulsion targets and scintillating fibre trackers is followed by a magnetic spectrometer, a calorimeter and a muon spectrometer for particle identification and momentum reconstruction. The most relevant τ-decay branches looked for by CHORUS are $\tau^- \to \mu^- \nu_\tau \bar{\nu}_\mu$ and $\tau^- \to \pi^- \nu_\tau$. However neither in the 1μ- nor in the 0μ-data sample analysed so far any ν_τ-candidate has been found.

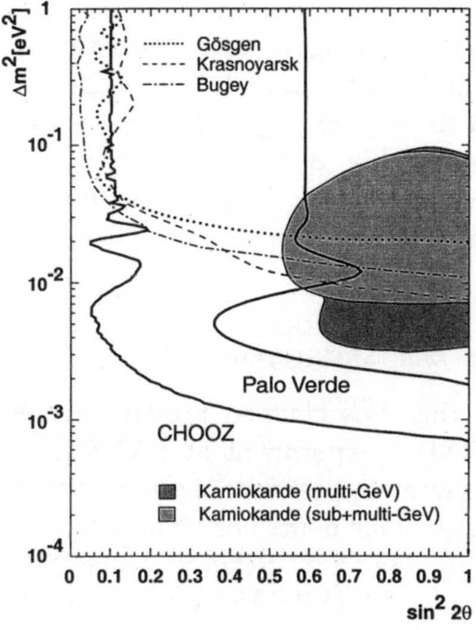

Figure 2. $\bar{\nu}_e \to x$-exclusion limits from CHOOZ and Palo Verde; hatched areas: evidence claims by Kamiokande.

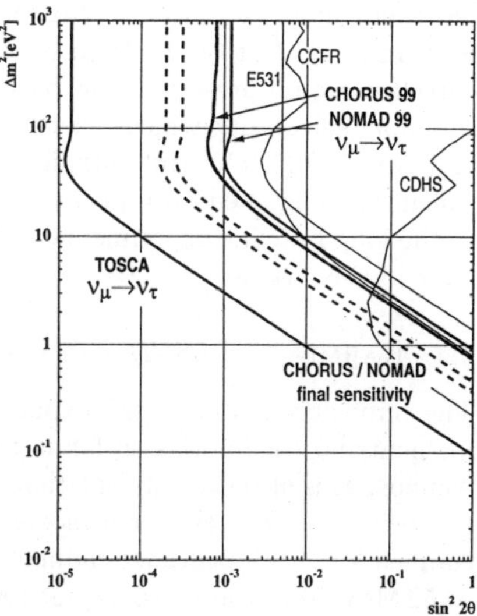

Figure 3. $\nu_\mu \to \nu_\tau$-exclusion limits from CHORUS and NOMAD; dashed lines: final sensitivities.

From this backgroundfree null result a 90 % CL upper limit for the oscillation probability $P_{\nu_\mu \to \nu_\tau} \leq 4 \times 10^{-4}$ has been deduced which is represented by the exclusion plot in Fig. 3.

NOMAD at the same neutrino beam is a fine grain magnetic spectrometer with a 2.7 t drift chamber neutrino target in a wide gap dipole magnet followed by transition radiation detectors, electromagnetic and hadronic calorimeters and muon chambers. Here ν_τ - or τ^- -identification resp. relies on the kinematics reconstruction from the outgoing particles particularily the determination of the missing transverse momentum in the two most relevant decay modes $\tau^- \to e^- \nu_\tau \bar{\nu}_e$ and $\tau^- \to \pi^- \nu_\tau$. With a small number of candidates being compatible with the expected background rate from ν_e -, ν_μ -CC and ν_μ -NC reactions the current 90 % CL exclusion limits with $P_{\nu_\mu \to \nu_\tau} \leq 0.6 \times 10^{-3}$ are slighty less stringent than those from CHORUS (see Fig. 3). However both experiments have still many data in the evaluation pipe expecting a final 90 % CL sensitivity of $\sin^2 2\Theta \geq 2 \times 10^{-4}$ or 3×10^{-4} resp. for large Δm^2 as indicated in Fig. 3 by the dashed lines.

In some kind of combination of the CHORUS and NOMAD techniques there had been a proposal for a new $\nu_\mu \to \nu_\tau$ -appearance experiment TOSCA with ten times higher sensitivity. Six emulsion-tracker-preshower modules with a total mass of 2.4 t of emulsion in the NOMAD magnet with hermetic muon detection would preserve a low background level of about 1 event. The corresponding sensitivity which reaches down to $\sin^2 2\Theta \geq 1 \times 10^{-5}$ (large Δm^2) and $\Delta m^2 \geq 10^{-1}\,\mathrm{eV}^2$ (full mixing) is shown in Fig. 3 too. Whether this experiment is going ahead will probably depend on the CERN policy with respect to the long baseline experiments (LBL) at Gran Sasso which will be discussed later in this article.

3.3 KARMEN *and* LSND — *the controversal* Δm^2 *-region*

The only accelerator based experiment that has claimed positive evidence for neutrino oscillations so far is the LSND [6] experiment at LANSCE, Los Alamos. It is also the only one that ever saw an appearance signal rather than a deficit as is the case for atmospheric and solar neutrinos. The evidence is claimed for the appearance channel $\bar{\nu}_\mu \to \bar{\nu}_e$ using a $\bar{\nu}_\mu$-beam with energies up to 52 MeV from the π^+-decay at rest sequence (DAR). Using the same DAR neutrino source and looking for the same $\bar{\nu}_\mu \to \bar{\nu}_e$-appearance the KARMEN [7] experiment at the spallation facility ISIS of the Rutherford Laboratory could not confirm the LSND result so far. Whereas Δm^2-values of the so called favoured region of LSND above 2 eV^2 are clearly excluded by the current 90 % CL exclusion limit of KARMEN there is a region of Δm^2 between 0.1 eV^2 and 1 eV^2 which need further investigation. To judge on the validity of these experiments and the significance of their results both experiments will be described in more detail below.

The LSND·Experiment

For LSND the 800 MeV proton beam of LANSCE is hitting a pion producing water target, followed by a high Z target and finally dumped in a water cooled Cu-beam dump. The energy spectrum of neutrinos ν_μ, ν_e and $\bar{\nu}_\mu$ from the stopped π^+-decay chain is shown in Fig. 4a. With only a tiny portion of simultaneously produced π^- not being captured before decaying there is otherwise no $\bar{\nu}_e$-contamination that could distort the search for $\bar{\nu}_e$ from a $\bar{\nu}_\mu \to \bar{\nu}_e$-oscillation. The $\bar{\nu}_e$-signature would be the inverse β-decay reaction $^1\mathrm{H}\,(\bar{\nu}_e, e^+)\,\mathrm{n}$ followed by p-n-capture $^1\mathrm{H}\,(n, \gamma)$ providing a delayed 2.2 MeV γ-signal. This signature is looked for in a 167 t hybrid oil Cherenkov detector with 1220 phototubes covering about 25 % of the detector surface. Whereas the Cherenkov cone of minimum ionizing particles like the e^+ will cause a prompt signal with a characteristic PM-pattern a small additive of scintillation fluor provides a slightly delayed, isotropic signal. Neutrons with no

Cherenkov light at all will cause, after thermalisation, the delayed and spatially correlated (n,γ)-signal with a capture time constant of about $188\,\mu s$. The Cherenkov fit, the time spread and the distance spread of the prompt signal thus provides a good e^+-'particle identifcation' (PID) clearly separated from that for neutrons. The 'correlated γ-identification', however, is provided by a ratio R of likelihoodfunctions L(corr)/L(uncorr) which are made from probability distributions for the time difference and the spatial correlation between the prompt and and the delayed signal as well as the number of γ-PMT hits for correlated and uncorrelated coincidences. These in turn are deduced from data samples of cosmic muon induced neutrons showing the typical $188\,\mu s$ time constant of a delayed signal. The R-distributions for correlated and uncorrelated events differ most significantly for large R-values.

LSND has taken data since 1993. In the data sample collected from 1993 to 1998 and asking for $R \geq 30$ there are 70 events during the beam-on period fulfilling these criteria where only 17.7 ± 1.0 events showed up in the corresponding beam-off period. As only 12.8 ± 1.7 events can be assigned to neutrino induced background there is a sample of 39.5 ± 8.8 "goldplated" excess events which LSND devote to $\bar{\nu}_\mu \to \bar{\nu}_e$-oscillations. The energy resolution as well as the statistics is not sufficient as to fix the Δm^2-value from these data.

Whereas the "goldplated" events are taken to demonstrate the effect of an $\bar{\nu}_e$-excess the oscillation analysis itself is based on the R-fit to the complete data set of 3049 events. From this fit a total excess of 90.9 ± 26.1 $\bar{\nu}_e$-events is found which relates to an oscillation probability of $P_{\bar{\nu}_\mu \to \bar{\nu}_e} = (0.33 \pm 0.09 \pm 0.05)\%$ (prelim. LSND result 93–98). This result is transformed into the LSND evidence plot for $\bar{\nu}_\mu \to \bar{\nu}_e$-oscillations of Fig. 6. The marked areas are called "favoured" regions as they are the 99 % and 90 % levels of an overall event based likelihood function. An analysis to deduce real confidence intervals according to the Unified Approach [8] is in preparation.

The LSND evidence claim for $\bar{\nu}_\mu \to \bar{\nu}_e$-oscillations is looking quite convincing however there are a few caveats which also need to be mentioned. The most important ν-induced background is caused by the reaction $\nu_e + {}^{12}C \to {}^{12}N_{g.s.} + e^-$ with electron energies up to $36\,\text{MeV}$ followed by the ${}^{12}N_{g.s.}$-decay. As not to be distorted by this background one might look only for events with energies above this value which have higher statistical significance. When in 1996 LSND published a first evidence of $\bar{\nu}_\mu \to \bar{\nu}_e$-oscillations based on 21.1 ± 6.3 "goldplated" candidate $\bar{\nu}_e$-events the ratio of events above $36\,\text{MeV}$ energy to those from 20 to $36\,\text{MeV}$ was 4.7:1. Taking the additional 18.4 ± 6.1. excess events that have been collected since then this high energy to low energy ratio is reversed to be about 1:2. In the fit to the oscillation

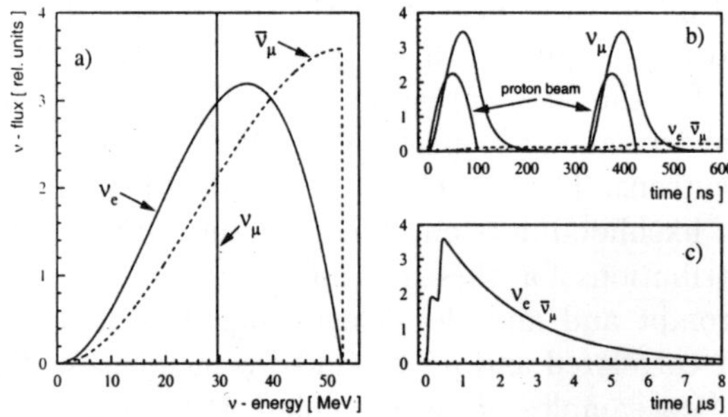

Figure 4. a) Energy spectra of the DAR neutrino source used by LSND and KARMEN. b) and c) time spectra of these neutrinos for the KARMEN experiment only, due to the ISIS proton beam.

parameters this favours lower Δm^2 -values compared to the 1996 publication. Moreover, although statistically still compatible it is amazing that this change of the energy spectrum coincides with a significant change of the composition of the neutrino production target. For all data since 1996 the water degrader had to be removed leaving only the classified high Z target in favour of the LANSCE demands for more neutrons. One might therefore ask 'Are the excess events really due to $\bar{\nu}_\mu \rightarrow \bar{\nu}_e$ -oscillations?' One would have liked to see this effect to be investigated more systematically.

The KARMEN experiment

A direct test of the LSND result is currently going on with the KARMEN experiment at the spallation facility ISIS of the Rutherford Appleton Laboratory, UK. With a high resolution segmented scintillation calorimeter of 56 t fiducial mass it looks for the same $\bar{\nu}_\mu \rightarrow \bar{\nu}_e$ -appearance oscillation using the very same DAR neutrino source of ν_μ, ν_e and $\bar{\nu}_\mu$ from stopped π^+ -decay (see Fig. 4). What is significantly different to LSND, apart from the much better energy resolution of the KARMEN detector is the distinct time structure of the ISIS neutrino beam shown in Fig. 4b, which is related to that of the 800 MeV proton beam of ISIS (2×100 ns, 50 Hz). This time structure has to be reflected in any measured time distribution of ν -induced events. Particularily ν_e - and $\bar{\nu}_\mu$ -events would have to follow the 2.2 μs slope of its parent muon decay thus providing a very characteristic and stringent signature (Fig. 4c).

Extensively exploiting these features, KARMEN, from 1990 to 1995 has performed an entire experimental programme of neutrino physics investigating

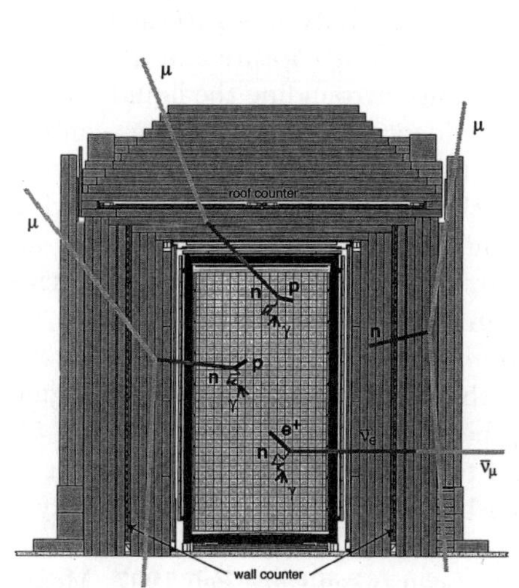

Figure 5. KARMEN2 detector, shielding and veto setup

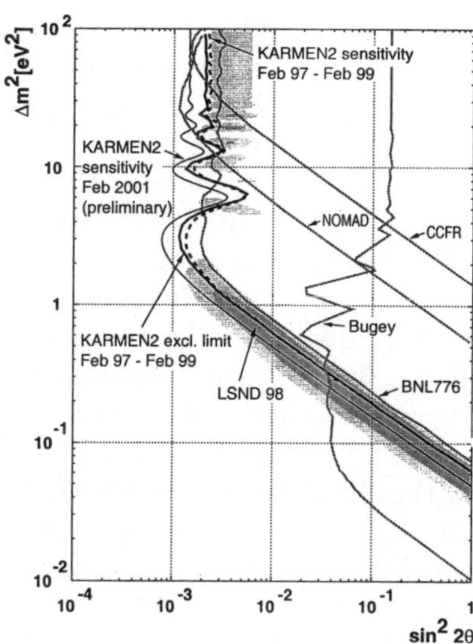

Figure 6. $\bar{\nu}_\mu \to \bar{\nu}_e$ 90 % CL exclusion limits of KARMEN2 and favoured evidence regions of LSND

quantitatively several charged current (CC) as well as neutral current (NC) neutrino nucleus interactions with implications on specific weak couplings, weak nuclear formfactors, exotic decay and interaction modes and others [9]. One of the main topics of the KARMEN experiment, however, has also been the search for neutrino oscillations particularly in the $\bar{\nu}_\mu \to \bar{\nu}_e$-appearance channel. An excess of $\bar{\nu}_e$, as at LSND, would have been detected via the reaction ^{1}H$(\bar{\nu}_e, e^+)$n followed by (n, γ)-capture but in this case not only by ^{1}H but also by Gd, implemented in the scintillator module boundaries and causing much higher γ-energies. Spatially correlated delayed coincidences within proper time and energy windows would be the candidates for possible neutrino oscillation events. For any $\bar{\nu}_e$-data sample due to oscillations the time as well as the energy spectrum of both the prompt and delayed signal is precisely known and would have been measured with high resolution. Although about 140 of those $\bar{\nu}_e$-like candidate events have been found in the 1990–95 data careful Maximum Likelihood (ML) analyses using this knowledge of time and energy spectral shape associated them to background and did not yield a positive evidence for any ν-oscillation.

However, the KARMEN1 exclusion limits on the oscillation parameters Δm^2

and $\sin^2 2\Theta$ could rule out only a small portion of the entire area of oscillation parameters of the LSND evidence. The sensitivity was governed by the background intensity caused by fast neutrons from μ-capture and μ-induced spallation processes in the 7000 t steel shielding surrounding the liquid scintillation calorimeter. Entering the detector unidentified, these neutrons cause a prompt signal by (n,p)-scattering followed by a (n,γ)-process of the scattered and thermalized neutrons thus providing the same signature as $\bar{\nu}_e$. Knowing the origin of this limiting background the problem could be cured. In an upgrade programme an additional layer of veto counters (5 cm plastic scintillator, 300 m^2) had been implemented in the shielding with still about 4.5 attenuation lengths of steel ($\lambda_n^{Fe} = 21.6\,cm$) towards the detector. Neutrons created within the enclosure of the veto shield and causing a detector signal are thus identified by the veto of having been induced by muons (see Fig. 5). This did reduce the relevant background to oscillation signals by a factor of 40 which means that the cosmic-induced background is now down to only 1 event/year.

With this new KARMEN2 setup data have been taken since Feb 1997. Meanwhile data for 4500 C protons on target of ISIS have been collected which is half of the data proposed to take until 2001. Within proper cuts on the energy, time and spatial correlation of the prompt and delayed signal 8 events compatible with a $\bar{\nu}_e$-signature have been found where 7.8 events of irreducible — now mainly ν-induced — background had been expected. As the number of events coincide with that expected from background there is no indication of any $\bar{\nu}_e$-excess that could be assigned to neutrino oscillations $\bar{\nu}_\mu \to \bar{\nu}_e$. A full Maximum Likelihood analysis following the Unified Approach of Feldman and Cousins [8] was applied to deduce correct confidence intervals. The corresponding exclusion curve shown in Fig. 6 represents the currently best limits for the $\bar{\nu}_\mu \to \bar{\nu}_e$-oscillation parameters. As both measured and expected background events are in good agreement the exclusion curve and the corresponding sensitivity (dashed line) fall almost on top of each other. With these exclusion limits together with those from the BUGEY $\bar{\nu}_e$-disappearance reactor experiment there is only a small portion of the LSND favoured parameter region left that on a 90 % confidence level could have caused the measured $\bar{\nu}_e$-like excess by oscillations.

KARMEN is assumed to run until 2001 when it will have doubled the current amount of data. The sensitivity it will have reached by then under very conservative assumptions is given by the thin solid line in Fig. 6. On a 90 % CL this would cover the entire 'favoured region' of LSND as published in 1996 [6] however the more recent LSND results (see above) shifted their evidence regions to somewhat lower Δm^2-values as indicated in Fig. 6. With the full data set in 2001 KARMEN2 will have observed about 15 background events. Provided the LSND oscillation probability of about 0.3 % is valid additional 11 to 15 excess events depending on Δm^2 will have to show up to confirm the evidence claim for $\bar{\nu}_\mu \to \bar{\nu}_e$ -oscillations.

*New Experiments to Resolve the LSND/*KARMEN *Controversy*

Obviously new experiments with higher significance are necessary to resolve the LSND/KARMEN controversy as to give theoretical ν -scenarios clear and definite experimental guidelines in this Δm^2 -region. At least one experiment — MiniBooNE [10] at FNAL — is definitely going ahead. It is particularily looking for an oscillation $\nu_\mu \to \nu_e$ using a neutrino beam based on the 8 GeV booster synchrotron of FNAL and providing ν_μ with energies from 0.25 to 2 GeV. ν_e will be detected by $^{12}\text{C}\,(\nu_e, e^-)\,^{12}\text{N}$ quasi elastic scattering in a large spherical tank with 445 t fiducial mass of mineral oil looked at by 1220 phototubes dismounted from the LSND experiment. The detector will be located at 500 m from the π -DIF area. The main problem will be a clear separation of e^-, μ^- and π^0 which should cause different Cherenkov patterns. It is assumed that one will see ~ 1000 ev/year if the LSND signal is due to neutrino oscillations. Data taking might begin in 2001. Provided MiniBooNE does see that signal a second phase — BooNE — will employ an additional detector at about 1000 m distance to fix the values of Δm^2 and $\sin^2 2\Theta$. Recently a CERN letter of intent, I216 [11] has proposed a similar $\nu_\mu \to \nu_e$ - oscillation experiment using the old but refurbished PS-neutrino beam line. Three identical tracking calorimeters of 130 t mass each — most probably a fine grain scintillator structure — will look for quasi elastic $n\,(\nu_e, e^-)\,p$ - and $n\,(\nu_\mu, \mu^-)\,p$ -scattering. One 'close' detector module will be located at 130 m and two 'far' modules at 885 m from the target. The difference Δ of the ratios N_{ν_e}/N_{ν_μ} for the far and the close detector will provide a rather flux independant signal for $\nu_\mu \to \nu_e$ -oscillations if Δ comes out to be significantly different from zero. The sensitivitiy of I216 after two years of data taking is shown in Fig. 7 compared to that of MiniBooNE after one year of running. As I216 is a self normalizing two distance experiment its sensitivity covers only the most interesting low Δm^2 -region of the LSND result. Currently a collaboration is becoming established working on the final detector design and getting the proposal approved. In view of the importance of the question the go ahead of competing experiments in this region of Δm^2 is highly desireable.

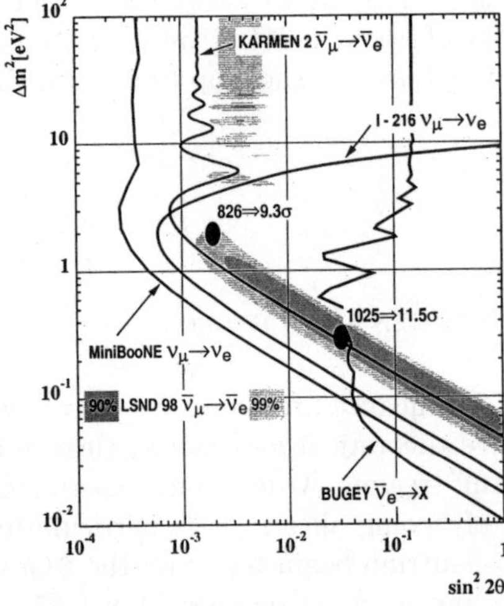

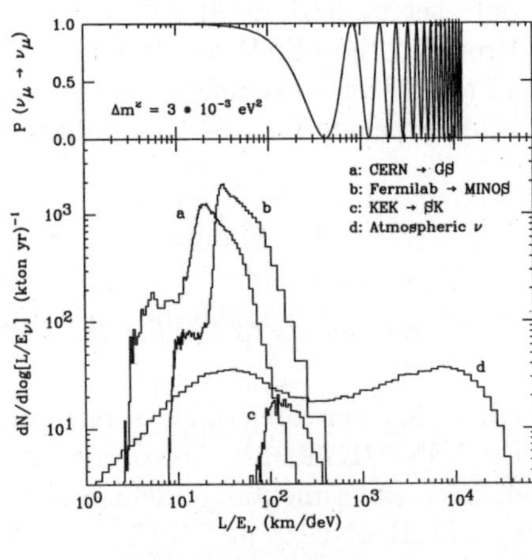

Figure 7. Sensitivities of future $\nu_\mu \to \nu_e$-oscillation experiments MiniBooNE and I216. Large dots: expected number of events for MiniBooNE if LSND evidence is valid.

Figure 8. Normalised ν_μ-CC event rates for LBL experiments at KEK, FNAL, CERN and for atmospheric neutrinos. Top: ν_μ disappearance probability for a given Δm^2 (adopted from [13]).

4 Long BaseLine Experiments LBL — the low Δm^2-region

With the proof of the L/E-dependence of the deficit of μ-type atmospheric neutrinos the SuperKamiokande experiment (SK) has provided the strongest evidence for neutrino oscillations so far [12]. The best fit to their data is achieved with the assumption of a $\nu_\mu \to \nu_\tau$-oscillation with $\Delta m^2 \approx 2 \times 10^{-3}\,\mathrm{eV}^2$ at maximum mixing $\sin^2 2\Theta = 1$. To test this region of Δm^2 with terrestrial ν_μ-sources requires large distances and/or small energies, however, if also $\nu_\mu \to \nu_\tau$-appearance is suggested to be proved, higher energies are compulsory.

Currently there are three places where long baseline neutrino experiments are performed, going ahead or being planned. For the K2K experiment in Japan the ν_μ-beam from the 12 GeV KEK machine is pointing towards the SK-detector at a distance of 235 km. In the US a ν-beam from the Main Injector of Fermilab will be looked for in the Soudan mine 734 km away from target by the MINOS project and in Europe a dedicated CERN ν-beam might be directed to the Gran Sasso Neutrino Laboratory LNGS at 732 km distance.

Rather than discussing the three projects in full detail the potential of these experiments is most generally reflected in the graph of Fig. 8 which has been adopted from [13]. Here the expected normalized event rate is plotted vs. the ratio L/E together with the oscillation pattern for $\Delta m^2 = 3 \times 10^{-3} \, eV^2$. Although for all three LBL experiments the oscillation probability will cut into the energy spectrum for values of $\Delta m^2 \geq 10^{-3} \, eV^2$ the corresponding oscillation sensitivity will only just cover the $\nu_\mu \rightarrow \nu_\tau$ -evidence region referring to the latest SK-results [12]. The most convincing signal of a clear oscillation minimum will probably be left to a new dedicated atmospheric neutrino experiment.

Whereas the CERN and FNAL ν-beam will also allow an explicit search for $\nu_\mu \rightarrow \nu_\tau$ -appearance the low energy beam of KEK limits the K2K [14] experiment to $\nu_\mu \rightarrow x$ or $\nu_\mu \rightarrow \nu_e$. However this project is the most advanced one and has started data taking in early 1999. Apart from the far detector — the 50 kt SK water Cherenkov detector — a smaller but otherwise identical close detector of 1 kt mass serves for comparison of the ν_μ-intensities. An additional detector system of a SciFi-H_2O tracker target, a shower detector and muon chambers provide precise monitoring of the ν_μ-beam. The sensitivity to $\nu_\mu \rightarrow x$ is limited to about $\Delta m^2 \geq 2 \times 10^{-3} \, eV^2$ whereas in the $\nu_\mu \rightarrow \nu_e$ -channel it is comparable to the CHOOZ [1] reactor experiment reaching down to $\Delta m^2 \geq 10^{-3} \, eV^2$ at full mixing. The first results from this experiment are eagerly awaited as they will be most decisive for the further development of the other LBL experiments.

The MINOS [15] project is based on the 120 GeV proton injector providing a wide band neutrino beam with energies up to 30 GeV with options for lower energy focussing. The 5.4 kt far detector in the Soudan mine is basicly a calorimeter of 1 inch magnetized iron plates interleaved with $25.800 \, m^2$ of active detector planes from scintillator strips of 1 cm thickness. A similar 1 kt near detector at 300 m serves for reference. Rates and energy spectra for ν_μ-CC and -NC events or NC-like events from ν_e and ν_τ as well as the ratio of NC/CC are used to determine oscillation modes and parameters provided they do exist within the sensitivities. ν_μ-disappearance and, indirectly, $\nu_\mu \rightarrow \nu_\tau$ and $\nu_\mu \rightarrow \nu_e$ can be tested down to $\Delta m^2 \approx 10^{-3} \, eV^2$ whereas direct ν_e -appearance might be comparable to the CHOOZ limit. A low energy version of the neutrino beam might even be able to cover the entire SK-evidence region in the ν_μ-disappearance channel. Exclusive ν_τ-appearance with the current detector design will be limited to much higher Δm^2-values. However with the employment of an additional emulsion hybrid detector of about 1 kt mass a few $\nu_\mu \rightarrow \nu_\tau$ -events per year can be expected. The MINOS project is currently under construction and is scheduled to take data in 2002.

The plans for LBL experiments at CERN are based on a neutrino beam from the 400 GeV SPS directing towards the Gran Sasso Neutrino Laboratory LNGS at a distance of 732 km. A variety of detectors are currently under discussion however the activties concentrate on three major issues [16]: the development of a neutrino beam with the goal of maximizing the ν_τ -CC interactions for a hypothetical $\nu_\mu \to \nu_\tau$ -oscillation, design of a dedicated $\nu_\mu \to \nu_\tau$ - appearance experiment with discovery potential for $\Delta m^2 \geq 2 \times 10^{-3}\,\mathrm{eV}^2$ and the development for a new atmospheric neutrino detector enabling an L/E precision experiment with sensitivties of $2 \times 10^{-4}\,\mathrm{eV}^2 \leq \Delta m^2 \leq 5 \times 10^{-3}\,\mathrm{eV}^2$ and the capability to simultaneously determine Δm^2 and $\sin^2 2\Theta$. Although still matter of optimisation a feasable neutrino beam with average energy $< E_{\nu_\mu} > = 30\,\mathrm{GeV}$ might provide a CC ν_τ -rate of about 7.5 ev/kt*year for $\Delta m^2 = 2 \times 10^{-3}\,\mathrm{eV}^2$. A $\nu_\mu \to \nu_\tau$ -search will thus always remain a low countrate experiment and therefore requires negligible, well controlled background. Among different concepts there is OPERA with a 0.75 kt lead-emulsion target looking for τ -decay kink reconstruction; ICARUS, a 2.4 kt LArTPC (probably installed at LNGS in 2003) with tracking and good energy resolution capabilities, might use the NOMAD experience of missing $p_\perp$ -reconstruction; and a 125 kt H_2O ring imaging Cherenkov detector, AQUARICH, has been proposed which might have particular sensitivity to quasi elastic $\nu_\tau\,\mathrm{n} \to \tau\mathrm{p}$ scattering followed by τ -muonic decay due to its good momentum, angle and time resolutions. All these experiments are expected to have a background of zero or one event for 4 years of data taking. Instead of giving any specific expected sensitivity of the different LBL experiments it might be fair to present a more general $\nu_\mu \to \nu_\tau$ -sensitivity curve in the summarizing plot of Fig. 9. If and what experiments are going ahead is currently a matter of discussion of a joint committe of CERN and LNGS.

Some of the dectector concepts mentioned above, might also suit for a new precision experiment on atmospheric neutrinos. To be superior to the SK experiment a clear oscillation pattern including the oscillation minimum is required to be seen. It means that the resolution in L/E has to be smaller than 1/2 of the oscillation length and this requires excellent angular and full calorimetric energy resolution for the atmospheric ν_μ -induced reactions. The ratio of track-mirrored up- and downgoing ν_μ could then provide an almost flux independand measure of the disappearance probability in the range of $2.5 \times 10^{-4}\,\mathrm{eV}^2 \leq \Delta m^2 \leq 5 \times 10^{-3}\,\mathrm{eV}^2$. With the corresponding ratio of 'no'- μ events it might even be possible to decide whether ν_μ has oscillated to ν_τ or into a sterile neutrino. Currently a dedicated atmospheric neutrino detector of 34 kt magnetized iron plates interleaved with active elements of 3 cm two dimensional pitch is under investigation to be placed at the Gran Sasso

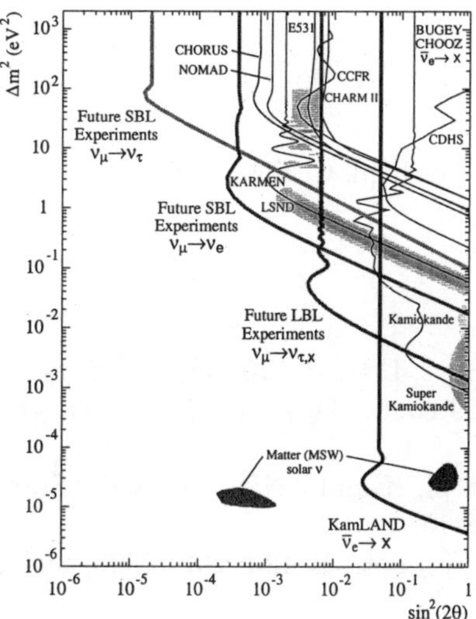

Figure 9. Summarizing 90 % CL exclusion and sensitivity limits for current and future SBL and LBL oscillation experiments.

underground laboratory[17].

5 Concluding Remarks

This paper tried to cover the status and the future prospects of reactor and accelerator based neutrino oscillation experiments. Evidences for oscillations claimed so far have initiated a variaty of neutrino mass and mixing scenarios which are still very speculative. At this stage, however, the most important request is an experimental one namely the unambiguous and conclusive experimental confirmation or disproof of all effects and hints that could be interpreted in terms of neutrino oscillations. The next generation of terrestrial neutrino experiments significantly exceed the current sensitivities as indicated in Fig. 9 for both the SBL and the LBL experiments . Moreover their results will also be most decisive. They will probe into certain solutions of the solar neutrino problem. They will give definite answers whether or not the LSND claim is valid and whether atmospheric neutrinos indeed oscillate and what are the corresponding oscillation parameters. These then will be the guidelines and the boundary conditions for realistic neutrino scenarios that are of greatest importance for our understanding of particle physics, astrophysics and cosmology.

Acknowledgments

The author would like to thank C. Eichner for his assistance in preparing this contribution.

References

1. M. Appollonio, Phys. Lett. B **420**, 397 (1998) and hep-ex/9907037
2. G. Gratta, WIN99 conf., Capetown hep-ex/9905011 and J. Wolf, privat communication
3. P. Alivisatos *et al*, KamLAND proposal, Stanford-HEP-98-03 and Tohoku-RCNS-98-15
4. E. Eskut *et al*, Phys. Lett. B **424**, 202 (1998) and CERN-EP/98-73 and hep-ex/9907015
5. P. Astier *et al*, Phys. Lett. B **435**, 169 (1999)
6. C. Athanassopoulos *et al*, Phys. Rev. C **54**, 2685 (1996) and E. Church, PANIC99 conf., Uppsala to be published in Nucl. Phys. B (conf. proc.)
7. B. Zeitnitz *et al*, Prog. Part. Nucl. Phys. **40**, 169 (1998) and G. Drexlin *et al*, Prog. Part. Nucl. Phys. **40**, 193 (1998) and R. Maschuw for the KARMEN coll., PANIC99 conf., Uppsala, to be published in Nucl. Phys. B (conf. proc.)
8. G. J. Feldman, R. D. Cousins, Phys. Rev. D **57**, 3873 (1998)
9. R. Maschuw *et al*, Prog. Part. Nucl. Phys. **40**, 183 (1998).
10. http://www.neutrino.lanl.gov/BooNE
11. CERN-SPSC/97-21 I216
12. Y. Fukuda *et al*, hep-ex/9807003, Y. Suzuki, WIN99 conf., Capetown and K. Kaneyuki, this conf.
13. G. Battistoni, P. Lipari, Proc. of 1998 Vulcano Workshop and hep-ph/9807475
14. K. Nishikawa *et al*, Nucl. Phys. B **59**, 289 (1997) (conf. proc.) and Y. Oyama, hep-ex/9803014
15. http://www.hep.anl.gov/NDK and NuMI note NuMI-L-337 Oct 1998
16. P. Picchi, F. Pietropaolo, hep-ph/9812222
17. MOMOLITH proposal, in preparation

CHARM OVERVIEW

STEFANO BIANCO

Laboratori Nazionali di Frascati
via E. Fermi 40, Frascati 00044, Italy
E-mail: bianco@lnf.infn.it

This paper is aimed at giving a complete review of the latest (post-1998 conference) experimental results on charm physics.

1 Introduction

The study of the c–quark generates about 200 papers and 800 measurements per year, for a total of 15 collaborations and 700 physicists involved. Such an effort has led us to a conflicting situation. On one hand, the advent of high-statistics, high-resolution experiments has turned c–quark physics into precision physics. On the other hand, the c–quark mass scale is too large for chiral symmetry methods, while theories based on expansions of heavy quark masses, such as Heavy Quark Effective Theory[1] (HQET) or Operator Product Expansion[2,3] (OPE), are often questionable because m_c may not be large enough. Nonetheless, remarkable agreement is often found when such theories make predictions, most notably on semileptonic decays, lifetimes, and spectroscopy.

The 1998-1999 scenario is full of results – from new experiments (SELEX and BES), experiments that have undergone significant upgrades (FOCUS, CLEO II.V, E835), and others that are planning upgrades for charm physics (HERMES), experiments at their peak publication rate (E791), and experiments (at LEP and HERA) that keep their charm working groups alive and vital. Interesting news comes from the neutrino (CCFR, CHARM II) and heavy-ion (NA50) groups.

In this paper I have tried to present a review of today's scenario rather than pursue the goal of detailed investigation. In setting the physics scenario for each item, I was guided by recent reviews[4,5,6,7,8], from which I borrowed copiously. With such a goal in mind, I apologize in advance to those whose work has been left unmentioned.

2 Production mechanisms

The QCD picture of charm production consists of parton-level hard scattering, which produces the $c\bar{c}$ pair, hadronization of the $c\bar{c}$ into a charmed hadron, and a final stage in which the charmed hadron travels through and interacts with the hadronic matter, followed by decay. Thanks to the generally large momentum transfers involved, the hard scattering is traditionally a good testing ground for perturbative QCD techniques, while the hadronization stage has the complication of matching experimental kinematical distributions of charm hadrons with predictions of a suitable dressing mechanism. Theoretical updates on both topics were given at HQ98 [9,10]. In general, any asymmetry is due to hadronization since $c\bar{c}$ asymmetries in NLO QCD are very small. For a seminal review see ref.[11,12].

A wealth of new results comes from Fermilab fixed-target experiments E791, SELEX and FOCUS (pion, hyperon, and photon beams respectively)[13]. Correlation studies of fully reconstructed hadroproduced $D\bar{D}$ pairs by E791[14] show less correlation than that predicted by the Pythia/Jetset Monte Carlo event generator for $D\bar{D}$ production, which, however, follows the experimental trend better than a pure NLO QCD parton level prediction (Fig.1). Coherently, $D\bar{D}$ correlation in photoproduction is much more pronounced because of a lack of hadronization contributions of the projectile remnants.

New results from E791[15] also report particle-antiparticle production asymmetries for $D^{\pm}$, $D_s^{\pm}$ and Λ_c, and a cross-section measurement[16] with a 500-GeV π^- beam on a nuclear target of $\sigma(D^0 + \bar{D}^0; x_F > 0) = 15.4^{+1.8}_{-2.3}\mu$b/nucleon. Asymmetry results show clear evidence for leading effects. Both measurements are in principle sensitive to m_c; in practice it is necessary to first nail down a few other parameters (factorization scale, intrinsic parton momentum k_t, etc). In general, E791 asymmetry results favor $m_c = 1.7\,\text{GeV}/c^2$, while the cross-section measurement is compatible with $m_c = 1.5\,\text{GeV}/c^2$. For the renormalization scale and scheme adopted for the above definitions of m_c the interested reader is referred to reff.[15,16]. SELEX preliminary results[17], which include comparison of Λ_c and $D^{\pm}$ asymmetries with proton, pion, and hyperon beams, also indicate large leading effects. FOCUS presented clear evidence[13] of Λ_c asymmetry in photoproduction (Tab.1). With a photon beam any asymmetry has to be attributed to the target, and the picture is consistent with $\Lambda_c(cud)$ production being more probable than $\bar{\Lambda}_c$. When comparing results from different photoproduction experiments, it should be also kept in mind how the asymmetry is in general $\sqrt{s}$-dependent, as well as dependent on the exact acceptance kinematics, i.e., the x_F range, etc.

Table 1. Compilation of $\Lambda_c/\bar{\Lambda}_c$ photoproduction asymmetry measurements (adapted from ref.[13]).

Experiment	Asymmetry
FOCUS prel.	0.14 ± 0.02
E687 (93)	0.04 ± 0.08
E691 (88)	0.11 ± 0.09
NA1 (87)	0.14 ± 0.12

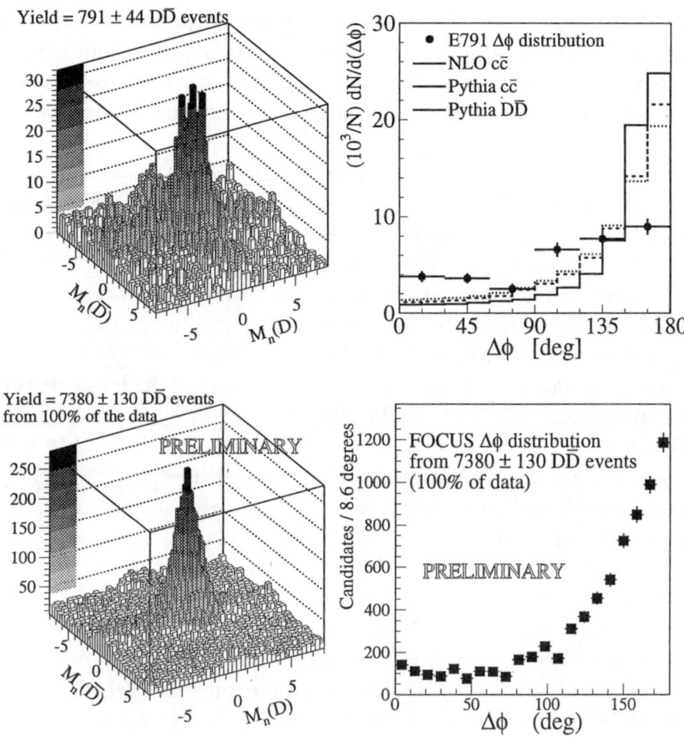

Figure 1. Mass correlation plots (left), and acoplanarity angle distributions (right) for hadro- (E791) and photo-produced (FOCUS) $D\bar{D}$ pairs.

Neutrino charm production also gives estimates of m_c. Results[18] of CCFR (Fermilab) and CHARM II (CERN) experiments provide analysis-dependent values: CCFR (next-to-leading order) and CHARM II find $m_c \sim 1.7\,\text{GeV}/c^2$, while CCFR (leading order) favors $m_c \sim 1.3\,\text{GeV}/c^2$. New data are expected very soon from the successor CCFR experiment, NuTeV.

The HERMES experiment at HERA presented[19] preliminary results (Fig.2) for open and hidden charm photoproduction at threshold. They expect to collect data in 1999 with an upgraded detector, and to provide a measurement of open charm cross section at threshold, with the ultimate goal of extracting the gluon momentum distribution $G(x)$. Theoretical predictions of cross sections at threshold suffer from major difficulties (the size of α_S, the role of higher order corrections, etc.). For the case of charmonium production at threshold, NLO predictions do exist[20,21]. New theory results may come from the utilization of novel methods (resummation of NLO logarithms) developed for the high-energy region [22,23].

New results on $G(x)$ have come from H1 and ZEUS. H1[24] determines the NLO gluon momentum distribution for $7.5 \cdot 10^{-4} < x < 4 \cdot 10^{-2}$ from DIS and direct detection of D^*'s photoproduced in the final state. Results on $G(x)$ agree with distributions found from scaling violations of the proton structure function. ZEUS[25] finds the D^* differential cross section well described by NLO QCD calculations, with *massless* calculations[26] performing better than *massive* calculations[27,28]. ZEUS also remeasured[29,30] with higher statistics the charm contribution $F_2^{c\bar{c}}$ to the proton structure function F_2, finding

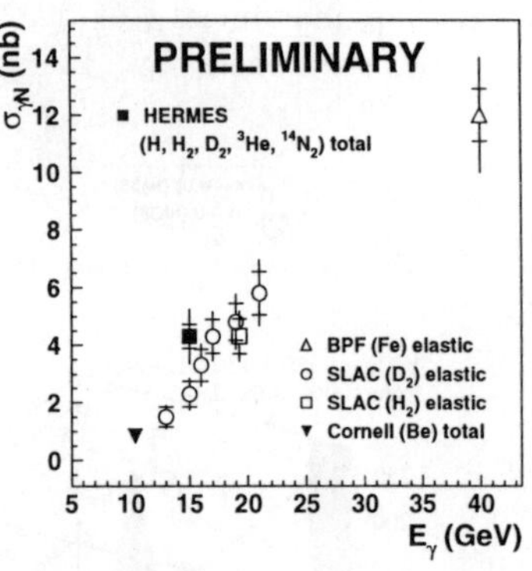

Figure 2. HERMES J/ψ photoproduction cross section.

that at low-x values a very large ($\sim 30\%$) fraction of DIS events contains open charm states, unlike EMC fixed target results at high-x. For a recent summary of HERA heavy quark results, see ref.[31]. Finally, a further observable that can be computed by NLO QCD is the probability $g_{c\bar{c}}$ of $c\bar{c}$ pair production by gluon splitting ($e^+e^- \rightarrow q\bar{q}g, g \rightarrow Q\bar{Q}$). New OPAL preliminary results[32] $g_{c\bar{c}} = 3.20 \pm 0.21 \pm 0.38) \times 10^{-2}$ are higher than theoretical estimates, as also recent measurements of $g_{b\bar{b}}$ by ALEPH and DELPHI.

In 1992 the CDF collaboration[33] discovered that the production of J/ψ and ψ' in $p\bar{p}$ collisions was enhanced by a factor of fifty with respect to predictions of the color-singlet model, which stated that produced $c\bar{c}$ pairs would dress into the observed charmonium state by keeping their quantum numbers, i.e., by rearranging their colors without gluon emission. To explain the result, a color-octet model was proposed in which the $c\bar{c}$ pair dressed in a charmonium hadron by emitting a soft gluon. The color-octect model predicts absence of polarization for charmonium production, since the initial polarization of $c\bar{c}$ pairs is destroyed by the radiation of gluons. Data by fixed target (WA92 at CERN), and neutrino (NUSEA) experiments on the polarization of J/ψ do confirm the color-octet prediction, while relevant polarization is observed in the HERA q^2 regimes[31], and at the collider (CDF), in agreement with the color-singlet model. The issue of charmonium production is dealt with in great detail in these proceedings[34].

Charmonium production is also investigated on the very distant field of relativistic heavy ion collisions, where the NA50 experiment[35] using 1996 data (158 GeV/nucleon Pb beams on Pb target) provided circumstancial evidence for charmonium suppression, which may be explained by the onset of a quark-gluon plasma regime. They measure J/ψ production relative to Drell-Yan pair production. After accounting for conventional nuclear absorption, their data (Fig.3) show evidence for a suddenly lower production, due to the attracting force between the $c\bar{c}$ quarks being screened by gluons, and fewer $c\bar{c}$ pairs hadronizing into J/ψ.

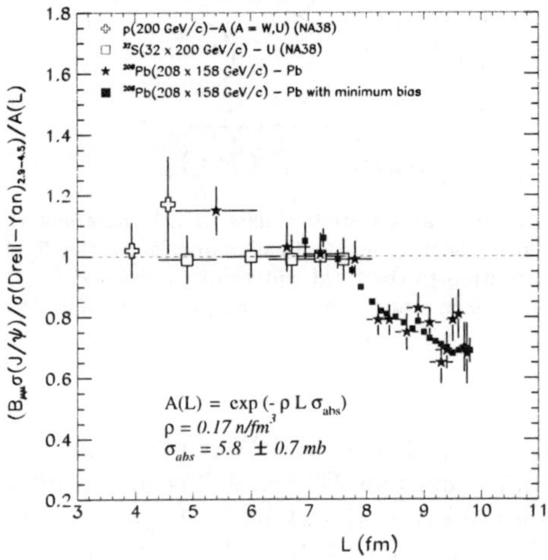

Those uncorrelated pieces of information taken together confirm the important role of gluons in the context of charmonium production dynamics.

Figure 3. NA50 J/ψ suppression in relativistic heavy ion collisions.

3 Lifetimes

If there were no other diagram but the spectator and no QCD effects causing charm hadrons to decay, we would have one lifetime for all states. The wide range of lifetimes measured (Fig.5) shows the extent to which this is not the case. The total width is written as a sum of the three possible classes of decays

$$\tau \equiv \frac{\hbar}{\Gamma_{Total}} \equiv \frac{\hbar}{\Gamma_{Semilept} + \Gamma_{Nonlept} + \Gamma_{Lept}}$$

The partial width $\Gamma_{Semilept}$ is universal (equal) for D^0 vs D^+ (an assumption experimentally verified within 10%) as a consequence of isospin invariance, and for D_s^+ vs D^0 on the basis of theoretical arguments. The partial width Γ_{Lept} is small due to the helicity suppression. Therefore, all differences experimentally found should be caused by $\Gamma_{Nonlept}$. Lifetimes are a window on decay dynamics: conventional explanations of differences among charm hadrons lie

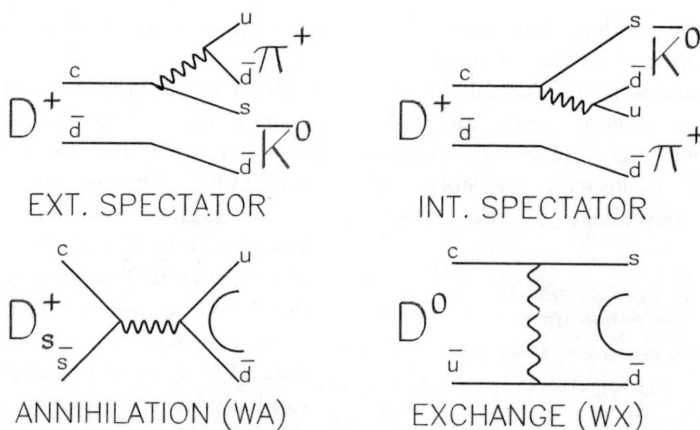

Figure 4. The two internal and external spectator decays are exclusive of D^+ mesons and, because of the destructive Pauli interference due to the identical $\bar{d}$ quarks in the final state, they enhance the lifetime. The contribution to the total widths of Cabibbo-Favored (CF) W-annihilation and W-exchange diagrams is, among mesons, unique to D_s^+ and D^0, respectively.

in the interplay among the spectator, W-exchange, and W-annihilation diagrams (Fig.4). The large difference in lifetimes for D^0 and D^+ is conclusively explained as being due to the presence of external and internal spectator diagrams. Instead, the D_S^+ lifetime as it appears in PDG98[36] is only different from D^0 at the 3σ level, i.e., $\tau_{D_s^+}/\tau_{D^0} = 1.12 \pm 0.04$.

The PDG98 measurements of charm lifetimes are dominated by old fixed-target photoproduction E687 experiment results. Besides new results from fixed-target experiments, a new player in the lifetime game in 1998-1999 was the e^+e^- experiment CLEO II.V. Their lifetime measurements[37] (relative to $3.7\,\mathrm{fb}^{-1}$, i.e., about 40% of their present data set) were made possible by the implementation of a double-sided Si vertex detector, which also has the beneficial effect of improving D^*-tagging by a better definition of the soft pion track. New results[42] are shown in Tab.2, with new world averages. Although CLEO II.V precision is at the level of E687, their continuous building up of statistics, a possible better understanding of the systematics of their new detector, and the planned CLEO III implementation of RICH particle ID may help them become competitive with fixed-target experiments in the future.

The most relevant new information comes from the E791[39] and FOCUS[40] measurements of D_s lifetime, which reduce the error on the ratio with the D^0 lifetime

$$R_\tau \equiv \tau_{D_s^+}/\tau_{D^0} = 1.22 \pm 0.02 \tag{1}$$

which is now ten standard deviations away from unity, indicating that although not dominant, the WA diagram is significant. In an approach based on Wilson's OPE[2] (where the interaction is factorized into three parts – weak interaction between quarks, perturbative QCD corrections, non-perturbative QCD effects), the decay rate is expanded in the heavy quark masses[3]

$$\Gamma(H_Q \to f) = \frac{G_F^2 m_Q^5}{192\,\pi^3} |KM|^2 \left[A_0 + \frac{A_2}{m_Q^2} + \frac{A_3}{m_Q^3} + \mathcal{O}(1/m_Q^4) \right] \tag{2}$$

Each term has a simple physical meaning: the leading operator A_0 contains the spectator diagram contribution; A_2 is the spin interaction of the heavy quark with light quark degrees of freedom inside the hadron; A_3, the PI, WA, WX contributions. A description of OPE goes beyond the scope of this review; interested readers are addressed to excellent review[41]. The OPE model predicts $R_\tau = 1.00 - 1.07$ if the WA operator does not contribute. If it does, the maximum effect predicted is $\pm 20\%$, i.e., $R_\tau = (0.8 - 1.27)$. The world average found is presently quite at the limits of the OPE predictions, and it could be used as a constraint to better define the WA operator, which also intervenes in semileptonic beauty decays[43].

In the baryon sector, the SELEX measurement[44] of the Λ_C lifetime disagrees with the PDG98 world average dominated by E687, which is instead preliminarily confirmed by FOCUS[40] new high-statistics measurement. Finally, a more precise measurement of Ω_c and Ξ_c^0 lifetimes is badly needed in order to confirm the lifetime pattern $\tau(\Omega_c^0) < \tau(\Xi_c^0) < \tau(\Lambda_c^+) < \tau(\Xi_c^+)$.

4 Nonleptonic weak decays

The highest impact new measurement is the CLEO II.V determination of the $\Lambda_c^+ \to pK^-\pi^+$ absolute branching fraction. This number, used to normalize all charm baryon branching ratios, consists of the PDG98 average of $(5.0 \pm 1.3)\%$ as an average of two model-dependent measurements, in mutual disagreement at the level of $2 - 3\,\sigma$. CLEO II.V tags charm events with the semielectronic decay of a D^*-tagged $\bar{D}$, and the Λ_c^+ production with a $\bar{p}$. Their final value is $B(\Lambda_c^+ \to pK^-\pi^+) = (5.0 \pm 0.5 \pm 1.5)\%$. A full discussion of the analysis technique is reported in these proceedings[81].

First observation[45] (confirmed shortly thereafter[46]) of Cabibbo-Suppressed (CS) $\Xi_c^+ \to pK^-\pi^+$ decay by the fixed-target, hyperon-beam experiment SELEX (Fig.6) provided information on the interplay of the external W-spectator decay and final-state interactions (FSI). These are interactions which occur in a space-time region where the final state particles have already been formed by the combined action of weak and strong forces, but

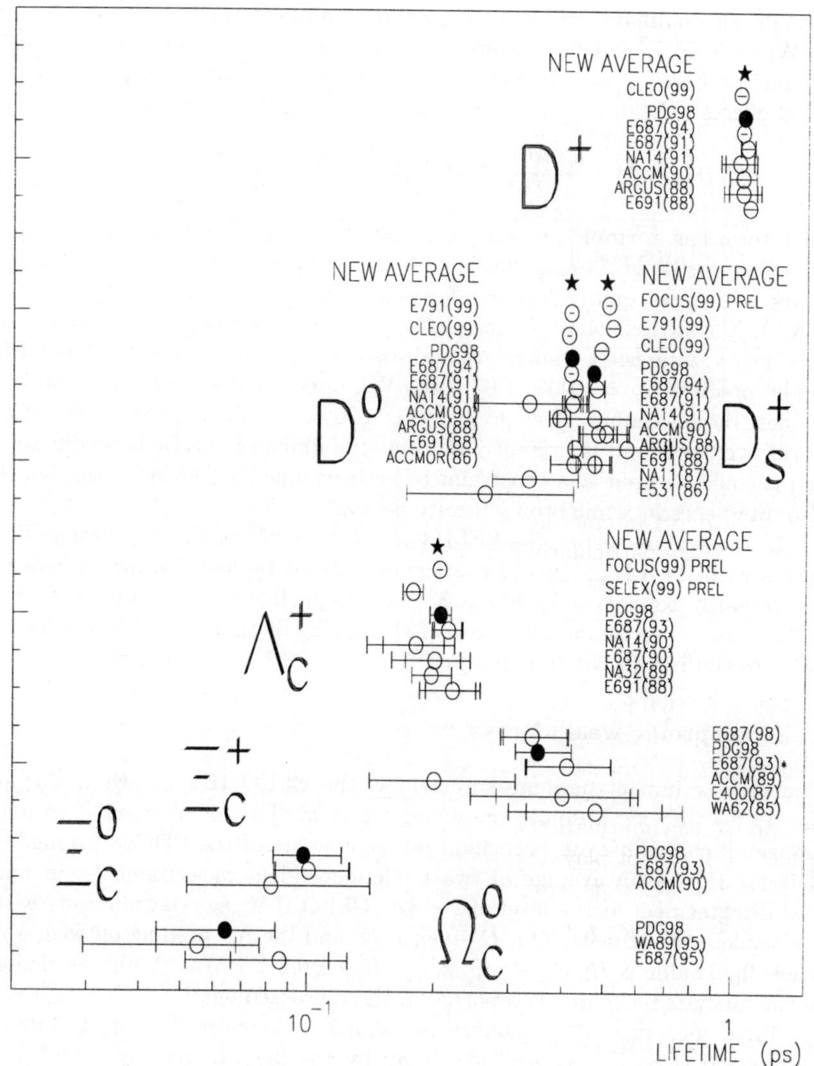

Figure 5. Compilation of charm meson and baryon lifetimes.

Table 2. Summary of new results in charm hadron lifetimes.

	Experiment	Lifetime (ps)	Events	Year	Technique
D^+	New Average	1.050 ± 0.015			
	CLEO	$1.0336 \pm 0.0221^{+0.0099}_{-0.0127}$	3777	98	e^+e^-
	PDG98	1.057 ± 0.015			
D^0	New Average	0.412 ± 0.003			
	E791	$0.413 \pm 0.003 \pm 0.004$	35k	99	Hadroprod
	CLEO	$0.4085 \pm 0.0041^{+0.0035}_{-0.0034}$	19k	98	e^+e^-
	PDG98	0.415 ± 0.004			
D_s^+	New Average	0.500 ± 0.007			
	FOCUS prel	$0.506 \pm 0.008(stat)$	5668	99	Photoprod
	E791	$0.518 \pm 0.014 \pm 0.007$	1662	99	Hadroprod
	CLEO	$0.4863 \pm 0.0150^{+0.0049}_{-0.0051}$	2167	98	e^+e^-
	PDG98	0.467 ± 0.017			
Λ_c^+	New Average	0.2019 ± 0.0031			
	FOCUS prel	$0.2045 \pm 0.0034(stat)$	8520	99	Photoprod
	SELEX prel	$0.177 \pm 0.010(stat)$	1790	99	Hyperons
	PDG98	0.206 ± 0.012			

are still strongly interacting while recoiling from each other. In charm meson decays, FSI are particularly problematic because of the presence of numerous resonances in the mass region interested[49]. The CS branching ratio, measured by SELEX relative to four–body CF decay

$$B(\Xi_c^+ \to pK^-\pi^+)/B(\Xi_c^+ \to \Sigma^+(pn)\,K^-\pi^+) = 0.22 \pm 0.06 \pm 0.03$$

is (once corrected for phase space) compatible with the branching ratio for the only other CS decay well measured, $\Lambda_c^+ \to pK^-K^+$, relative to three–body CF decay $\Lambda_c^+ \to pK^-\pi^+$. This is different from the charm meson case, where branching ratios depend heavily on the multiplicity of the final state, and is interpreted as confirmation of the fact that for charmed baryons, contrary to mesons, FSI do not play a relevant role. Finally, first evidence of DCS decay $D^+ \to K^+K^-K^+$ was reported[47] by FOCUS (Fig.6c), which measures

$$\Gamma(D^+ \to K^+K^-K^+)/\Gamma(D^+ \to K^+\pi^-\pi^+) = (1.41 \pm 0.27) \times 10^{-4} \qquad (3)$$

Such a decay cannot proceed via a spectator diagram, since the $\bar{d}$ initial state quark disappears in the final state. Possible mechanisms are pure WA, or Long-Distance (LD) processes including a light meson which strongly couples to KK. In either case, a Dalitz analysis would be of extreme interest to possibly investigate the decay resonant structure. For a DCSD, in the simplest picture one has $\Gamma_{DCSD}/\Gamma_{CF} \propto \tan^4\theta_C \simeq 2 \times 10^{-3}$. Any deviation from this value is due to effects such as interference, hadronization, FSI, etc.

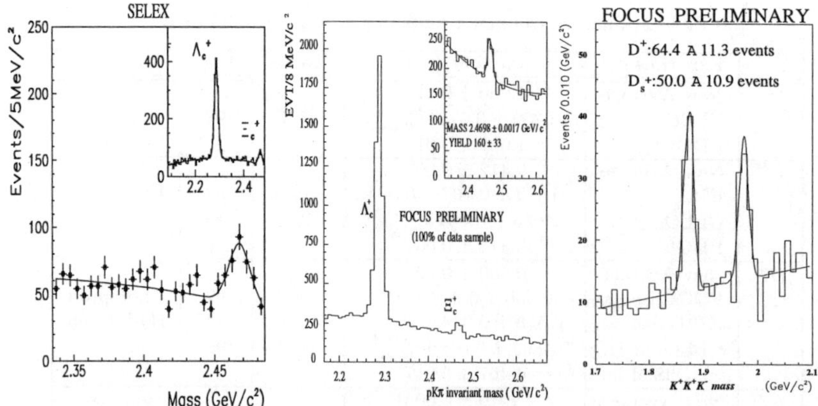

Figure 6. Observation of CS decay $\Xi_c^+ \to pK^-\pi^+$ (left and center), and DCS decay $D^+ \to K^+K^-K^+$ (right).

Although in principle accessable (with an important *caveat* being the treatment of FSI) by means of lattice methods, nonleptonic decays lack an organic theoretical framework rigorously descending from first principles, while the most interesting (two-body) decays are still largely undetected[49]. A theoretical approach that has been pointed to as comprehensive is ref.[50],[51], which has the merit of fully incorporating FSI in the prediction of two-body nonleptonic decays and also formulates CP-violation (CPV) asymmetries and CP-eigenstate lifetime differences.

5 Semileptonic decays

Comprehensive older reviews of leptonic and semileptonic decays are in [52,53]. Form factors describe dressing of $Q\bar{q}$ into a daughter hadron at the hadronic W-vertex of the spectator decay (Fig.7a). In the simplest case of a charmed pseudoscalar meson decaying to a light pseudoscalar meson, lepton, and antineutrino, the differential decay rate is

$$\frac{d\Gamma}{dq^2} = \frac{G_F^2|V_{cq}|^2P^3}{24\pi^3}\left\{|f_+(q^2)|^2 + |f_-(q^2)|^2\mathcal{O}(m_\ell^2) + ...\right\} \qquad (4)$$

where P is the momentum of the pseudoscalar meson in the reference frame of the charmed meson, and the $Wc\bar{q}$ vertex is described by only two form factors $f_\pm(q^2)$. Parameterizations for $f_\pm(q^2)$ form factors inspired respectively by a

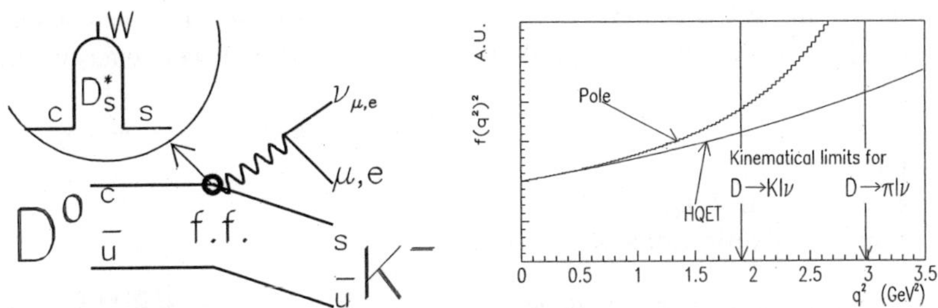

Figure 7. Semileptonic decays of the D^0 meson, illustrating coupling of a virtual $D_s^*(c\bar{s})$ vector state which originates the nearest-pole dominance model (left); $|f_+(q^2)|^2$ as a function of q^2 for pole and HQET parameterizations. The kinematic limits for $K\ell\nu$ and $\pi\ell\nu$ are drawn with vertical lines (adapted from ref.[8])

pole dominance model and HQET[54] are

$$f_\pm(q^2) = f_\pm(0)(1 - q^2/M_{pole}^2)^{-1} \qquad \text{(Pole)} \qquad (5)$$

$$f_\pm(q^2) = f_\pm(0)e^{\alpha q^2} \qquad \text{(HQET)} \qquad (6)$$

The value for M_{pole} is somehow arbitrarily chosen such as to be the closest $Q\bar{q}$ state with the same J^P as the hadronic weak current (Fig.7a). Unfortunately (Fig.7b from ref.[8]), there is little or no difference between a pole or an exponential form in the range of small q^2 accessible by CF $K\ell\nu$ decays, while maximal sensitivity is allowed for CS decay $\pi\ell\nu$. FOCUS should be able to finally measure the q^2 dependance by making use of the collected sample of 5,000 $\pi\ell\nu$ semileptonic decays.

In the case of a pseudoscalar-to-vector decay there are four form factors $(V, A_{1,3})$, q^2-dependent. After assuming a nearest-pole dominance model, they are customarily expressed via the ratios $r_V \equiv V(0)/A_1(0)$, $r_2 \equiv A_2(0)/A_1(0)$, with $A_3(0)$ becoming negligible in the (questionable) limit of zero lepton mass. E791 has presented new measurements[55] of form-factor ratios for $D_s^+ \to \phi\ell^+\nu_\ell$, with $(\ell = e, \mu)$, which investigate the extent to which the SU(3) flavor symmetry is valid by comparing form factors with what was measured in $D^+ \to \bar{K}^{*0}\ell^+\nu_\ell$ previously, where a spectator $\bar{d}$ quark is replaced by a spectator $\bar{s}$ quark. Measurements are based on a sample of 144 electron decays and 127 muon decays: r_V is consistent with the expected SU(3) flavor symmetry between D_s and D^+ semileptonic decays, while r_2 appears inconsistent (Fig.8).

In the e^+e^- sector, OPAL has recently produced the first measurement[56,57] of the semileptonic branching ratio of charm hadrons produced in $Z^0 \to c\bar{c}$ decays, finding $B(c \to \ell) = 0.095 \pm 0.006^{+0.007}_{-0.006}$, in good agreement with the ARGUS lower energy data.

6 Decay constants f_D and f_{D_s} in charm leptonic decays

No experimental results have emerged on leptonic decays. Theoretical activity on the main reasons for these studies (i.e., the pseudoscalar decay constants f_D and f_{D_s}) is very intense[58]. Lattice calculations have now converged to $f_{D_S} \sim 220 \pm 15\,\mathrm{MeV}$, to be compared with the world average $254 \pm 31\,\mathrm{MeV}$. On the contrary, the lattice result $f_D \sim 195 \pm 15\,\mathrm{MeV}$ can only be compared with the 1988 MARK III

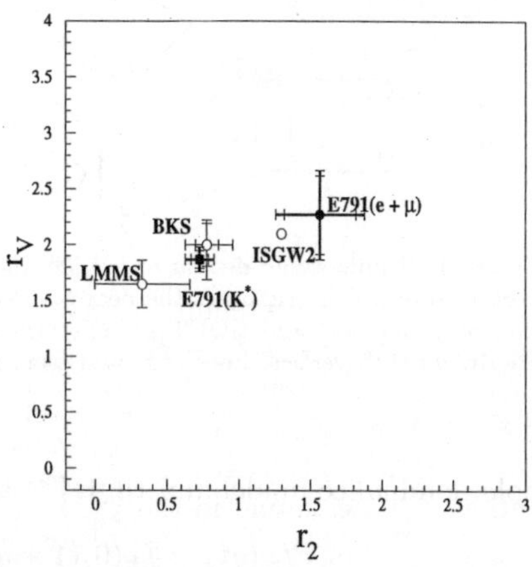

Figure 8. E791 semileptonic vector form-factor ratios for D^+ and D_s^+, and comparison with predictions.

limit of $f_D < 290\,\mathrm{MeV}$. An experiment able to study the challenging decay $D \to \ell\nu_\ell$ is badly needed. An alternative model-dependent technique[59] relates the $D^{*+} - D^{*0}$ mass isosplittings to f_D, via the wavefunction at the origin $|\psi(0)|^2$. The value inferred from the best isosplit measurement [60] is $f_D = (290 \pm 15)\,\mathrm{MeV}$, very distant from the lattice computation.

7 Rare and forbidden decays, CP violation.

In the charm sector, Flavor-Changing Neutral Current processes such as $D^+ \to h^+\mu^+\mu^-$, $D^0 \to \mu^+\mu^-$, etc, are suppressed in the SM via the GIM mechanism, with predictions spanning an enormous range $10^{-9} - 10^{-19}$. Lepton Family Number Violating $D^+ \to h^+\ell_1^+\ell_2^-$, and Lepton Number Violating $D^+ \to h^-\ell_1^+\ell_{1,2}^+$ processes are instead strictly forbidden. This is why charm rare decays can provide unique information. E791 has presented [62] a set of new limits that improve the PDG98 numbers by a factor of 10, reaching approximately the 10^{-5} region.

CP-violation asymmetries in D decays are expected to occur via the interplay of weak phases stemming from penguin, Single-Cabibbo-Suppressed diagrams, and a (strong) FSI phase, and are predicted at the 10^{-3} level[50].

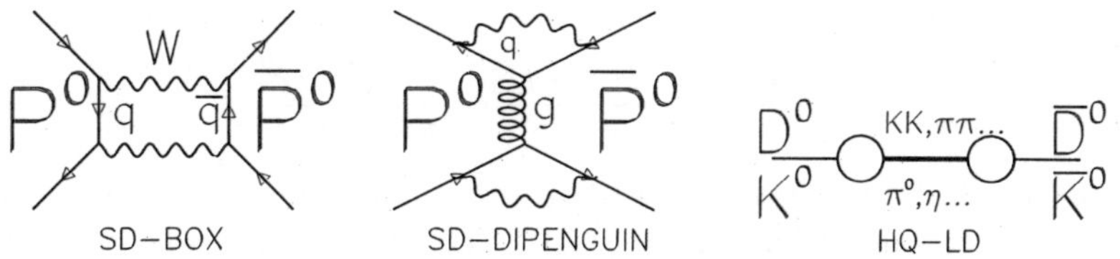

Figure 9. Box (proportional to $(m_q^2 - m_u^2)/m_W^2$), penguin, and long-distance diagrams for mixing.

FOCUS presented preliminary results[63] on CPV asymmetries in the two most accessible modes ($D^+ \to K^- K^+ \pi^+$ and $D^0 \to K^- K^+$), which improve the current limits down to the 10^{-2} level. CPV in the charm sector still has to be discovered. The availability of large clean samples of fully reconstructed $D^+ \to K^- K^+ \pi^+$ decays entitles one to investigate CPV by comparing phases and amplitudes found in the two CP conjugate Dalitz plots[47].

8 $D^0 \bar{D}^0$ mixing

Important new results have been presented on $D^0 \bar{D}^0$ mixing (for an updated review see [64]). It is useful to recall the key features of particle-antiparticle mixing [65]. Because of weak interactions, flavor $f = s, c, b$ of a generic pseudoscalar neutral meson P^0 is not conserved. Therefore it will try and decay with new mass eigenstates $P^0_{1,2}$ which no longer carry definite flavor f: they are new states with different mass and lifetime $|P^0_{1,2}\rangle \propto (p|P^0\rangle \pm q|\bar{P}^0\rangle)$ where complex parameters p and q account for any CPV. The time evolution of $|P^0(t)\rangle$ is given by the Schrödinger equation. After a time t the probability of finding the state P^0 transformed into $\bar{P}^0$ is

$$|\langle \bar{P}^0|P^0(t)\rangle|^2 \propto \left|\frac{q}{p}\right|^2 e^{-\Gamma_1 t}[1 + e^{\Delta\Gamma t} + 2e^{\frac{\Delta\Gamma}{2}t}\cos(\Delta m t)] \tag{7}$$

with definitions $\Delta m \equiv m_1 - m_2$, $\Delta\Gamma = \Gamma_1 - \Gamma_2$ and $\bar{\Gamma} \equiv (\Gamma_1 + \Gamma_2)/2$. The two states will oscillate with a rate expressed by Δm and $\Delta\Gamma$, which are naturally expressed when calibrated by the average decay rate within the parameters $x \equiv \Delta m/\bar{\Gamma}$ and $y \equiv \Delta\Gamma/(2\bar{\Gamma})$.

In the case of charm mesons[66], because of the Cabibbo-favored decay mechanism and the large phase space available for their decay, decay widths are very similar ($y \ll 1$), and the time-integrated ratio of mixed and nonmixed rates is

$$r \equiv \frac{\Gamma(D^0 \to \bar{D}^0 \to \bar{f})}{\Gamma(D^0 \to f)} = \left|\frac{q}{p}\right|^2 \frac{x^2 + y^2}{2} \qquad (8)$$

Theoretical estimates of x fall into two main categories, short distance (SD) and heavy quark/long distance (HQ-LD): the former arise from the box diagram[67] (Fig.9a), with GIM mechanism suppressing the charm case (Tab.3) or the dipenguin diagram[68], the latter come from QCD diagrams[69] and FSI[67] such as rescattering of quarks with known intermediate light states (Fig.9c). An important comment was made recently[70] on the possibility of measuring y separately from x. Indeed, $x \neq 0$ means that mixing is genuinely produced by $D^0 \bar{D}^0$ transitions (either SD or HQ-LD, or both), while $y \neq 0$ means that the fast-decaying component D_1^0 quickly disappears, leaving the slow-decaying component D_2^0 behind, which is a mixture of D^0 and $\bar{D}^0$. Infinite discussion is active on the extent

Table 3. Box diagram contributions to mixing.

P^0	q	$x \equiv \frac{\Delta m}{\Gamma}$
K^0	c	0.48
D^0	s	$\ll 1$
B_d^0	t	0.73
B_s^0	t	> 14

to which the three contributions are dominant: consensus seems to exist on the HQ–LD being, in the case of charm mesons, larger than the SD, and in any case utterly small. Standard Model predictions are[72]

$$x, y < 10^{-7} - 10^{-3} \qquad r^{SM} < 10^{-10} - 10^{-4} \qquad (9)$$

still below the PDG98 limit[75] $r < 5 \times 10^{-3}$. Any observation of $D^0 \bar{D}^0$ mixing above the predicted level, once HQ–LD effects are understood, is a signal that new physics contributions are adding to the box diagrams[71]. Traditionally, $D^0 \bar{D}^0$ mixing is searched for by means of event-counting techniques, while advances in event statistics now allow studies of the y parameter.

8.1 Wrong sign vs right sign counting

Mixing is searched for in the decay chains

$$D^{*+} \to \pi^+, \qquad D^0 \to \bar{D}^0 \to K^+ \pi^-, K^+ \pi^- \pi^+ \pi^-, K^+ \ell^- \bar{\nu}_\ell \qquad (10)$$

with the particle/antiparticle nature of D^0 at production and at decay given by the sign of π^+ and K^- respectively.

In the case of a hadronic final state, life is complicated by pollution of the mixing by the Doubly-Cabibbo-Suppressed Decay $D^0 \to K^+ \pi^-$, proportional to $\tan^4 \theta_C$. The measurable r_{WS} – the rate of wrong-sign events – has therefore contributions[73] from DCSD, interference, and mixing

$$r_{WS} = \frac{\Gamma(D^0 \to f)}{\Gamma(\bar{D}^0 \to f)} = \frac{e^{-\bar{\Gamma}t}}{4}|\langle f|H|D^0\rangle|^2_{CF}\left|\frac{q}{p}\right|^2(X + Yt + Zt^2) \tag{11}$$

$$X \equiv 4|\lambda|^2 \quad Y \equiv 2\Re(\lambda)\Delta\Gamma + 4\Im(\lambda)\Delta m \quad Z \equiv (\Delta m)^2 + (\Delta\Gamma)^2/4 \tag{12}$$

$$\lambda \equiv \frac{p}{q}\frac{\langle f|H|D^0\rangle_{DCS}}{\langle f|H|\bar{D}^0\rangle_{CF}} \tag{13}$$

The X term (pure DCS) is characterized by an exponential decay time behavior, unlike the Z term (pure mixing), and this feature can in principle be used to suppress the DCS pollution. The Y (interference) term receives contributions from $\Im(\lambda)$, which can be nonzero if a) CPV is present, thus introducing a phase φ in p/q; and/or b) a strong phase δ is present, due to different FSI in the DCS and CF decays. By assuming CP conservation, i.e., $|p/q| = 1$, defining

$$e^{i\varphi} \equiv \frac{p}{q} \qquad e^{i\delta}\sqrt{r_{DCS}} \equiv \frac{\langle f|H|D^0\rangle_{DCS}}{\langle f|H|\bar{D}^0\rangle_{CF}} \tag{14}$$

and measuring t in units of $\bar{\Gamma}$ one can write[70] a simpler expression for r_{WS}

$$r_{WS} \propto e^{-t}[r_{DCS} + t^2(r/2) + t\sqrt{2rr_{DCS}}\cos\phi] \tag{15}$$

where the interference angle is given by $\phi = \arg(ix + y) - \varphi - \delta$. Equation 15 shows how a meaningful quote of the r result must specify which assumptions where made on the CPV and strong angles φ and δ. If one assumes CP invariance ($\varphi = 0$), then

$$r_{WS} \propto e^{-t}\{r_{DCS} + (r/2)t^2 + (y'\sqrt{r_{DCS}})t\} \tag{16}$$

$$y' \equiv y\cos\delta - x\sin\delta \qquad x' \equiv x\cos\delta + y\sin\delta \tag{17}$$

The alternative option in counting techniques is the use of semileptonic final states $K\ell\nu$, which do not suffer from DCSD pollution but are harder experimentally.

8.2 Lifetime difference measurements

The y parameter can be determined directly by measuring the lifetimes of CP=+1 and CP=−1 final states, assuming CP conservation, i.e., that D_1^0 and D_2^0 are indeed CP eigenstates. This would allow in principle, along with an independent measurement of r, limits to be set on x. The experimentally most accessible CP-eigenstates are K^+K^- and $\pi^+\pi^-$ (CP=+1), $K_S\phi$ (CP=−1), and $K^-\pi^+$ (mixed CP).

Table 4. Synopsis of recent mixing results. CPV phase is φ, strong phase is δ, interference angle is $\phi = \arg(ix + y) - \varphi - \delta$.

	Assumptions	Mode	N_{RS}	Result (%)
ALEPH[74]	No mix	$K\pi$	1.0k	$r_{DCS} = 1.84 \pm .59 \pm .34$
(95% CL)	$\varphi = 0, \cos\phi = 0$			$r < 0.92$
	$\varphi = 0, \cos\phi = +1$			$r < 0.96$
	$\varphi = 0, \cos\phi = -1$			$r < 3.6$
E791[75]	$\varphi = 0$	$K\ell\nu$	2.5k	$r = 0.11^{+0.30}_{-0.27}$ $(r < 0.50)$
(90% CL)				
E791[76]	No mix	$K\pi$	5.6k	$r_{DCS} = 0.68^{+0.34}_{-0.33} \pm 0.07$
(90% CL)	No mix	$K3\pi$	3.5k	$r_{DCS} = 0.25^{+0.36}_{-0.34} \pm 0.03$
	$\varphi \neq 0$ in Y			$r = 0.39^{+0.36}_{-0.32} \pm 0.16$
				$(r < 0.85)$
		$K\pi$		$r_{DCS} = 0.90^{+1.20}_{-1.09} \pm 0.44$
		$K3\pi$		$r_{DCS} = -0.20^{+1.17}_{-1.06} \pm 0.35$
	None			$r(\bar{D}^0 \to D^0) = 0.18^{+0.43}_{-0.39} \pm 0.17$
	None			$r(D^0 \to \bar{D}^0) = 0.70^{+0.58}_{-0.53} \pm 0.18$
	No Y			$r = 0.21^{+0.09}_{-0.09} \pm 0.02$
E791[77]	$\varphi = 0, \delta = 0$	KK	6.7k	$\Delta\Gamma = 0.04 \pm 0.14 \pm 0.05\,\mathrm{ps}^{-1}$
(90% CL)		$K\pi$	60k	$(-0.20 < \Delta\Gamma < 0.28)\,\mathrm{ps}^{-1}$
				$y = 0.8^{+2.9}_{-1.0}$ $(-4 < y < 6)$
CLEO[78]	$\varphi = 0, \delta = 0$	$\pi\pi$	475	
(90% CL)		KK	1.3k	$y = -3.2 \pm 3.4$ $(-7.6 < y < 1.2)$
$5.6\,\mathrm{fb}^{-1}$		$K\pi$	19k	
CLEO[79]	No mix	$K\pi$	16k	$r_{DCS} = 0.34 \pm 0.07 \pm 0.06$
(95% CL)	$\varphi = 0, \delta \neq 0$			$r_{DCS} = 0.50^{+0.11}_{-0.12} \pm 0.08$
$9\,\mathrm{fb}^{-1}$	$\varphi = 0, \delta \neq 0, x' = 0$			$y' = -2.7^{+1.5}_{-1.6} \pm 0.2$
				$(-5.9 < y' < 0.3)$
	$\varphi = 0, \delta \neq 0, y' = 0$			$x' = 0 \pm 1.6 \pm 0.2$
				$|x'| < 3.2$ $(r < 0.05)$

8.3   Mixing results and projections

The most recent mixing results in Tab.4 are compiled from ALEPH (out of 4×10^6 hadronic Z decays) and E791 (2×10^5 reconstructed decays) recently published results, as well as CLEO II.V preliminary results (lifetime difference ($5.6\,fb^{-1}$) and hadronic counting ($9.0\,fb^{-1}$)). Once compared with the same assumptions, measurements are consistent, with the exception of a mild discrepancy in the ALEPH measurement of r_{DCS}. CLEO II.V's best limit on y' corresponds to $r < 0.05\%$ if $x' = 0$. Progress should come from CLEO II.V (semileptonic counting and lifetime differences including the $K_S\phi$ decay mode) and FOCUS, which both project sensitivities around $y' < 1\%$, corresponding to $r < 0.005\%$. A compilation of predictions on $D^0\bar{D}^0$ mixing was recently presented[80], and it was pointed out how the CLEO II.V limit already rules out a lot of them.

It is clear that, in order to further reduce the limit on r, one should await results from B-factories. Mixing and CPV are the very topics which sorely experience the lack of a concrete project of τ–charm factory.

9   Spectroscopy

A specialized review of the many new results available and their implications for Heavy Quark Symmetry predictions was given at the conference[81], so I shall present only an overall picture, referring the interested reader to details in D. Besson's paper.

With the high-statistics, high-mass resolution experiments attaining maturity, focus has been shifted from the ground state (0^- and 1^-) $c\bar{q}$ mesons and ($1/2^+$ and $3/2^+$) cqq baryons to the orbitally- and, only very recently, radially-excited states[a]. An organic and consistent theoretical framework for the spectrum of heavy-light mesons is given by the ideas of Heavy Quark Symmetry (HQS), later generalized by Heavy Quark Effective Theory in the QCD framework. The basic idea (mediated from the JJ coupling in atomic physics) is that in the limit of infinite heavy quark mass: a) the much heavier quark does not contribute to the orbital degrees of freedom, which are completely defined by the light quark(s) only; and b) properties are independent of heavy quark flavor. The extent to which the infinite heavy quark mass limit is appropriate for charm hadrons is the subject of infinite discussion; however, things seem to work amazingly well. For recent experimental reviews see [87,88,89].

9.1   Mesons

Heavy Quark Symmetry provides explicit predictions on the spectrum of excited charmed states[82,83]. In the limit of infinite heavy quark mass, the spin of the heavy quark $\mathbf{S_Q}$ decouples from the light quark degrees of freedom (spin $\mathbf{s_q}$ and orbital $\mathbf{L}$), with $\mathbf{S_Q}$ and $\mathbf{j_q} \equiv \mathbf{s_q} + \mathbf{L}$ the conserved quantum numbers. Predicted excited states are formed by combining $\mathbf{S_Q}$ and $\mathbf{j_q}$. For $L = 1$ we have $j_q = 1/2$ and $j_q = 3/2$ which, combined with S_Q, provide prediction for two $j_q = 1/2$ (J=0,1) states, and two $j_q = 3/2$ (J=1,2) states. These four states[b] are named respectively D_0^*, $D_1(j_q = 1/2)$, $D_1(j_q = 3/2)$ and D_2^*. Fi-

---

[a]In the past, these excited states were called generically and improperly D^{**}.

[b]Common nomenclature for excited states is $D_J^{*(','',...)}$, where J is the total angular momentum (spin+orbital), $(','',...)$ indicates radial excitations, and $*$ indicates natural

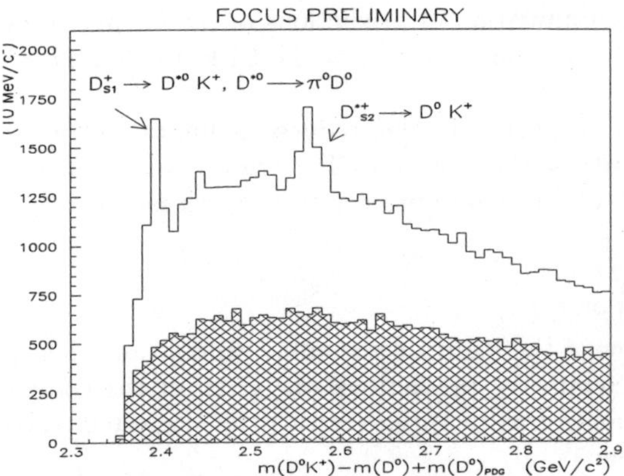

Figure 10. The $D^0 K^+$ invariant mass spectrum, showing the D_{s2}^* peak along with the $D_{s1}(j_q = 3/2)$ reflection peak due to an undetected π^0. Wrong-sign ($D^0 K^-$) background is shown shaded.

Table 5. Experimental status of (L=1, n=1) and (L=0, n=2) $c\bar{q}$ and $c\bar{s}$ mesons (adapted from ref.[87]). Not well established states are shown in bold. Units are MeV/c^2. Spin-parity assignment for D_{s2}^* is not pinned down yet, therefore this state is quoted as $D_{sJ}(2573)$ in PDG98. Theory predictions are taken from ref.[81,93].

j_q	1/2	1/2	3/2	3/2	1/2	1/2
J^P	0^+	1^+	1^+	2^+	0^-	1^-
L	1	1	1	1	0	0
n	1	1	1	1	2	2
	$\mathbf{D_0^*}$	$\mathbf{D_1}$	D_1	D_2^*	$\mathbf{D'}$	$\mathbf{D^{*\prime}}$
Mass exp.		$2461^0\,{}^{+42}_{-35}$	$2422^0, 2427^\pm$	$2459^{0,\pm}$		2637 ± 6
Mass th.	2400	2490	2440	2500	2580	2640
Width exp.		$290^0\,{}^{+104}_{-83}$	$19^0, 28^\pm$	$23^0, 25^\pm$		< 15
Width th.	> 170	> 250	$20 - 40$	$20 - 40$		$40 - 200$
Decay Mode	$D\pi$	$D^*\pi$	$D^*\pi$	$D\pi, D^*\pi$		$D^*\pi\pi$
	$\mathbf{D_{s0}^*}$	$\mathbf{D_{s1}}$	D_{s1}	D_{s2}^*	$\mathbf{D_s'}$	$\mathbf{D_s^{*\prime}}$
Mass exp.			$2535^\pm$	$2573^\pm$		
Mass th.	2480	2570	2530	2590	2670	2730
Width exp.			$< 2.3^\pm$	$15^\pm$		
Width th.			< 1	$10 - 20$		
Decay Mode			$D^* K$	DK		

nally, parity and angular momentum conservation force the ($j_q = 1/2$) states to decay to the ground states via S-wave transitions (broad width), while

---

$(0^+, 1^-, 2^+, ...)$ J^P assignment.

$(j_q = 3/2)$ states would decay via D-wave (narrow width).

Such a pattern was recently beautifully borne out by the CLEO evidence for the $D_1(j_q = 1/2)$ broad state[79]. An open question remains the 1997 DELPHI observation[91] of first radial excitation $D^{*\prime+}$ in the $D^{*+}\pi^-\pi^+$ final state, not confirmed either by OPAL [92] or by CLEO[93], and questioned by theory predictions[94]. The situation of our present knowledge of excited D mesons is shown in tab.5. A major confirmation for HQS would also be the observation of the missing $(L = 1, j_q = 1/2, J^P = 0^+)$ D_0^* broad state. Finally, very little is known on excited $(c\bar{s})$ states (Fig.10).

9.2 Baryons

In the framework of SU(4) at ground state we expect nine cqq $J^P = 1/2^+$ baryons (all of them detected after the recent CLEO II.V observation of Ξ_c') and six cqq $J^P = 3/2^+$ baryons (Σ_c^{*+} and Ω_c^{*0} remain undetected; they are expected to decay via the experimentally difficult channels $\Sigma_c^{*+} \to \pi^0\Lambda_c^+$ and $\Omega_c^{*0} \to \gamma\Omega_c^0$). All the ccq and ccc states remain undiscovered[95]. In the HQS framework, cq_1q_2 baryons are considered as a system made of a heavy quark and a light diquark. Conserved numbers are the heavy quark spin $\mathbf{S_Q}$ and the diquark angular momentum $\mathbf{j}_{q_1q_2} \equiv \mathbf{L} + \mathbf{s}_{q_1q_2}$, where $\mathbf{L}$ is the diquark orbital momentum and $\mathbf{s}_{q_1q_2} \equiv \mathbf{s}_{q_1} + \mathbf{s}_{q_2} + \mathbf{l}_{12}$ the diquark total momentum[c].

During 1998, CLEO[96] presented evidence for the two $1/2^+$ missing states $\Xi_c^{0\prime}$ and $\Xi_c^{+\prime}$, respectively $c\{sd\}$ and $c\{su\}$, through the radiative decay $\Xi_c\gamma$. Masses measured are compatible with predictions[95,97].

Only a few of the orbitally excited P-wave $(L = 1)$ baryons have been observed. The first doublet $\Lambda_{c1}(2593)$ $(1/2^-)$ and $\Lambda_{c1}^*(2625)$ $(3/2^-)$ was ob-

Table 6. Isospin mass splittings.

	$M(\Sigma_c^{++} - \Sigma_c^0)$ MeV
FOCUS prel.	$0.28 \pm 0.31 \pm 0.15$
PDG98	0.57 ± 0.23
E791	$0.38 \pm 0.40 \pm 0.15$
CLEO II	$1.1 \pm 0.4 \pm 0.1$
CLEO	$0.1 \pm 0.6 \pm 0.1$
ARGUS	$1.2 \pm 0.7 \pm 0.3$

served several years ago by CLEO II.V, ARGUS, and E687. In 1999 CLEO II.V presented evidence[98] for the charmed-strange baryon analogous to $\Lambda_{c1}^*(2625)$, called $\Xi_{c1}^*(2815)$, in its decay to $\Xi_c\pi\pi$ via an intermediate Ξ_c^* state. The presence of intermediate Ξ_c^* instead of Ξ_c' is an indication of $3/2^-$ assignment, while HQS explicitly forbids a direct transition to the $\Xi_c\pi$ ground

---

[c]I adopt for excited baryon states the nomenclature in [12]. Thus, members of 3/2 multiplets are given a $(*)$, the subscript is the orbital light diquark momentum $\mathbf{L}$, and $(\prime)$ indicates symmetric quark wavefunctions $c\{q_1q_2\}$ with respect to interchange of light quarks, opposed to antisymmetric wavefunctions $c[q_1q_2]$.

state, because of angular momentum and parity conservation.

The last topic of this section is the mass difference between isospin states of charmed baryons (isosplits). Lately, interest in this issue was revamped[99,100] led by the consideration that, while for all well-measured isodoublets one increases the baryon mass by replacing a u–quark with a d–quark, the opposite happens in the case of the poorly measured $\Sigma_c^{++}(cuu) - \Sigma_c^0(cdd)$ isosplit. Besides the u/d quark mass difference, isosplits are in general sensitive to em effects, and spin–spin hyperfine interactions. New FOCUS results[101] are consistent with E791, and mildly inconsistent with CLEO II (Tab.6). Finally, first measurement of the Σ_c width was presented at this conference – full details in D. Besson's review.

In the baryon sector, the discovery of double-charm states would be of fundamental importance. A recent theoretical work[102] shows how a (nearly) model–independent approach based on the Feynman-Hellman theorem is able to compute heavy-flavor hadron masses in very good agreement with experiments, and to make predictions for double-charm states. As an example, the mass of the Ξ_{cc}^+ is predicted at $\sim 3.7\,\mathrm{GeV}/c^2$, and dominant decay modes are $D^+\Sigma^+$ and $D^+\Lambda^0$. Such a discovery does not seem within reach of either CLEO, or present fixed-target experiments.

9.3 Charmonium

Charmonium states are produced at e^+e^- storage rings and through $p\bar{p}$ annihilation. The e^+e^- annihilation proceeds, at first order, via a virtual photon and only $J^{PC} = 1^{--}$ states can be directly formed. Nonvector states such as the χ_J states can be observed only via two–step processes as $e^+e^- \to \psi' \to (c\bar{c}) + \gamma$, or higher order processes. On the contrary, in $p\bar{p}$ annihilations all J^{PC} quantum numbers are accessible.

New data come from e^+e^- BES[103] and $p\bar{p}$ E835[104]. Results include measurements of masses and widths of χ_{c0}, χ_{c2}, η_c. A 3σ disagreement remains between their η_c mass determinations. Moreover, the width $\Gamma(\eta_c \to \gamma\gamma)$ measured by E835 is nearly a factor of two narrower than the PDG98 world average, with which, instead, recent data from L3, OPAL, and BES agree.

The most intriguing puzzle in charmonium physics seems to remain the case for the pseudoscalar radial excitation $\eta_c'(2^1S_0)$. Claimed in 1982 by Crystal Barrel at $3.594\,\mathrm{GeV}/c^2$, it was not confirmed by either DELPHI or E835 extensive searches ($30\,\mathrm{pb}^{-1}$ in the range 3.666 to $3.575\,\mathrm{GeV}/c^2$). The E835 data-taking period at the end of 1999 sees the search for the η_c' approved as prioritary.

10 Conclusions and outlook

The major news this year in charm physics comes from the advances in lifetime measurement technology. The D_s^+ lifetime is now conclusively measured as being larger than D^0, and the lifetime ratio can be used (along with the new measurements of DCSD branching ratios) to constrain the sizes of the WA and WX operators. New limits on $D^0\bar{D}^0$ mixing also come mainly from the novel capability of measuring the lifetimes of opposite CP eigenstates. The other field where impressive news comes from is spectroscopy, with HQS predictions being spectacularly confirmed by the observation of an excited broad meson state, while the puzzles of meson radial excitations and the very existance of the η_c' still elude us. At the opening of the third millenium, besides results from E791, FOCUS, SELEX, CLEO II.V, E835, and BES, we should expect first data on $D^0\bar{D}^0$ mixing from the BaBar and BELLE B-factories.

The future of post-Y2K charm physics is less clear. No new fixed-target data-taking periods for charm studies are planned either at Fermilab, or at CERN for the near future (COMPASS commissioning is scheduled to begin in 2000, with a busy physics programme which includes charm muonproduction, DIS spin-physics, gluon structure functions, and light-quark hadronic physics). Despite the significant upgrade planned, it is not clear whether CLEO III will retain competitivity with respect to B-factories. Future experiments at high-energy hadron machines (Hera-B, BTeV, LHC-B) will need to tame huge backgrounds in order to contribute to charm physics. Intense workshop activity[105] on a e^+e^- τ-charm factory has not translated into an actual proposal — perhaps the operational experience coming from low-energy, high-luminosity ($\approx 10^{33}$cm^{2}s^{-1}) machines such as Frascati DAΦNE is needed. Interesting ideas come from a proposal for a $p\bar{p}$ collider operating at the open charm threshold[106]. On the other hand, a few specialized efforts have been approved, in the form of experiments undertaking upgrades for specific study of open charm physics (HERMES, NA50). The distant future will probably see charm ν-production from muon storage rings. Next year, the Lisbon conference will be an appropriate time to verify whether such a scenario had evolved.

Acknowledgments

This review would not have been possible without the unconditioned help of the spokespersons and analysis coordinators of the experiments which provided me with recent data: ALEPH, DELPHI, OPAL, BES, E791, L3, SELEX, E835, ZEUS, H1, NA50, HERMES, NUSEA. I especially enjoyed discussions with D. Besson, D. Asner, A. Zieminski, and M.L. Mangano. I thank all my FOCUS colleagues for continuous help, especially E.E. Gottschalk, P. Sheldon, M. Hosack, D. Menasce, M. Merlo, K. Stenson, H. Cheung, A. Zallo, H. Mendez, C. Riccardi, G. Boca, S.P. Ratti, L. Moroni, E. Vaandering. I should like to thank I. Bigi, J. Cumalat, D. Pedrini, and J. Wiss for fundamental comments and my colleagues Franco L. Fabbri and Shahzad Sarwar for daily discussions on charm physics. Finally, I should like to thank and congratulate the International Advisory Committee and the Local Organizing Committee for a completely successful conference.

References

1. N. Isgur and M.B. Wise, Phys. Lett. **B232**, 113 (1989); Phys. Lett. **B237**, 527 (1990).
2. K.G. Wilson, Phys. Rev. **179**, 1499 (1969).
3. I.I. Bigi, N.G. Uraltsev and A.I. Vainshtein, Phys. Lett. **B293**, 430 (1992); for a review of the subject, I.I. Bigi *Proc. Heavy Quark at Fixed Target, St. Goar (Germany) 1996, Frascati Physics Series vol. 7, (Köpke Ed.)* hep-ph/9612293.
4. C. Quigg, *Perspectives on Heavy Quark 98*, 4th Work. on Heavy Quarks at Fixed Target (HQ 98), Batavia, IL, 10-12 Oct 1998, AIP Conf. Proc. **459** (Cheung, Butler Eds.) p.485.
5. E. Golowich, hep-ph/9701225 (1997).
6. J. Butler, Proc. First Latin-American Symp. on HEP, Merida, Mexico 1996, AIP Conf. Proc. **400** (D'Olivo, Klein-Kreisler, Méndez Eds.) p.3.
7. J.P. Cumalat, *ibidem* p.91.
8. J.E. Wiss, Proc. I.S.P., Varenna (Italy) 1998, published by IOS-OHMSHA (Bigi, Moroni Eds), p.39.
9. F.I. Olness, ref.[4], p.238, hep-ph/9812270.
10. E. Norrbin, ref.[4], p.228, hep-ph/9812460
11. J.A. Appel, *Ann. Rev. Nucl. Part. Sci.* **42**, 367 (1992).
12. J.A. Appel, FERMILAB-CONF-93-328, Nov. 1993.
13. K. Stenson, XXXIV Renc. de Moriond (QCD and H.E. Had. Inter.), Les Arcs 1800, March 20-27, 1999.
14. E.M. Aitala *et al.* [E791 Coll.], hep-ex/9809029.
15. K. Stenson [E791 Coll.], ref.[4], p.159.
16. E.M. Aitala *et al.* [E791 Coll.], hep-ex/9906034.
17. E. Gottschalk, Fermilab DOE99 review, May 5, 1999; A. Kushnirenko [SELEX Coll.], ref.[4], p.168.
18. T. Adams et al., ref.[4], p.198.

19. E.C. Aschenauer [HERMES Coll.], ref.[4], p.189.

20. F. Maltoni, et al., Nucl. Phys. **B519**, 361 (1998).

21. A. Petrelli, et al., Nucl. Phys. **B514**, 245 (1998).

22. M.L. Mangano, private communication.

23. R. Bonciani, et al., Nucl. Phys. **B529**, 424 (1998).

24. C. Adloff *et al.* [H1 Coll.], Nucl. Phys. **B545**, 21 (1999).

25. O. Deppe et al., [ZEUS Coll.], Proc. PHOTON99 23-27 May 1999 , Freiburg, Germany.

26. B.A. Kniehl, et al., Z. Phys. **C76**, 689 (1997).

27. S. Frixione, et al., Phys. Lett. **B348**, 633 (1995).

28. S. Frixione, et al., Nucl. Phys. **B454**, 3 (1995).

29. J. Breitweg *et al.* [ZEUS Coll.], hep-ex/9908012.

30. D. Bailey, Proc. NUCLEON99, 7-9 Jun 1999, Frascati, Italy.

31. B. Naroska, ref.[4], p.207.

32. G. Abbiendi *et al.* [OPAL Coll.], hep-ex/9908001.

33. F. Abe *et al.* [CDF Coll.], Phys. Rev. Lett. **69**, 3704 (1992).

34. A. Zieminski, *these proceedings*.

35. M.C. Abreu *et al.* [NA50 Coll.], CERN-EP-99-013.

36. C. Caso *et al.*, Eur. Phys. J. **C3**, 1 (1998).

37. G. Bonvicini et al. [CLEO Coll.], *Phys. Rev. Lett.* **82**, 4586 (1999)

38. P.L. Frabetti et al., [E687 Coll.] *Phys. Lett.* **B427**, 221 (1998)

39. E.M. Aitala *et al.* [E791 Coll.], *Phys. Lett.* **B445**, 449 (1999)

40. H.W.K. Cheung, Proc. Heavy Flavours 8, Southampton, UK, July 1999.

41. G. Bellini, I.I. Bigi, P.J. Dornan, *Phys. Rept.* **289**, 1 (1997).

42. H.W.K. Cheung, ref.[4], p.291.

43. I.I. Bigi, private communication.

44. A.Y. Kushnirenko [SELEX Coll.], ref.[4], p.168.

45. S.Y. Jun *et al.* [SELEX Coll.], hep-ex/9907062.

46. C. Riccardi [FOCUS Coll.], Proc. PANIC 99, June 1999, Uppsala (Sweden).

47. L. Moroni [FOCUS Coll.], Proc. EPS 99, Tampere, August 1999.

48. P. Dini [FOCUS Coll.], Proc. WHS 99, March 1999, Frascati (Italy), (Bressani, Feliciello, Filippi Eds.), ISBN88-86409-19-2, p.59.

49. T.E. Browder, K. Honscheid, D. Pedrini, *Ann. Rev. Nucl. Part. Sci.* **46**, 395 (1996).

50. F. Buccella et al., *Phys. Rev.* **D51**, 3478 (1995).

51. F. Buccella, M. Lusignoli and A.Pugliese, *Phys. Lett.* **B379**, 249 (1996).

52. J.G. Körner and G.A. Schuler, Z. Phys. **C46**, 93 (1990).

53. J.D. Richman and P.R. Burchat, *Rev. Mod. Phys.* **67**, 893 (1995).

54. D. Scora and N. Isgur, Phys. Rev. **D39**, 799 (1989); Phys. Rev. **D52**,

2783 (1995).

55. E.M. Aitala *et al.* [E791 Coll.], Phys. Lett. **B450**, 294 (1999).

56. P. Gagnon [OPAL Coll.], Nucl. Phys. Proc. Suppl. **75B**, 216 (1999).

57. G. Abbiendi *et al.* [OPAL Coll.], Eur. Phys. J. **C8**, 573 (1999).

58. A.S. Kronfeld, ref.[4], p.355.

59. J. Amundson et al. PR **D47** (1993) 3059; J. Rosner et al., PR **D47** (1993) 343.

60. D. Bortoletto et al. PRL **69** (1992) 2046.

61. G. Burdman, ref.[4], p.461.

62. E.M. Aitala *et al.* [E791 Coll.], hep-ex/9906045.

63. D.Pedrini [FOCUS Coll.], Proc. KAON 99, Chicago, June 1999.

64. P. Sheldon, Proc. Heavy Flavours 8, Southampton, UK, July 1999.

65. E. Leader and E. Predazzi, An introduction to gauge theories and modern particle physics, Cambridge 1996, ISBN 0 521 57742 X.

66. E. Golowich, hep-ph/9505381.

67. J.F. Donoghue, et al., Phys. Rev. **D33**, 179 (1986).

68. A.A. Petrov, Phys. Rev. **D56**, 1685 (1997).

69. H. Georgi, Phys. Lett. **B297**, 353 (1992); T. Ohl, et al., Nucl. Phys. **B403**, 605 (1993).

70. T. Liu, CHARM 2000, hep-ph/9408330; FCNC 97, Santa Monica, February 1997 hep-ph/9706477.

71. J.L. Hewett, T. Takeuchi and S. Thomas, hep-ph/9603391.

72. G. Burdman, FERMILAB-Conf-94/190; S. Pakvasa, FERMILAB-Conf-94/190; J.L. Hewett, SLAC-PUB-95-6821; I. Bigi, FERMILAB-Conf-94/190; T.A. Kaeding, hep-ph/9505393.

73. I.I. Bigi, *23rd Int. Conf. on High Energy Physics, Berkeley, CA, Jul 16-23, 1986* (SLAC-PUB-4074), S.C. Loken (Ed.), World Scientific, 1986, p.857 ; G. Blaylock, A. Seiden and Y. Nir, Phys. Lett. **B355**, 555 (1995).

74. R. Barate *et al.* [ALEPH Coll.], Phys. Lett. **B436**, 211 (1998).

75. E.M. Aitala *et al.* [E791 Coll.], Phys. Rev. Lett. **77**, 2384 (1996).

76. E.M. Aitala *et al.* [E791 Coll.], Phys. Rev. **D57**, 13 (1998).

77. E.M. Aitala *et al.* [E791 Coll.], Phys. Rev. Lett. **83** (1999) 32.

78. M. Selen [CLEO Coll.], APS99, Atlanta (USA), May 1999.

79. M. Artuso *et al.* [CLEO Coll.], hep-ex/9908040.

80. H.N. Nelson, hep-ex/9908021.

81. D. Besson, *these proceedings*.

82. S. Godfrey and N. Isgur, Phys. Rev. **D32**, 189 (1985).

83. S. Godfrey and R. Kokoski, Phys. Rev. **D43**, 1679 (1991).

84. N. Isgur and M.B. Wise, Phys. Rev. Lett. **66**, 1130 (1991).

85. E.J. Eichten, C.T. Hill and C. Quigg, Phys. Rev. Lett. **71**, 4116 (1993)

86. J. Bartelt and S. Shukla, *Ann. Rev. Nucl. Part. Sci.* **45**, 133 (1995).

87. F.L. Fabbri, ref.[48], p.627.

88. S.P. Ratti, ref.[48], p.337.

89. S. Paul, hep-ph/9903311.

90. S. Anderson *et al.* [CLEO Coll.], hep-ex/9908009.

91. P. Abreu et al., (DELPHI Coll.), *Phys. Lett.* **B426**, 231 (1998).

92. G. Abbiendi et al. [OPAL Coll.], Opal Conf. Rep. PN352/ICHEP9ε-1037 (1998).

93. D. Besson, these proceedings.

94. D. Melikhov and O. Pene, Phys. Lett. **B446**, 336 (1999).

95. J.G. Körner, M Krämer, D. Pirjol, *Prog. in Part. Nucl. Phys.* **33**, 787 (1994).

96. C.P. Jessop *et al.* [CLEO Coll.], Phys. Rev. Lett. **82** (1999) 492.

97. E. Jenkins, Phys. Rev. **D55**, 10 (1997).

98. J.P. Alexander *et al.* [CLEO Coll.], hep-ex/9906013.

99. J.L. Rosner, Phys. Rev. **D57**, 4310 (1998).

100. K. Varga et al., Phys. Rev. **D59**, 014012 (1999).

101. E.W. Vaandering [FOCUS Coll.], APS 99, Atlanta, (USA).

102. E. Predazzi ref.[48], p.265.

103. F.A. Harris [BES Coll.], EPS 99, Tampere (Finland), August 1999, hep-ex/9910027.

104. D. Bettoni , Proc. EPS 99, Tampere (Finland), August 1999.

105. M.L. Perl and P.C. Kim, *Summary of the Tau-charm Physics Workshop, Stanford, CA, 6-9 Mar 1999.*

106. HESR at GSI (Germany) http://www.ep1.ruhr-uni-bochum.de/gsi/

Heavy Quark Spectroscopy: Transitions and Decays

DAVE Z. BESSON*
*Physics Department, University of Kansas,
Lawrence, KS USA 66045-2151*

(Representing the CLEO collaboration)

1. Introduction

The study of charm hadron decays and widths typically begins with the long-standing pattern of charm hadron lifetimes. In the most naive picture, in which the only accessible decay diagram is the simple external W-emission spectator diagram (W_{ext}), the lifetime of all the charm particles should be identical and calculable from the mass of the charm quark. The fact that there are large discrepancies between the various charm lifetimes is an indication that other diagrams contribute. The pattern of lifetimes, as well as the primary diagrams expected to contribute to charm decay, is presented below, including recent contributions to this conference from CLEO (D^+, D^0 and D_s) and SELEX (Λ_c):

Particle	τ (ps)	Main Diagrams
D^+	1.057 ± 0.015^1	$|W_{ext} - W_{int}|^2, W_{annihilation}$
	$1.033\pm0.022^{+0.0102}_{-0.013}$	
D_s	0.467 ± 0.017^1	$W_{ext}, W_{annihilation}$
	$0.486\pm0.015 \pm 0.005^2$	
D^0	0.415 ± 0.004^1	$W_{ext}, W_{exchange}$
	$0.408 \pm 0.004 \pm 0.003^2$	
Λ_c	0.206 ± 0.012^1	$W_{ext}, W_{int}, W_{exchange}$
	$0.177 \pm 0.010 \pm sys^3$	
Ξ_c^+	$0.35^{+0.071}_{-0.04}$	W_{ext}, W_{int}
Ξ_c^0	$0.098^{+0.0231}_{-0.015}$	$W_{ext}, W_{int}, W_{exchange}$
Ω_c^0	0.064 ± 0.020^1	W_{ext}, W_{int}

The 'standard' explanation for this pattern, at least in the case of charged and neutral D-mesons, is that destructive interference between the internal and external W-emission diagrams which can contribute to the same final state (e.g., $D^+ \to \overline{K^0}\pi^+$) dilates the D^+ lifetime.[4] Nonetheless, it is likely that other diagrams, such as annihilation and exchange, also modify the D-meson lifetimes – the fact that the D_s and D^0 lifetimes deviate (whereas in the naive quark model, their lifetimes would be identical) is fairly compelling evidence that additional non-external spectator effects must

---

*Special thanks go to Peter Brabant, Matt Cervantes, Armando Daniel, Nathan DeLee, Russell P. Stutz

be present. We note here that the pre-CLEO99 measurement of the D^0 and D_s lifetimes gave values that deviated by approximately 3σ; the CLEO99 measurements give lifetimes that are irreconcilable at the 4.5σ level.

Note that, for baryon decays, the presence of a third quark can mitigate the color/helicity suppression. The Λ_c is typically modeled as a system containing a heavy charm quark surrounded by an inert (ud) diquark, bound in an S=0 state, with L=0 relative to the heavy charm quark. The (ud) system is in an isospin=0 state, unlike the Σ_c, which has I=1. The simplest picture of Λ_c decays is then as the weak decay of the charm quark to the strange quark by external W-emission, such that the (ud) diquark is entirely decoupled from the $c \rightarrow s$ transition. In such a naive picture, $\mathcal{B}(\Lambda_c \rightarrow \Lambda X)$=100%. However, the tabulated value for this inclusive decay is $27 \pm 9\%$[1]; moreover, the simplest spectator decay $\mathcal{B}(\Lambda_c \rightarrow \Lambda \pi^+) = 0.58 \pm 0.16\%$[1]; by comparison the comparable decay in the charmed meson sector: $D^0 \rightarrow K^- \pi^+$ has a branching ratio about 8 times as large. This, coupled with the measurement of a Λ_c lifetime half that of the D^0 and the D_s, indicates that there is a large hadronic width proceeding by diagrams other than the simple external W-emission diagram. To some extent, this is to be expected in the baryonic sector, where exchange decays ($\Lambda_c \rightarrow \Sigma^+ \phi$, e.g.) are not subject to the same helicity suppression operative in charmed meson decays. The fact that $\Lambda_c \rightarrow pK\pi$ is so large also indicates that internal W-emission could play a substantial role in charmed baryon decays.

2. Λ_c lifetime

The Λ_c lifetime has recently been remeasured by the SELEX Collaboration. SELEX is a fixed target, charmed baryon production experiment which uses an extracted hyperon beam at Fermilab to enhance the baryonic (and charmed baryonic) fraction of the final state particles. They combine silicon detection and good particle identification (based on Čerenkov counters) to achieve precise vertex resolution and also good signal to noise in final states including kaons and protons ($\Lambda_c \rightarrow pK^-\pi^+$, e.g.). To select a clean sample of Λ_c events, they require that the $\Lambda_c \rightarrow pK\pi$ candidate be well separated from the primary interaction point. This gives them a sample of approximately 1790 Λ_c with a S:N=3:1 at the peak. SELEX obtain a preliminary Λ_c lifetime of 177 ± 10 fs (statistical errors only),[3] to be compared with the presently tabulated PDG value of 206 ± 12 fs.[1]

3.   Inclusive $\Lambda_c \rightarrow \Lambda X$ Production

This confirms that the Λ_c lifetime is markedly smaller than the lifetime of the charmed mesons. It is important, then, to attempt to quantify the branching fractions of the most accessible Λ_c decay modes and verify their values. As mentioned perviously, given the smallness of the Λ_c lifetime, we might expect $\Lambda_c \rightarrow \Lambda X <<$100%. CLEO have recently reported on the inclusive rate for charmed baryons to decay into Λ particles using charm-event tagging. They select $e^+e^- \rightarrow c\bar{c}$ events which have a clear charm tag and measure the Λ content in the hemisphere opposite the tag (charge conjugate modes are implicit). This allows to estimate the product branching fraction: $\mathcal{B}_\Lambda = \mathcal{B}(c \rightarrow \Theta_c) \cdot \mathcal{B}(\Theta_c \rightarrow \Lambda + X)$, where the r.h.s. represents a sum over all charmed baryons Θ_c produced in e^+e^- fragmentation at $\sqrt{s}$=10.5 GeV, given the CLEO-specific tag mixture.

In brief, inclusive measurements of charmed baryon decay products provide essential information on the relative contributions of different decay processes (e.g., external W-emission, internal W-emission, W-exchange) to the weak $c \to sW^+$ transition in baryons. Difficulties in distinguishing direct charm decay products from jet fragmentation particles have hampered such inclusive measurements. An example is the measurement of $\mathcal{B}(\Lambda_c \to \Lambda + X)$. Using the total Λ yield at $\sqrt{s} = 10$ GeV/c to measure $\mathcal{B}(\Lambda_c \to \Lambda + X)$ requires separating the Λ component due to light quark fragmentation from production via $c \to \Lambda_c \to \Lambda + X$. The difficulty of separating fragmentation Λ's from those resulting from Λ_c decays can be overcome by using a tagged sample of $e^+e^- \to c\bar{c}$ events. CLEO have used charm-event tagging to measure the product of the likelihood for a charm quark to materialize as a charmed baryon Θ_c times the branching fraction for a charmed baryon to decay into a Λ: $\mathcal{B}(c \to \Theta_c) \cdot \mathcal{B}(\Theta_c \to \Lambda + X)$. (From JETSET 7.3 Monte Carlo simulations,[5] using the default LEP-tuned control parameters, we expect that $\sim$88% of Θ_c particles produced in $e^+e^- \to c\bar{c}$ fragmentation at $\sqrt{s}$=10 GeV will be Λ_c's.)

Previous measurements[6-8] of $\mathcal{B}(\Lambda_c \to \Lambda X)$ have either measured the step in the Λ production rate as the $e^+e^- \to \Lambda_c \overline{\Lambda}_c$ threshold is crossed, Λ production from topologically tagged Λ_c decays, or derived a value for $\mathcal{B}(\Lambda_c \to \Lambda X)$ based on measurements of baryon production in B-meson decay. Knowing, for example, that $\mathcal{B}(B \to p/\overline{p}(direct) + anything) = 5.5 \pm 0.5\%$,[1] $\mathcal{B}(B \to \Lambda/\overline{\Lambda} + anything) = 4.0 \pm 0.5\%$, and assuming that $\overline{B} \to \Lambda_c \overline{N} X$ dominates baryon production in B-decay, with $\overline{N}$ equally likely to be $\overline{p}$ or $\overline{n}$, one can estimate: $\mathcal{B}(\Lambda_c \to \Lambda X) \sim \frac{4}{5.5 \times 2}$=36%. This simple-minded estimate, however, needs to be modified to take into account many corrections, among them the recent result that $\frac{\mathcal{B}(\overline{B} \to \overline{\Lambda} X)}{\mathcal{B}(\overline{B} \to (\Lambda X)}) = 0.43 \pm 0.09 \pm 0.07$.[9] That result implies that only $\sim$2/3 of the inclusive ($\Lambda + \overline{\Lambda}$ yield in $\overline{B}$-decay come from decays of charmed baryons; the remainder presumably due to associated production or decay of anticharm baryons.

A sample $c\bar{c}$ event is schematically shown below, showing some of the particles relevant to our measurement. In the event depicted below, the fragmentation of the original $c\bar{c}$ quark-antiquark results in a Λ recoiling in one hemisphere opposite the charm tag (either the soft pion or the electron) in the other hemisphere. For this analysis, event "hemispheres" are defined using the thrust axis. Note that both the soft pion and the electron are charged opposite to the p daughter of the Λ. In addition to the tags depicted below, CLEO also tag $c\bar{c}$ events with fully reconstructed $\overline{D}^0 \to K^+\pi^-$ or $D^- \to K^+\pi^-\pi^-$ events, in which the $\overline{D}$ daughter kaon contains an $\bar{s}$-quark, in contrast to the Λ. For all four tags, we will therefore refer to our signal as an opposite hemisphere, opposite sign (OH/OS) correlation.

The product branching fraction $\mathcal{B}(c \to \Theta_c) \cdot \mathcal{B}(\Theta_c \to \Lambda + X)$ can be derived from the fraction of times that an event containing a charm tag in one hemisphere contains a

Λ in the opposite hemisphere. This, effectively, is the fraction of Λ particles per $c\bar{c}$ event and should equal the probability of a c quark fragmenting to produce a Θ_c multiplied by the probability of the Θ_c decaying to a Λ multiplied by the efficiency for detection of a Λ in our charm-tagged event sample. In equation form, defining $\frac{N(\Lambda)}{c\bar{c}}$ as the ratio of the number of reconstructed Λ's in tagged $c\bar{c}$ events to the total number of $c\bar{c}$ event tags, we have:

$$\frac{N(\Lambda)}{c\bar{c}} = \mathcal{B}(c \to \Theta_c) \cdot \mathcal{B}(\Theta_c \to \Lambda + X) \cdot \epsilon_\Lambda \tag{1}$$

for both data and Monte Carlo. Assuming that the Monte Carlo simulation accurately reproduces the efficiency for finding a Λ in a tagged event, the yield of non $c\bar{c}$ tags ($<4\%$), and the fraction of non-signal $\Lambda-tag$ correlations ($<5\%$), one can then calibrate the observed value of Λ's per $c\bar{c}$ in data to Monte Carlo:

$$\frac{\frac{N(\Lambda)}{c\bar{c}}^{Data}}{\frac{N(\Lambda)}{c\bar{c}}^{MC}} = \frac{\mathcal{B}(c \to \Theta_c + X) \cdot \mathcal{B}(\Theta_c \to \Lambda + X)^{Data}}{\mathcal{B}(c \to \Theta_c + X) \cdot \mathcal{B}(\Theta_c \to \Lambda + X)^{MC}}; \tag{2}$$

the Monte Carlo values for $\mathcal{B}(c \to \Theta_c + X)$ and $\mathcal{B}(\Theta_c \to \Lambda + X)$ are 0.0667 and 0.369, respectively. A recent measurement by the ALEPH collaboration[10] at $\sqrt{s}=90$ GeV has determined $\mathcal{B}(c \to \Lambda_c + X) = 0.079 \pm 0.008 \pm 0.004 \pm 0.020$.

Using four different $e^+e^- \to c\bar{c}$ tags, CLEO measure the product branching ratio $\mathcal{B}_\Lambda = \mathcal{B}(c \to \Theta_c + X) \cdot \mathcal{B}(\Theta_c \to \Lambda + X)$:

Tag	$\mathcal{B}_\Lambda$
electron	$(1.62 \pm 0.10 \pm 0.32)\%$
π_{soft} from $D^{*+} \to \pi_{soft}D^0$	$(1.53 \pm 0.06 \pm 0.30)\%$
$\overline{D}^0$	$(2.12 \pm 0.09 \pm 0.30)\%$
D^-	$(2.09 \pm 0.13 \pm 0.42)\%$;

these results sum over the charmed baryons Θ_c produced at $\sqrt{s}=10$ GeV. Separating common from independent systematic errors, we combine these four numbers to obtain a weighted product branching fraction:

$$\mathcal{B}_\Lambda = (1.87 \pm 0.03 \pm 0.2)\%$$

We can convert this result into a contour in the plane: $\mathcal{B}(\Theta_c \to \Lambda X)$ vs. $\mathcal{B}(c \to \Theta_c)$, Using the Monte Carlo value for $\mathcal{B}(c \to \Theta_c + X)$ of 6.67% (this is consistent with the tabulated product cross-section $(e^+e^- \to c\bar{c}) \cdot \mathcal{B}(c \to \Lambda_c) \cdot \mathcal{B}(\Lambda_c \to pK^-\pi^+)$ using $\mathcal{B}(\Lambda_c \to pK^-\pi^+)=5.0\%$), and taking the results from our four tags, we can infer a weighted average value:

$$\mathcal{B}(\Theta_c \to \Lambda + X) = (28 \pm 1 \pm 4)\%$$

It is important to note that this measurement is independent of the $\Lambda_c \to pK^-\pi^+$ normalization but is dependent on the Monte Carlo estimated value for $\mathcal{B}(c \to \Theta_c)$.

This measurement is the first of its kind at $\sqrt{s} = 10$ GeV. We also note that the difference between the inferred value for $\mathcal{B}(\Theta_c \to \Lambda + X)$ and the sum of the exclusive Λ_c modes to Λ's suggests that as-yet undiscovered $\Lambda_c \to \Lambda X$ exclusive decay modes are not large.

In the simplest picture, a charmed baryon such as a Λ_c decays weakly through external W-emission. Neglecting fragmentation at the lower vertex, this produces either a Σ^0 or, if isospin does not change, a Λ. Since all Σ^0's decay into Λ, we therefore expect that $\Lambda_c \to \Lambda + X \approx 100\%$ if the external spectator diagram dominates. This simple-minded prediction is expected to be obeyed in semileptonic decays; i.e., $\frac{\mathcal{B}(\Lambda_c \to \Lambda l \nu)}{\mathcal{B}(\Lambda_c \to X l \nu)} \to 1$. Present data, however, give a value of approximately 50% for this ratio, albeit with large errors.[1]

By contrast, internal W-emission or W-exchange can lead to NKX final states ($\Lambda_c \to pK_s^0$, e.g.). The fact that the Λ_c lifetime is only half that of the D^0 meson suggests that internal W-emission and W-exchange processes may comprise a large fraction of the total Λ_c width. Although internal W-emission may be suppressed in decays of charmed mesons due to the color-matching requirement (which would predict $\frac{\mathcal{B}(D \to W_{int} X)}{\mathcal{B}(D \to W_{ext} X)} = \frac{1}{9}$), the larger number of degrees of freedom in baryon decays may mitigate this suppression, leading to a potentially large fraction of pKX final states. In the case of the Λ_c, exchange decays can produce either Λ's or NK in the final state, depending on the quark configuration. The naive expectation that the absence of exchange diagrams in Ξ_c^+ decays will lead to a longer lifetime for Ξ_c^+ compared to Ξ_c^0 and Λ_c is consistent with current experimental data (although this argument seems to break down in the case of the Ω_c).

The current world average for $\mathcal{B}(\Lambda_c \to \Lambda + X)$[1] is consistent with the notion that the simple-minded external W-emission picture does not saturate Λ_c decays. Exclusive $\Lambda_c \to \Lambda + X$ channels have also been measured; normalized to an estimate of $\mathcal{B}(\Lambda_c \to pK^-\pi^+) = (5.0 \pm 1.3)\%$,[1,11] the sum of the observed exclusive modes account for approximately half of the presently tabulated inclusive $\Lambda_c \to \Lambda + X$ rate $\left(\frac{\Sigma \mathcal{B}(\Lambda_c \to \Lambda + X)_{exclusive}}{\mathcal{B}(\Lambda_c \to pK^-\pi^+)} \sim 5,\ \text{where the sum includes a contribution of } \frac{\Sigma \mathcal{B}(\Lambda_c \to \Sigma^0 + X)_{exclusive}}{\mathcal{B}(\Lambda_c \to pK^-\pi^+)} \sim 1.5\right)$. The value for $\mathcal{B}(\Lambda_c \to \Lambda + X)$ therefore has implications for the external vs. internal spectator fractions in charmed baryon decay.

If charmed baryons produced in e^+e^- events are predominantly Λ_c's, and if the JETSET expectation for $f(c \to \Lambda_c)$ is accurate, then our results are in approximate agreement with the current world average for $\mathcal{B}(\Lambda_c \to \Lambda + X)$. Our results are therefore qualitatively consistent with a possibly substantial internal spectator contribution to charmed baryon decay. We note that the methodology of this analysis differs substantially from the previous CLEO analysis,[8] which relied on a model of charmed baryon production in B-decay to derive $\mathcal{B}(\Lambda_c \to \Lambda + X)$.

Naively, one might expect fragmentation and decay of charmed baryons to be similar to bottom baryons. Using vertex tagging techniques, the OPAL collaboration has determined $\mathcal{B}(b \to \Theta_b)\mathcal{B}(\Theta_b \to \Lambda + X) = (3.50 \pm 0.32 \pm 0.35)\%$.[12] Perhaps the simplest way to reconcile the two numbers is to assume (*ad hoc*) that $\mathcal{B}(b \to \Theta_b)$ at $\sqrt{s} \sim 90$ GeV is approximately twice as large as $\mathcal{B}(c \to \Theta_c)$ at $\sqrt{s}=10$ GeV, and that $\mathcal{B}(\Lambda_b \to \Lambda_c X) \approx 1.0$. However, the fact that the Λ_b lifetime is only 2/3 that of the B-mesons,[1] coupled with the fact that Λ_b has already been observed through $\Lambda_b \to \psi\Lambda$ imply that $\mathcal{B}(\Lambda_b \to \Lambda_c X) < 1$. Correspondingly, we expect an enhancement of $\mathcal{B}(b \to \Theta_b)$ at OPAL relative to $\mathcal{B}(c \to \Theta_c)$ at CLEO.

We stress that this final central value for $\mathcal{B}_\Lambda$ averages over the specific mix of charm-tags that we use in this analysis. The composition of the $\overline{D}$-meson tags will not be the same as the composition of the electron tags, insofar as the lepton-tagged sample represents a weighted sum of $\overline{\Theta}_c \to lX$, $D_s^- \to lX$, $\overline{D}^0 \to lX$ and $D^- \to lX$. The quoted final result can be general only if the two hemispheres in an $e^+e^- \to c\bar{c}$ event fragment independently. Although not yet measured, it is possible that there may be correlated $\overline{\Theta}_c\Theta_c$ production, in which case the likelihood of observing $\Theta_c \to \Lambda X$ would be larger for $\overline{\Theta}_c$ tags than for $\overline{D}$ tags, and the assumption of independent fragmentation would be invalid. A study of correlated $\overline{\Theta}_c\Theta_c$ production still awaits to be done.

4. Re-measurement of the $\Lambda_c \to pK\pi$ branching fraction

Of the four fundamental normalization branching fractions of charmed hadrons ($\mathcal{B}(D^0 \to K^-\pi^+)$, $\mathcal{B}(D^+ \to K^-\pi^+\pi^+)$, $\mathcal{B}(D_s^+ \to \phi\pi^+)$, and $\mathcal{B}(\Lambda_c \to pK^-\pi^+)$), the $\Lambda_c \to pK^-\pi^+$ branching fraction is the least well-known, and presently the most controversial. There have been two basic methods used to measure this branching fraction. The first uses, as input, the ratio of branching fractions: $\frac{\Lambda_c \to \Lambda Xl\nu}{\Lambda_c \to pK^-\pi^+}$ [13,14] and the well-measured Λ_c lifetime. One can deduce a total semileptonic branching fraction for Λ_c decays

$$\mathcal{B}(\Lambda_c \to Xl\nu_l) = \frac{\Gamma(\Lambda_c \to Xl\nu_l)}{\Gamma^{\text{tot}}(\Lambda_c)}.$$

, assuming that the total semileptonic width is the same in Λ_c decays as in $D_s \to Xl\nu$, $D^0 \to Xl\nu$, and $D^+ \to Xl\nu$ (the approximate equality of the semileptonic widths for all the charmed mesons lends credence to this assumption, although mass and phase space effects in semileptonic decays may be significant[15]), and assuming $\frac{\Lambda_c \to \Lambda Xl\nu}{\Lambda_c \to Xl\nu} \approx 1.0$.[16–18] Knowing the fraction of semileptonic Λ_c decays which produce Λ's (as opposed to, e.g., $\Lambda_c \to N\overline{K}Xl\nu$), one can therefore estimate the absolute branching fraction for $\Lambda_c \to \Lambda Xl\nu$, and, correspondingly, the absolute branching fraction for $\Lambda_c \to pK^-\pi^+$ from the measured $\frac{\Lambda_c \to \Lambda Xl\nu}{\Lambda_c \to pK^-\pi^+}$ ratio. Such a procedure yields values of $\mathcal{B}(\Lambda_c \to pK^-\pi^+)$ in the range of 6-8%.[18]

In the second approach, one uses the fact that baryon number must be conserved in B-decay and that $\mathcal{B}(b \to c) \approx 1.0$. Under the assumption that baryon production in B-decay occurs through $\overline{B} \to \Lambda_c\bar{p}W$, the total number of observed $\overline{B} \to \bar{p}X$ events can be related to $\overline{B} \to \Lambda_c X$. Measurements of the $\Lambda_c \to pK^-\pi^+$ yield in such events therefore allow a determination of the absolute $\Lambda_c \to pK^-\pi^+$ branching fraction.[16,17]* The Particle Data Group uses a combination of this technique and D and Λ_c charm semileptonic measurements to estimate $\mathcal{B}(\Lambda_c \to pK^-\pi^+) = 5.0 \pm 1.3\%$.[1]

CLEO use a new technique, again based on flavor-tagging, to determine $\mathcal{B}(\Lambda_c \to pK^-\pi^+)$ using e^+e^- annihilation continuum events. An unbiased, Λ_c-enriched sample

---

*Unfortunately, a more recent study of flavor-tagged baryon production in B-decay indicates that diagrams other than $\overline{B} \to \Lambda_c\bar{p}W$ may contribute substantially to Λ_c, $\overline{\Lambda}_c$, and $p/\bar{p}$ production in B-decay.[9]

is isolated by selecting events containing a $\overline{D}$ in the hemisphere (as determined by the event thrust axis) opposite an antiproton. Such events are expected to have a Λ_c in the hemisphere opposite the $\overline{D}$ (we use parentheses to designate opposite hemisphere correlations: "$(\Lambda_c\overline{D})$") and in the same hemisphere as an antiproton (using square brackets to designate same hemisphere correlations: "$[\overline{p}\Lambda_c]$") to conserve both baryon number and charm; the $pK^-\pi^+$ yield in these events therefore gives an implied value for $\Lambda_c \to pK^-\pi^+$.

The analysis comprises two techniques – in one, CLEO constructs a three-particle correlation to determine the $\Lambda_c \to pK^-\pi^+$ branching fraction, and in the second, a two-particle correlation is sufficient to infer $\mathcal{B}(\Lambda_c \to pK^-\pi^+)$. The $\Lambda_c \to pK^-\pi^+$ yield is thus measured in a sample of two-jet continuum events containing both a charm tag ("$\overline{D}$") as well as an antiproton ($e^+e^- \to \overline{D}\overline{p}X$), with the antiproton in the hemisphere opposite the $\overline{D}$ (measurement of charge conjugate modes is implicit throughout). Under the hypothesis that such selection criteria tag $e^+e^- \to \overline{D}\overline{p}\Lambda_cX$ events, the $\Lambda_c \to pK^-\pi^+$ branching fraction can be determined by measuring the $pK^-\pi^+$ yield in the same hemisphere as the antiprotons in our $\overline{D}\overline{p}X$ sample. Three types of $\overline{D}$ charm tags are used - π^-_{soft} (from $D^{*-} \to \overline{D}^0\pi^-_{\text{soft}}$), electrons (from $\overline{D} \to Xe\nu_e$), and fully reconstructed $\overline{D}^0 \to K^+\pi^-$ or $D^- \to K^+\pi^-\pi^-$ or $D_s^- \to \phi\pi^-$.

For both tags, we can quantify the ratio of tagged events containing a Λ_c decaying into $pK^-\pi^+$ to all tagged events. This ratio is equal to:

$$
R = \frac{\mathcal{Y}(e^+e^- \to \overline{D} + \overline{p} + (\Lambda_c \to pK^-\pi^+) + X)}{\mathcal{Y}(e^+e^- \to \overline{D} + \overline{p} + X)}
\tag{1}
$$

where $\mathcal{Y}$ stands for "yield in e^+e^- annihilation", "$\overline{D}$" designates any one of our three charm tags, and the $\overline{p}$ and the $\overline{D}$ are in opposite hemispheres with respect to the thrust axis of the event.

There are corrections which need to be made for backgrounds. For all tags, fake antiprotons (primarily kaons) need to be subtracted from the denominator of our ratios. All tags must also subtract events containing an antiproton opposite a $\overline{D}$-meson which do not tag $\Lambda_c\overline{p}\overline{D}$ events, viz. $D\overline{D}N\overline{p}$ events. Both of these subtractions are 10-15% corrections, and are derived (as much as possible) from the data itself. Additionally, for the double correlation sample, there must be a correction due to $\Lambda_c\overline{p}$ events which do not contain a charmed meson (a 4% correction, obtained from the data; these are predominantly four-baryon events such as $\Lambda_c\overline{\Lambda}_c\overline{p}N$, e.g.).

5. Results

Results, showing the yields $\mathcal{Y}$, efficiencies ϵ, and backgrounds, in both data and Monte Carlo, are tabulated in Table 1. The weighted average of the three techniques corresponds to $\mathcal{B}(\Lambda_c \to pK^-\pi^+) = 5.0 \pm 0.5\%$ (statistical error only).

Including systematic errors, then combining the results obtained from the various charm tags, we obtain $\mathcal{B}(\Lambda_c \to pK^-\pi^+) = (5.0 \pm 0.5 \pm 1.2)\%$.

Table 1. Event yields for signal and backgrounds, in data and Monte Carlo. As before, same hemisphere correlations are designated with brackets [], and opposite hemisphere correlations are designated with parenthesis (). Background yields which are subtracted from the numerator or denominator are indicated with a minus sign.

Double Correlations	MC	Data
$\mathcal{Y}[\Lambda_c\bar{p}]$ (Numerator)	1656±65	1093±47
$\mathcal{Y}(\overline{D}^0\bar{p})$ (Denominator)	2725±84	1369±55
$\mathcal{Y}(D^-\bar{p})$ (Denominator)	1501±113	963±71
$\mathcal{Y}(D_s^-\bar{p})$ (Denominator)	111±19	51±11
$\epsilon_{\Lambda_c}/\epsilon_{D^0}/\epsilon_{D^-}/\epsilon_{D_s^-}$	26.5%/43.7%/32.2%/14.5%	
$\mathcal{Y}(\overline{\Lambda}_c\bar{p})$ (Bkgnd to Num.)	−84±49	−75±39
$\mathcal{Y}[D^0\bar{p}]$ (Bkgnd to Den.)	−268±40	−68±23
$\mathcal{Y}[D^+\bar{p}]$ (Bkgnd to Den.)	−417±67	−152±39
$\mathcal{Y}[D_s^+\bar{p}]$ (Bkgnd to Den.)	−26±11	−1±6
fake $\bar{p}$ in $[D\bar{p}]$ evts. (Bkgnd. to Den.)	−272±31	−298±36
f_2 ($\Lambda_c\overline{\Lambda}_c\bar{p}N$ correction)	5.1%	6.9%
f_3 ($\overline{D}\bar{p}DN$ correction)	17.5%	10.6%
$\mathcal{B}(\Lambda_c \to pK^-\pi^+)$	4.3% (input)	(4.9±0.5)%
$(\pi_{\text{soft}}^-\bar{p})$ Triple Correl.		
$\mathcal{Y}(\pi_{\text{soft}}^-\bar{p})$ (Denominator)	34222±1092	14553±485
fake $\bar{p}$ in $(\pi_{\text{soft}}^-\bar{p})$	−3318±310	−1867 ± 261
$\mathcal{Y}([\Lambda_c\bar{p}]\pi_{\text{soft}}^-)$ (Numerator)	202.8±27.8	101.6±20.6
$\mathcal{B}(\Lambda_c \to pK^-\pi^+)$	4.3% (input)	(5.2±1.3)%
$(e^-\bar{p})$ Triple Correl.		
$\mathcal{Y}(e^-\bar{p})$ (Denominator)	4178±65	1739±47
fake $\bar{p}$ + fake e^-	−382±39	−272 ± 41
$\mathcal{Y}([\Lambda_c\bar{p}]e^-)$ (Numerator)	20.1±5.2	10.3±3.8
$\mathcal{B}(\Lambda_c \to pK^-\pi^+)$	4.3% (input)	(5.6±2.5)%

280

5.1. *Discussion*

At present, this technique is limited by the precision of the non-signal backgrounds; presumably, more data would allow a greater understanding of those backgrounds. Our result is roughly consistent with the determination of $\mathcal{B}(\Lambda_c \to pK^-\pi^+)=7\pm2\%$ suggested by Dunietz,[18] based on the measured ratio for $\frac{\Lambda_c \to \Lambda X l \nu_l}{\Lambda_c \to pK^-\pi^+}$ and assuming that the semileptonic charmed baryon width is the same as the semileptonic charmed meson width. It is also consistent with the value of $5.0 \pm 1.3\%$ derived by the Particle Data Group.[1] We now disuss the implications of this result and its consistency with related measurements.

The product branching fraction: $\mathcal{B}(B \to (\Lambda_c X \ or \ \overline{\Lambda}_c X)) \cdot \mathcal{B}(\Lambda_c \to pK^-\pi^+)$ can be directly determined by simply measuring the efficiency-corrected $\Lambda_c \to pK^-\pi^+$ yield in $B\overline{B}$ events. CLEO has performed such a measurement, finding a value of $\mathcal{B}((B + \overline{B}) \to \Lambda_c) \cdot (\Lambda_c \to pK^-\pi^+) \sim (2 \pm 0.2) \times 10^{-3}{}^{22}$ for this product branching fraction. Given that, $\mathcal{B}(\Lambda_c \to pK^-\pi^+) = 0.05$ implies that $\mathcal{B}(B \to (\Lambda_c \ or \ \overline{\Lambda}_c)) \sim 3.6\%$. This can be compared to the Particle Data Group value of $\mathcal{B}(B \to p \ or \ \overline{p}) \sim 8.0\%$.[1] Our result therefore implies that $B \to baryons$ may be occurring at a substantial rate through modes such as $\overline{B} \to DN\overline{N}X$,[21] $\overline{B} \to \Xi_c \overline{Y}X$, or $\overline{B} \to \Xi_c\overline{\Lambda}_c$. CLEO has recently published evidence for the latter of these.[19]

We can also place bounds on the $\Lambda_c \to pK^-\pi^+$ branching fraction by using the measured CLEO $e^+e^- \to$ hadrons cross-section, assuming that the $c\bar{c}$ fraction is 40% of the total hadronic cross-section. CLEO has measured $\mathcal{B}(\Lambda_c \to pK^-\pi^+) \cdot \sigma(e^+e^- \to (\Lambda_c + \overline{\Lambda}_c)) = 10 \pm 1$ pb. That measurement simply determines the total yield of either Λ_c or $\overline{\Lambda}_c$ in e^+e^- annihilations; i.e., it determines the sum of $c \to \Lambda_c$ plus $\bar{c} \to \overline{\Lambda}_c$. Our value of $\mathcal{B}(\Lambda_c \to pK^-\pi^+)=0.05$ implies that $\sigma(e^+e^- \to (\Lambda_c + \overline{\Lambda}_c)) =200$ pb. Using the recent CLEO measurement of $R \equiv \frac{\sigma(e^+e^- \to q\bar{q})}{\sigma(e^+e^- \to \mu^+\mu^-)}$,[20] which corresponds to a value of $\sigma(e^+e^- \to q\bar{q}) \sim 3.3$ nb, and using a Monte-Carlo inferred estimate of $\mathcal{B}(c \to \Lambda_c) \sim 0.07$, we have: $\sigma(e^+e^- \to q\bar{q}) \times \frac{c\bar{c}}{q\bar{q}} \times (c \to \Lambda_c + \bar{c} \to \overline{\Lambda}_c) = 3300$ pb$\times 0.4 \times 0.07 \times 2 = 185$ pb, in good agreement with our measurement above.

Lastly, since the presently tabulated exclusive Λ_c decays are all normalized to $\mathcal{B}(\Lambda_c \to pK^-\pi^+)$, we conclude that only $\sim 50\%$ of the exclusive decays of Λ_c have been measured. There is, correspondingly, a large fraction of the Λ_c width which is unaccounted for. Since the Λ_c lifetime is only $\sim 40\%$ of the D^0/D_s lifetime, it has long been realized that diagrams such as exchange diagrams, and/or final states including neutrons, are likely to be large contributors to Λ_c decay and may produce final states different than the 'usual' states expected from simple $\Lambda_c \to \Lambda W_{\text{external}}$ diagrams. Measurement of such decays await additional data and analysis.

6. Charm Hadron Spectroscopy Overview

The simplest mesons to describe, from a theoretical standpoint, are bound states consisting of two heavy quarks (the ψ and Υ states). Because of the heaviness of the quarks involved, the relativistic corrections are small, and the mesons can be treated as slowly moving particles in a strong potential well. The spectroscopy of these bound states can then be derived from a perturbative treatment, once a form for the potential $V(r)$ as a function of interquark spacing r is specfied.[23-34]

In the perturbative approach, the interquark Hamiltonian is written as an unperturbed Hamiltonian H_0 plus the perturbative corrections:

H=p^2 + V(non-rel.) (H_0)

- +V(spin-ind. relativistic corrections)

- +V(spin-spin) $(\bigtriangledown^2 V_V \mathbf{S_1} \cdot \mathbf{S_2})/3m^2 r$

  - -splits $^1S_1$ (η_b) from $^3S_{J=0,1,2}$ (Υ);

- +V(spin-orbit) $(3V_V' - V_S')/(3m^2 r)$

- +V(tensor) $(V_V'/r - V_V'')(3(\mathbf{S_1} \cdot \mathbf{r})(\mathbf{S_2} \cdot \mathbf{r}) - \mathbf{S_1} \cdot \mathbf{S_2}))/2m^2 r)$

  - -the last two perturbations split states with $L > 0$ by J (χ_b:$^3 P_{J=0,1,2}$, e.g.)

where these corrections have been written in terms of a piece which Lorentz transforms as a vector (V_V) and a piece which transforms as a scalar (V_S). Theorists give different estimates for V_V and V_S; typical corrections are approximately 10 MeV for bottomonium.

A simple prescription for V, which obeys the long-distance (confinement) and short-distance (single-vector exchange) boundary behavior is: $V_0(r) = \alpha r$ (confinement) + β/r (single-gluon exchange at high q^2). We can then use the available experimental data from charmonium to determine α, β and pin the ground state energy[39] and build up the entire expected mass level diagram. So far, the S- and P- states agree rather well with expectation, the D-states have yet to be observed.[†]

The spectroscopy of the b$\bar{\text{b}}$ system, which indicates the magnitudes of the splitting in a very non-relativistic system, are shown in Figure 1. The data typically are in agreement with theoretical predictions for the locations of mass levels to better than 0.1%, or 10 MeV on the scale of 10 GeV.

7. From Heavy-Heavy→Heavy-Light systems

For heavy quark - light quark mesons, we expect that the formalism will be very similar to what we have written for heavy-heavy mesons, however the magnitude of the corrections to the unperturbed Hamiltonian are now substantially larger. This results from the larger mean velocity of the light quark and correspondingly larger spin-dependent corrections. For example, the splittings between the vector Υ(1S) and pseudoscalar η_b are markedly smaller ($\sim$30 MeV) than for the comparable case of the D^* and the D ($\Delta M \sim$140 MeV), as the relativistic corrections are larger for the heavy-light system.

In principle, Heavy-Light J(PC) spectroscopy can be built up analogous to Heavy-Heavy[‡]:

[†]There are, however, two anomalies in bottomonia - first that the $\Upsilon(4S)$-$\Upsilon(3S)$ mass splitting is generally low relative to predictions, and second that the relative splitting of the J=2,1,0 sub-states of the χ_b'/χ_b triplets also are at variance with prediction (r(2P)<r(1P)), although recent CLEO data favor $r(2P) > r(1P)$.

[‡]Note that the L=1 states are always even parity as they have: $P = -1^\ell P_q P_{\bar{q}}=+1$).

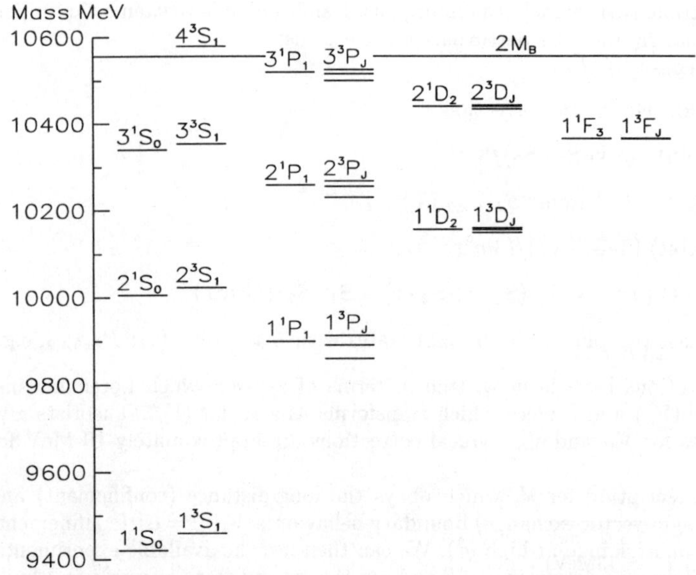

Fig. 1. Mass spectrum of b$\bar{\text{b}}$ bound states below the $B\bar{B}$ threshold predicted by one of the potential models.[37]

q$\bar{\text{q}}$ content	J(PC)=0(− +)	J(PC)=(1- -)§	J=0,1,2; L=1
b$\bar{\text{b}}$	η_b	Υ	χ_b
c$\bar{\text{u}}$	D	D^*	D_J
c$\bar{\text{s}}$	D_s	D_s^*	$D_{s,J}$

The spectroscopy of heavy-quark plus light diquark (integral spin) baryonic systems does not carry over as easily; the "ground state" charmed baryon (Λ_c) has no mesonic analog - since $S_{diquark}{=}0$ in a Λ_c, the spin-spin corrections present in the mesonic sector are not present for Λ_c's. In that sense, the baryonic analogs of the D and D^* are the Σ_c (3/2)- and the Σ_c^* (3/2)-.

8. Heavy-Light Spectroscopy in HQET

Just as the it is the light electron, and not the heavy, stationary nucleus which determines atomic spectroscopy, one might expect that the light quark quantum numbers in heavy-light systems of strongly interacting quarks determine the properties of those heavy-light systems; alternately, the "light degrees of freedom decouple". In the usual picture of spectroscopy, for hadrons with $L \neq 0$ (D_J, e.g.), we define the quantum number J=L+S as the total angular momentum of: (spin s_Q of the heavy quark Q) + (spin s_q of the light quark q) + $L_{relative}$. For (L=1) D_J states, e.g., the case of $S_{total} = S_{qQ} = 1$ gives us a triplet with L+S=J; $J = 2, 1, 0$ and, in the case of $S_{qQ} = 0$, a singlet with L+S=J=0, for a total of four states.¶In the heavy quark picture,[38] we can recast this picture alternately, where we rewrite the four possible spin states, incorporating L into the definition of total light quark angular momentum j_q. This gives $j_q = L \pm s_q$=1/2 or 3/2. In this scheme, there are now two doublets which result when we combine the total light quark angular momentum j_q with the possible $\pm 1/2$ spin projections of heavy quark: (j_q, s_Q).

Ordered by mass, we obtain a scheme as follows:

J(P) (Decays)	j_q	s_Q	Allowed Decays
0+	1/2	1/2	$D\pi$ (S-Wave) $D^*\pi$ (S-Wave)
1+	1/2	1/2	$D^*\pi$ (S-Wave or D-Wave)
1+ ($\rightarrow$1-0-)	3/2	1/2	$D^*\pi$ (S-Wave or D-Wave)
2+ ($\rightarrow$0-0-/1-1-)	3/2	1/2	$D\pi$, $D^*\pi$ (D-Wave)

Although the mass ordering seems contrary to what one would find in bottomonium, this pattern is approximately followed in the strange sector, where we have[1]:

J(P)	j_{light}	s_{heavy}	Allowed Decays (Measured BR's)
0+ ($K_0^*(1430)/\Gamma = 287 \pm 23$MeV)	1/2	1/2	$K\pi$ (S-Wave) (93$\pm$10%) $K^*\pi$ (S-Wave)
1+ ($K_1(1400)/\Gamma = 174 \pm 13$MeV)[40]	1/2	1/2	$K^*\pi$ (S-Wave or D-Wave) (94$\pm$6%)
1+ ($K_1(1270)/\Gamma = 90 \pm 20$MeV)	3/2	1/2	$K\rho$ (S-Wave or D-Wave) (42$\pm$6%)
2+ ($K_2^*(1430)/\Gamma = 105 \pm 5$MeV)	3/2	1/2	$K\pi$ (50$\pm$1.2%), $K^*\pi$ (25$\pm$2%) (D-Wave)

There are two observations that can be made in extrapolating from strange spectroscopy to charm spectroscopy:

1. In principle, the two J(P)=1+ states can mix. In the strange sector, the K_1 states are actually mixed versions of $^3P_1(1260)$ and $^1P_1(1235)$ with a mixing angle that can, with enough data, be derived from τ decay. Large mixing between these states would probably have the effect of markedly broadening the narrower j_q=3/2 state (J(P)=1+) relative to the other j_q=3/2 state (i.e., the J(P)=2+ state). We will use this principle later to argue that the mixing between the analogous states in the charm sector is small.

2. The small mass splitting between the two L=1, (1/2)+ states observed in the strange sector (the $K_0^*(1430)$ and the $K_1(1400)$) leads us to believe that the as-yet undiscovered (1/2)+ charm doublet is also approximately degenerate in mass.

The allowed D- and S- wave amplitudes of the 1+ can also mix. This latter possibility can be (and has been) experimentally investigated in charm by examination of the possible helicity angles in these transitions. In the HQET limit $m_Q \rightarrow \infty$, S_Q and the spin of the light degrees of freedom are separately conserved by the strong

---

¶In fact, before the advent of HQET, this classification scheme was realized almost a decade ago.[35]

interactions[41]; both of the J=1 (3/2)+ states therefore are expected to decay D-wave.$^{\parallel}$ Although the $K_2^*(1430)$ must decay D-wave by angular momentum conservation, the J=1+, $j_q = 3/2$ state can decay either S-wave or D-wave. The D-wave decay is expected to have a suppressed width due to the angular momentum barrier which must be overcome in this case. The narrowness of the $K_1(1270)$ j=3/2 state relative to the j=1/2 state and the fact that it is approximately equal in width to the $K_2^*(1430)$ is consistent with the D-wave decay to $K\rho$ being dominant.

Splittings in heavy-light systems can provide essential information on the nature of the interquark potential. Perhaps the simplest splitting to consider is the mass splitting between the L=0 pseudoscalar and the L=0 vector states. The hyperfine splitting ($D^* - D$, e.g.): $\uparrow\uparrow - \uparrow\downarrow$ results primarily from the chromomagnetic interaction between the heavy and light quark ($\sim \frac{e\hbar}{2mc}$ in the classical approximation), although the electromagnetic interaction also contributes. The splitting therefore depends directly on the wave function overlap of the two quarks.** The energy difference between the higher energy J=1 vector state ($\uparrow\uparrow$) and the lower energy J=0 pseudoscalar state ($\uparrow\downarrow$), in the hyperfine interaction picture, can be written in terms of the charges and the spinors of the two quarks i and j as:

$$\Delta E_{ij}^{dipole,dipole} \sim \frac{q_i q_j}{m_i m_j} |\Psi(0)|^2 \sigma_{\mathbf{i}} \cdot \sigma_{\mathbf{j}}$$

In this simple model, the mass splitting therefore varies as: $\Delta_m \sim 1/m_Q$ or $\frac{m_{B^*} - m_B}{m_{D^*} - m_D} \sim \frac{m_c}{m_b}$. The splitting between the $B^* - B$ should therefore be only about 1/3 as large as the splitting between the $D^* - D$.

Table 2 displays the vector-pseudoscalar mass splittings for charmed and beauty mesons. The similarity of the D and D_s splittings, and the reduction of the mass splitting observed for the B system are in accord with the qualitative picture given above.

Vector-Pseudoscalar states	Mass Splitting (MeV)
$D^{*0} - D^0$	142.12±0.07
$D^{*+} - D^+$	140.64±0.10
$D_s^* - D_s$	144.22±0.60
$B^* - B$	46.0±0.6
$B_s^* - B_s$	47.0±2.6

Table 2. Vector-pseudoscalar mass splittings for light-heavy systems. Shown is the data on the $D^* - D$, $D_s^* - D_s$, and $B^* - B$.

$^{\parallel}$The $J = 1$ $D_s J(2536)$ (which is very close to threshold) has a substantial S-wave admixture since the D-wave decays are $p^{2\ell+1}$ suppressed.

**With enough theoretical/experimental precision, one can perhaps infer $\Psi(0)$ based on hyperfine splittings, and therefore estimate decay constants.

It is of interest to compare precisely the vector-pseudoscalar mass splitting for the $c\bar{s}$ system with that of the $c\bar{d}$ system. Two factors in the expression for the mass difference depend on the mass of the light quark: (i) the chromomagnetic effect is expected to be smaller for the $c\bar{s}$ system due to the heaviness of the strange quark compared to the down quark, but (ii) the square of the wave function overlap at the origin is expected to be larger because of the larger reduced mass of the strange quark. The present data for the $D_s^* - D_s$ mass difference[1] indicate a larger vector-pseudoscalar splitting in the $c\bar{s}$ system than in the $c\bar{d}$ system, indicating that wave function overlap is the dominant effect.

8.1. Mass splittings in systems with L=1

For orbitally excited states, the mass splitting between J=2 and J=1 states scales similarly to the vector-pseudoscalar case considered above. There is then an equality expected among the orbitally excited states, as indicated below (with the calculated world average of the available experimental measurements shown in parenthesis):

- $\Delta M(D^{*0}(2+)) - M(D^0(1+))$ (36.6±2.3 MeV)

- $= \Delta M(D^{*+}(2+)) - M(D^+(1+))$ (37.0±5.0 MeV)

- $= \Delta M(D_s^*(2+)) - M(D_s^*(1+))$ (38.1±1.9 MeV)

- $= \Delta M(\Lambda_{c,J}(2630)) - M(\Lambda_{c,J}(2593))$ (35.0±2.0 MeV)

- $\approx (m_c/m_b)\Delta M(B^*(2+)) - M(B^*(1+))$

Note that the magnitude of splitting for the L=1 states is substantially smaller than the vector-pseudoscalar hyperfine splitting between the L=0 meson states. This follows directly from the fact that the wave function overlap at the origin ($|\Psi(0)|^2$) is smaller for orbitally excited mesons.

In the HQET limit, the $\Lambda_{c,J}$ splittings are identical to the D_J splittings since, in this picture, the splitting results directly from the up-down or up-up configuration of the orbital angular momentum relative to the light quark spin (or diquark spin, in the case of the $\Lambda_{c,J}$). The HQET expectations of equal mass splittings are obeyed remarkably well, as shown by the experimental data; for the orbitally excited mesons, the world average mass for the $D_1^{*0}(2420) = 2422.2\pm1.52$ MeV, and the world average mass for the $D_2^{*0}(2460) = 2458.8\pm1.70$ MeV.

Although the strongly decaying 1/2 states (S-wave) are difficult to see, owing to the large expected width for these decays, they should be observable in, e.g. $D_{sJ} \to DK$. If these states have masses unexpectedly low and below strong decay threshold, they will be narrow - the D_{sJ} states corresponding to the $j_{light}=1/2$ doublet would decay to either $D_s\pi^0$, $D_{sJ} \to D_s\pi^+\pi^-$, or $D_{s(J=0)} \to D_s^*\gamma$. CLEO have recently made the first observation of the S-wave D_J,[44] using the fact that $B \to D_J\pi$ allows them to kinematically isolate a sample of D_J decays through B-reconstruction. Furthermore, the fact that the spin-parity of the B and π are known allows them to exploit angular distributions to further refine their candidate D_J sample. They observe the D_J through its decay into $D^{*+}\pi^-$; the final state is therefore $D^{*+}\pi^-\pi^-$ proceeding through $D_J^0 \to$

$D^{*+}\pi^-$. From their analysis, they find: $M(D_1^0(j = \frac{1}{2}) = 2.461^{+0.041}_{-0.034} \pm 0.010 \pm 0.032$ GeV and $\Gamma(D_1^0(j = \frac{1}{2}) = 290^{+101}_{-79} \pm 26 \pm 36$ MeV; these states are therefore wide as expected.

The L3 collaboration have also recently made measurements of the spectroscopy of orbitally excited B-mesons,[43] also in the decay $B_J \to B\pi$. The daughter B-meson is only partially reconstructed, approximating a subset of the energy flow along the jet-axis as the B four-momentum. By combining the "pseudo-B" with a candidate transition pion, the B_J can be reconstructed. In practice, it is experimentally simpler to calculate the mass difference: $\Delta M = M(B_{pseudo}\pi) - M(B_{pseudo})$, which should peak at the true B_J mass. Of course, the mass difference is extremely smeared due to the fact that the B is not fully reconstructed. Nevertheless, L3 see a signal above the background expected from uncorrelated B-random pion combinations. The four B_J states contributing to this excess are separated statistically by fitting the excess to a sum of four states. From this analysis, L3 determine: $M_{B_2^*} = (5770 \pm 6 \pm 4)$ MeV, $\Gamma_{B_2^*} = (23 \pm 26 \pm 15)$ MeV, $M_{B_1^*} = (5675 \pm 12 \pm 4)$ MeV, $\Gamma_{B_2^*} = (76 \pm 28 \pm 15)$ MeV.

We can now compare the results from the B-system to those from the D-system, using the recent measurements of L3 and CLEO.[44] We construct ratios of mass differences and widths: $\frac{\Delta M(B_2^*-B_1^*)}{\Delta M(D_2^*-D_1^*)} = \frac{95\pm13\pm4}{(-2^{+41}_{-34}\pm33)MeV}$. The fact that the measured D_1^{*0} mass is actually higher than the measured D_2^{*0} mass is, obviously, somewhat surprising, since the angular momentum correction to the D_2^{*0} mass should be larger. Additionally, $\frac{\Delta M(B_2^*-B^0)}{\Delta M(D_2^*-D^0)} = \frac{490\pm7 MeV}{590\pm2 MeV}$. In this case, the pattern is consistent with the naive expectation that the relativistic corrections should be smaller in the case of the heavier B. Additionally, although the B_2^* width is approximately equal to the D_2^* width, the measured CLEO width for the D_1^* is approximately four times the width observed by L3 for the B_1^*, albeit with roughly 30% errors. In this case, although the B-states are, indeed, expected to be narrower, the factor of four is rather surprising.

References

1. C. Caso *et al.* (The Paricle Data Group), Euro. Phys. Jour. **C3**, 1 (1998)
2. G. Bonvicini *et al.* (The CLEO Collaboration), Phys. Rev. Lett. **82**, 4586 (1999)
3. "Charm Physics results from SELEX", presented by Alexander Kushnirenko at Heavy Quarks at Fixed Target, Fermilab, Oct. 10-12, 1998
4. M. Bauer *et al.*, Z. Phys. **C34**, 103 (1987).
5. S. J. Sjostrand, LUND 7.3, CERN Report No. CERN-TH-6488-92, 1992.
6. Abe *et. al.* , Phys. Rev. D **33**, 1 (1986).
7. Adamovich *et. al.* , EPL **4**, 887 (1987).
8. CLEO Collaboration, G. Crawford *et. al.* , Phys. Rev. D **45**, 752 (1992).
9. R. Ammar et al., (The CLEO Collaboration), Phys. Rev. **D55**, 13, 1997.
10. R. Barate *et al.* (The ALEPH Collaboration), hep-ex/9909032, CERN/EP 99-094 (1999).
11. M. Selen, presented at the 1999 American Physical Society meeting, Atlanta, GA.
12. The OPAL Collaboration, G. Abbiendi *et al.*, CERN-EP/98-186, accepted by Eur. Phys. J. **C**.
13. H. Albrecht *et al.* (The ARGUS Collaboration), Phys. Lett. **B269**, 234 (1991).
14. T. Bergfeld *et al.* (The CLEO Collaboration), Phys. Lett. **323**, 219 (1994).
15. M. A. Ivanov *et al.*, hep-ph/9910342

16. G. Crawford et al. (The CLEO Collaboration), Phys. Rev. **D45**, 752 (1992), and M. Procario et al., Phys. Rev. Lett. **73**, 1472 (1994).

17. H. Albrecht *et al.* (The ARGUS Collaboration), Zeit. Fur. Phys. **C58**, 191 (1993), and H. Albrecht *et al.* (The ARGUS Collaboration), Zeit. Fur. Phys. **C42**, 191 (1989).

18. Isi Dunietz, hep-ph/9805287, and Phys. Rev. **D58** (1998).

19. B. Barish *et al.* (The CLEO Collaboration), Phys. Rev. Lett. **79**, 3599 (1997).

20. R. Ammar *et al.* (The CLEO Collaboration), Phys. Rev. **D57**, 1350 (1998). The cross-section quoted here is, of course, approximate since we do not specify the infrared cutoff for initial state radiation.

21. Isi Dunietz, FERMILAB -CONF-97-230-T, Mar 1997., hep-ph/9708252

22. M. Zoeller, Ph. D. thesis (SUNY, Albany), unpublished (1994).

23. Godfrey and Isgur, Physical Review **D32**, 189-231 (1985).

24. Fulcher L. P., **Phys. Rev.** D42, 2337 (1990).

25. Lichtenberg D.B. et al., **Z. Phys.** C41, 615 (1989).

26. Fulcher L. P., **Phys. Rev.** D37, 1259 (1988).

27. Fulcher L. P., **Phys. Rev.** D44, 2079 (1991).

28. Gupta S. N.,Repko W. W.,Suchyta III C. J., **Phys. Rev.** D39, 974 (1989).

29. Buchmuller W., Grunberg G., Tye S.-H., **Phys. Rev. Lett.** 45, 103 (1980), **Phys. Rev.** D24, 132 (1981).

30. Moxhay P.,Rosner J. L., **Phys. Rev.** D28, 1132 (1983).

31. Richardson J. L., **Phys. Lett.** 82B, 272 (1979).

32. Gupta S. N.,Radford S. F.,Repko W. W., **Phys. Rev.** D26, 3305 (1982), **Phys. Rev.** D30, 2424 (1984), see also ref..[33]

33. Gupta S. N., Radford, S. F., Repko, W. W., **Phys. Rev.** D34, 201 (1986).

34. Bhanot G., Rudaz S.,**Phys. Lett.** 78B, 119 (1978).

35. G. Garvey, **Comm. Nucl. Part. Phys. 3** 109 (1986).

36. K. Hikasa *et al.*, The Particle Data Group, Phys. Rev. D45, 1 (1992).

37. Kwong W., Rosner J. L., **Phys. Rev.** D38, 279 (1988).

38. Gupta S., and Johnson J., HEPPH-9408405, presented at 1994 Meeting of the American Physical Society, 2-6 Aug. 1994, Albuquerque, NM; B. E. Palladino, IFT-P-018-94, May 1994,; A. Das and V. Mathur, Phys. Rev. D49, 2508 (1994).; A. G. Grozin, BUDKERINP-92-97, Dec. 1992; T. Ito and T. Morii, Z. Phys. C59, p. 57 (1993); F. E. Close, Phys. Lett. B289, p. 143, (1992); A. Falk, Ph. D. Thesis, UMI-91-31945, Harvard U., (1991) (unpublished); J. Körner and G. Thompson, DESY-91-119, Oct. 1991

39. D. Besson and T. Skwarnicki, Ann. Rev. Nucl. Part. Sci. 43, 333 (1993).

40. The very broad $K_1^*(1410)$ is likely a radial excitation of the lower vector K^*.

41. N. Isgur and M. Wise, **Phys. Rev. Lett.**, 66, 1130 (1991)

42. Y. Kubota et al., (The CLEO Collaboration), Phys. Rev. Lett. 72, 1972 (1994).

43. M. Acciarri et al. (L3 Collaboration), Phys. Lett. B. accepted. CERN-EP/99-116, 5 August 1999. L3 preprint 181.

44. The CLEO Collaboration, contributed to this conference.

Extracting CKM matrix elements from semileptonic b decays

Pauline Gagnon

Centre for Research in Particle Physics, Ottawa, Canada
mailing address: EP Division, CERN, Genève 23, CH 1211, Suisse
E-mail: pauline.gagnon@cern.ch

The fundamental parameters of the Standard Model such as the CKM matrix elements, are not predicted by the model and must be measured from experimental data. I outline here how two of these matrix elements, V_{ub} and V_{cb}, can be obtained from the study of semileptonic b hadron decays. I also show how a precise determination of these two elements provide stringent although indirect constraints on the complex phase of the CKM matrix, which is needed within the Standard Model to allow CP-violation.

1 Introduction

Since the Standard Model was first postulated three decades ago, tremendous experimental effort has been invested in verifying its predictions. So far, no deviations from theoretical expectations have been found. Despite this impressive performance, the Standard Model has its limitations and cannot predict the fundamental parameters needed by the model, such as the Cabibbo-Kobayashi-Maskawa matrix elements. These must be determined from experimental data. The CKM matrix relates the mass eigenstates d, s, b (i.e. the physical quarks themselves) to the weak eigenstates d', s', b' (the states that couple through the weak interaction), namely:

$$\begin{pmatrix} d' \\ s' \\ b' \end{pmatrix} = \begin{pmatrix} V_{ud} & V_{us} & V_{ub} \\ V_{cd} & V_{cs} & V_{cb} \\ V_{td} & V_{ts} & V_{tb} \end{pmatrix} \begin{pmatrix} d \\ s \\ b \end{pmatrix}$$

Although the Standard Model does not predict the values of the CKM matrix elements, it requires that the matrix be unitary, imposing constraints between the elements such as:

$$V_{ud}V_{ub}^* + V_{cd}V_{cb}^* + V_{td}V_{tb}^* = 0$$

which reduces to:

$$V_{ub}^* + V_{td} = s_{12}V_{cb}^*$$

when V_{ud} and V_{tb} are assumed to be unity, and s_{12} is the Cabibbo angle. This unitarity constraint is usually represented by a triangle in the complex plane, as shown in Figure 1. Unitarity means that the sides of the triangle must cross at the apex such as to form a closed triangle.

After imposing the unitarity constraints, the nine CKM matrix elements can be expressed in terms of only four unknown parameters. A common

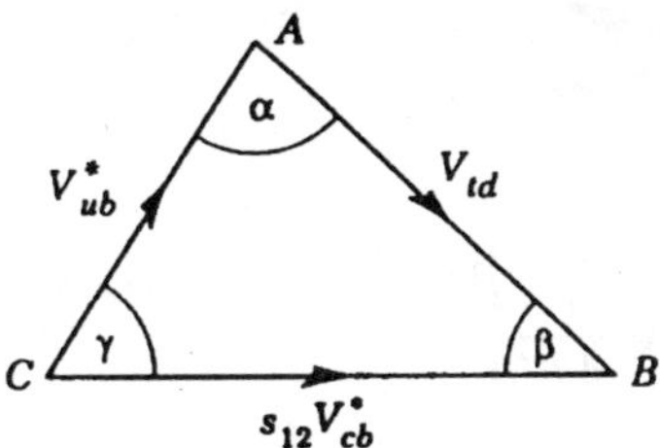

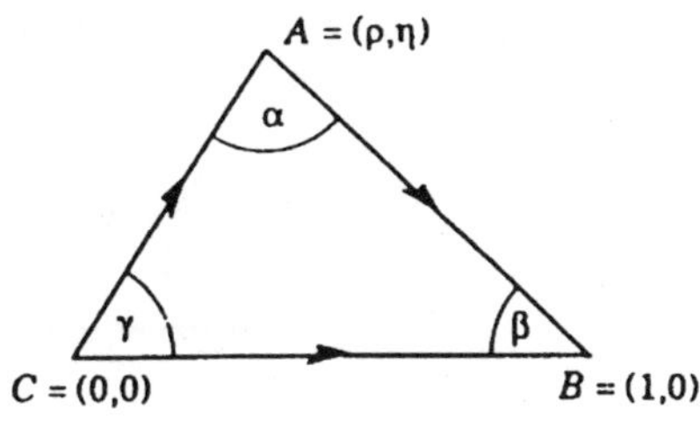

Figure 1: The unitarity triangle in the complex plane expressed in terms of the CKM matrix elements V_{ub}^*, V_{td} and V_{cb} and, equivalently, the Wolfenstein parameters ρ and η.

parametrisation is the Wolfenstein parametrisation which uses the following four parameters: $\lambda \equiv s_{12}$, the Cabibbo angle, $A = |V_{cb}|^2/\lambda^2$, ρ and η, where η corresponds to a complex phase. In the Wolfenstein parametrisation, the CKM matrix reduces to:

$$\begin{pmatrix} 1 - \lambda^2/2 & \lambda & A\lambda^3(\rho - i\eta) \\ -\lambda & 1 - \lambda^2/2 & A\lambda^2 \\ A\lambda^3(1 - \rho - i\eta) & -A\lambda^2 & 1 \end{pmatrix} + \mathcal{O}(\lambda^4).$$

The complex phase η is necessary to explain CP-violation within the Standard Model. A non-zero phase allows CP-violation to take place since a particular decay could have a different decay rate than its CP-conjugate. This can occur when two amplitudes with different phases interfere such that the magnitude of the decay rate changes when taking the CP-conjugate. The implication is that matter and anti-matter may have different decay rates. Hence, it is particularly interesting to determine if indeed η is non-zero (to test the possibility of CP-violation within the Standard Model) and to check that the unitarity triangle is closed (to check the Standard Model requirement that the CKM-matrix must be unitary).

To achieve these two goals unambiguously, one must measure the three sides and the three angles α, β and γ of the unitary triangle shown in Figure 1. This will be best done at the B factories and future hadron colliders, but meanwhile, stringent although indirect constraints [1,2] can already be obtained through precision measurements of a few parameters such as V_{ub} and V_{cb}, as discussed in Section 4.

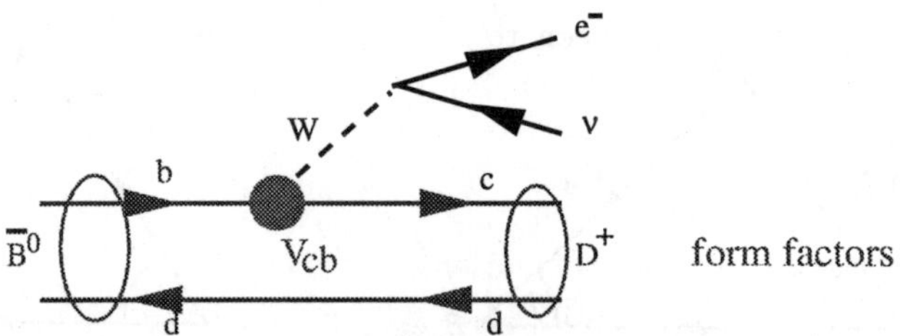

Figure 2: Schematic semileptonic decay.

1.1 Extracting V_{ub} and V_{cb} from semileptonic decays

Since the CKM matrix elements are not predicted by the theory, they must be determined experimentally. Semileptonic decays are particularly suited for this task since they can be represented as the product of two currents, as shown schematically in Figure 2: a leptonic current, which is exactly known, and a hadronic current, which can be parametrised in terms of form factors and contains a CKM matrix element.

In the following discussion, all LEP data refers to data collected near the Z^0 resonance, with subsequent decay into $b\bar{b}$, that is, $e^+e^- \to Z^0 \to b\bar{b}$. CLEO and ARGUS data refer to $e^+e^- \to \Upsilon(4S) \to b\bar{b}$ data.

Two general approaches exist to extract V_{cb} and V_{ub} from semileptonic decays. One method consists of getting the semileptonic widths, $\Gamma(b \to X_c\ell)$ and $\Gamma(b \to X_u\ell)$, from the measured inclusive decay rates, $BR(b \to X_c\ell)$ and $BR(b \to X_u\ell)$, and the lifetime, τ_b, using the relations:

$$\Gamma(b \to X_c\ell) = BR(b \to X_c\ell)/\tau_b \quad \text{and} \quad \Gamma(b \to X_u\ell) = BR(b \to X_u\ell)/\tau_b$$

In principle, the semileptonic width can be calculated theoretically although there is some disagreement on the method, as mentioned in the next section. The other method relies on the calculation of the decay rate of exclusive decay channels, such as $B \to \rho\ell\nu$ for V_{ub}, and $B \to D^*\ell\nu$ or $B \to D\ell\nu$ for V_{cb}, in terms of a parametrisation of the form factors. Since the errors on the theoretical inputs are largely uncorrelated between these two methods, it is possible to achieve a greater precision by combining the results. Details and results from both the "inclusive" and "exclusive" methods follow.

1.2 Producing LEP averaged results

In 1998, two new LEP working groups were formed for the purpose of combining V_{cb} and V_{ub} results from the four LEP experiments. Recently, members from the CLEO and SLD collaborations have also joined these groups. One important task has been to seek some agreement on the theoretical inputs needed for the extractions of V_{cb} and V_{ub}. Several theorists have performed calculations of the inclusive decay rates for b → u and b → c transitions, as well as of the form factors needed for exclusive modes. The methods involved for these calculations differ and so do the opinions on the size of the errors to be assigned to these calculations.

To attempt to clarify this situation, a mini workshop was held at CERN at the end of May 99, where interested theorists and experimentalists gathered to exchange ideas on the subject and produced a document summarising the outcome of those discussions. The results presented here are derived using the theoretical inputs described in this document[3]. Small differences between the results presented here and those shown at the conference reflect the fact that the theoretical inputs were not finalised at the time of the conference.

2 Measuring V_{ub} from the inclusive semileptonic rate

The first step required for the extraction of V_{ub} is the determination of the inclusive semileptonic branching fraction BR(b → $X_u\ell$). This rate can be related to the semileptonic width using the average b hadron lifetime that is, BR(b → $X_u\ell$) = $\tau_b\Gamma$(b → $X_u\ell$) where Γ(b → $X_u\ell$) is to be obtained from theoretical calculations. The results presented in this section have been updated to reflect the results derived by the LEP V_{ub} Working Group for the 1999 summer conferences[4] and released soon after the Physics in Collision 99 conference.

2.1 Measuring BR(b → $X_u\ell$)

The main difficulty in this measurement consists in differentiating b → u from the much more prominent b → c decays. This can be achieved by exploiting the fact that b → u transitions are characterised by both a low hadronic mass M_X and a high lepton energy E_ℓ.

In $b → c$ transitions, the minimum hadronic mass is limited to 1.87 GeV/c^2 by kinematics whereas 90% of $b → u$ transitions are found below this limit, as shown in the upper left plot of Figure 3. In principle, the lepton energy spectrum also provides excellent separation power since no leptons from b → c transitions can have energy beyond 2.3 GeV. This variable is more problematic though since it strongly depends on the decay modelling. These are the two most discriminant variables used to separate b → u from the b → c decays.

Analyses have been performed at LEP by DELPHI[5], L3[6] and ALEPH[7]. Typically, information on the lepton momentum, the hadronic mass and the neutrino momentum (evaluated from the missing momentum in the event) is fed to a neural network trained to distinguish b → u from the b → c decays.

292

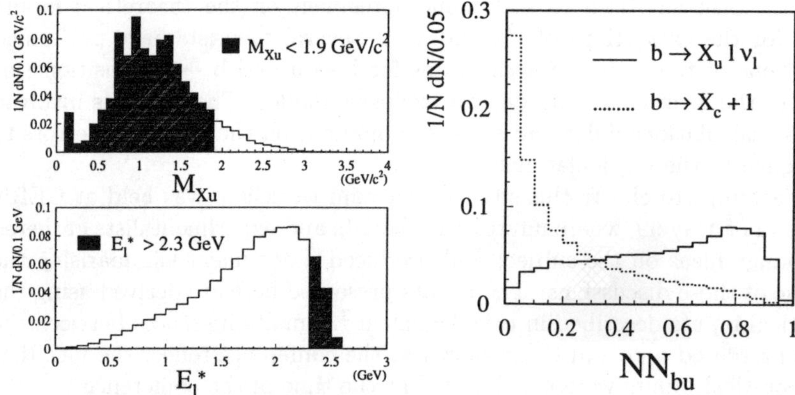

Figure 3: Hadronic mass distribution (upper left) and lepton energy spectrum (lower left) for b → u transitions. The dark histograms show events with $M_X < 1.87$ GeV and $E_\ell > 2.3$ GeV. The rightmost plot illustrates the discrimination power of the ALEPH neural network in separating b → u from b → c decays using information from M_X, E_ℓ and ν momentum.

The DELPHI analysis also includes information from the secondary vertex and kaon identification. The output of the neural network used by ALEPH is shown on the right plot of Figure 3, illustrating the separation power when applied to two Monte Carlo samples of b → u and b → c decays of equal size. The neural network output is shown for data in Figure 4. A clear excess of b → u events can be seen on the right hand plot when the data is compared to a Monte Carlo simulation that does not contain any b → u events.

The results for BR(b → $X_u\ell$) from three LEP experiments are:

$$ALEPH = (1.73 \pm 0.56 \pm 0.51 \pm 0.21) \times 10^{-3}$$
$$DELPHI = (1.53 \pm 0.46 \pm 0.44 \pm 0.22) \times 10^{-3}$$
$$L3 = (3.3 \pm 1.3 \pm 1.4 \pm 0.5) \times 10^{-3}$$
$$LEP = (1.67 \pm 0.35 \pm 0.38 \pm 0.20) \times 10^{-3},$$

where the first error corresponds to statistics and detector systematics combined, and the two others to b → c and b → u transitions modelling, respectively. All correlations have been taken into account.

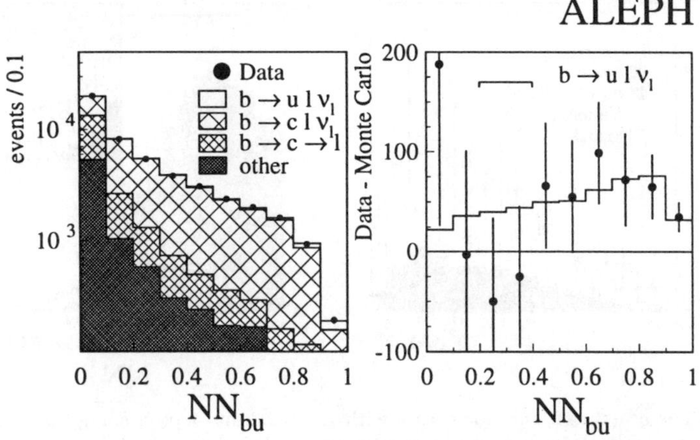

Figure 4: Distribution of the neural network output variable as seen in the ALEPH data. A clear excess of event is shown in the rightmost plot, showing the extra events found with respect to a Monte Carlo simulation that did not contain any b → u transitions.

2.2   Determining V_{ub} at LEP

Theoretical calculations for $\Gamma(b \to X_u\ell)$ [8,9] give $|V_{ub}|$ to be

$$0.00445 \left(\frac{\mathrm{BR}(b \to X_u\ell)}{0.002} \right)^{1/2} \left(\frac{1.55}{\tau_b} \right)^{1/2} \times (1 \pm 0.020 \ (pert.) \pm 0.035 \ (m_b))$$

where the errors come from perturbative QCD corrections and the estimate of the b quark mass.

With the average b hadron lifetime $\tau_b = (1.564 \pm 0.14)$ ps [10], the value of V_{ub} can be extracted from the equation above. The errors on $\mathrm{BR}(b \to X_u\ell)$ and τ_b are taken to be Gaussian and a probability function is formed to extract the value of V_{ub} at LEP, yielding:

$$|V_{ub}| = (4.05 \pm^{+0.39}_{-0.46} \ (exp.)^{+0.51}_{-0.60} \ (theory)) \times 10^{-3} \ (LEP \ average)$$

as derived by the LEP V_{ub} Working Group [4].

2.3   CLEO measurement for V_{ub}

The CLEO analysis looks at exclusive channels. Since the B mesons are produced nearly at rest, the B and $\bar{B}$ decay products overlap in space and cannot be disentangled. Hence, the hadronic mass of the recoiling system cannot be reconstructed and the CLEO analysis relies mostly on the lepton energy

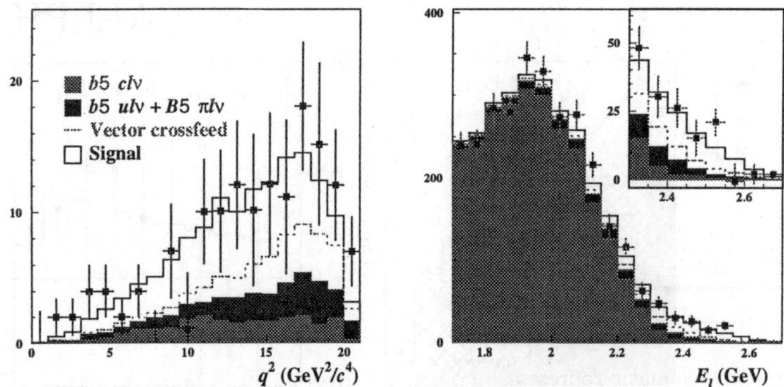

Figure 5: The q^2 (shown only for events with $E_\ell > 2.3$ and lepton energy distributions for $B \to \rho\ell\nu$, b $\to$ c and other b $\to$ u events in the CLEO data. A clear excess of events is visible in the high energy region, which can be attributed to $B \to \rho\ell\nu$ events.

spectrum, looking for an excess of events past the b $\to$ c transition end-point. Information on q^2, the virtual W mass squared, is also used to reduce the model-dependence. The analysis is performed from five different exclusive modes [11]: $B^0 \to \rho^-\ell^+\nu$ and $\pi^-\ell^+\nu$, and $B^+ \to \omega\ell^+\nu$, $\rho^0\ell^+\nu$, and $\pi^0\ell^+\nu$. The q^2 distribution and the lepton energy spectra are shown in Figure 5, where an excess of events in the high energy region is clearly visible. By averaging with a previous result [12], they obtain:

$$|V_{ub}| = (3.25 \pm 0.14 \text{ (stat.)}^{+0.21}_{-0.29} \text{ (syst.)} \pm 0.55 \text{ (theory)}) \times 10^{-3} \text{ (CLEO)}.$$

2.4 Combined results for V_{ub}

Combining the results given in Sections 2.2 for LEP and 2.3 for CLEO assuming the theoretical errors to be uncorrelated between the inclusive method used at LEP and the exclusive method used at CLEO, but full correlation for the decay modelling errors, my average is:

$$|V_{ub}| = (3.53 \pm 0.20 \text{ (exp.)} \pm 0.13 \text{ (theory)} \pm 0.54 \text{ (models)}) \times 10^{-3}.$$

3   Measuring V_{cb}

The results from $B \to D^*\ell\nu$ presented in this section have been updated since the conference to reflect the summer 99 average presented by the LEP V_{cb} Working Group in early July. Results for the inclusive V_{cb} measurement have been kept as presented at the conference although updates are now available. They

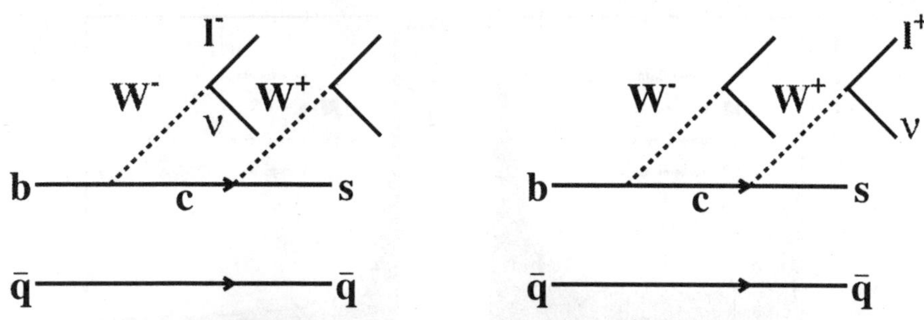

Figure 6: Schematic representation of b → ℓ (left) and b → c → ℓ decays (right).

can be found in[3] for the theoretical inputs needed for the inclusive channel and in [13] and [14] for the experimental results at LEP. There is no official average of the $B \to D\ell\nu$ channel available at this point, so my own average is presented here.

3.1 V_{cb} from the inclusive method

The DELPHI and OPAL collaborations had results on BR(b → Xℓ) reviewed at this conference. Both use similar methods to select the events: each $Z^0 \to b\bar{b}$ event is divided in two hemispheres and a lifetime b-tagging technique is used on each hemisphere to identify b events. Each hemisphere opposite a b-tagged hemisphere is searched for a lepton. The difficulty in this measurement is to separate leptons coming from b → ℓ from b → c → ℓ decays.

The diagrams in Figure 6 show the "direct" b → ℓ decays and the "cascade" b → c → ℓ decays. In direct decays, the lepton comes from the bottom quark decay, whereas for cascade decays, the lepton comes from the subsequent decay of the charm quark. Since b quarks are much heavier, leptons coming from direct b quark decays tend to have more transverse momentum with respect to the b jet than leptons coming from cascade decays. In addition, the leptons in these two types of decays have a different sign correlation with respect to the original b quark charge, which can be reconstructed using a jet charge technique or related to the charge of a second lepton found in the event. The lepton total and transverse momenta, and the sign correlations are the most powerful variables used to distinguish b → ℓ from b → c → ℓ decays, although using the lepton momentum introduces large uncertainties since it is strongly model-dependent.

DELPHI [15] uses the transverse momentum distribution for all events se-

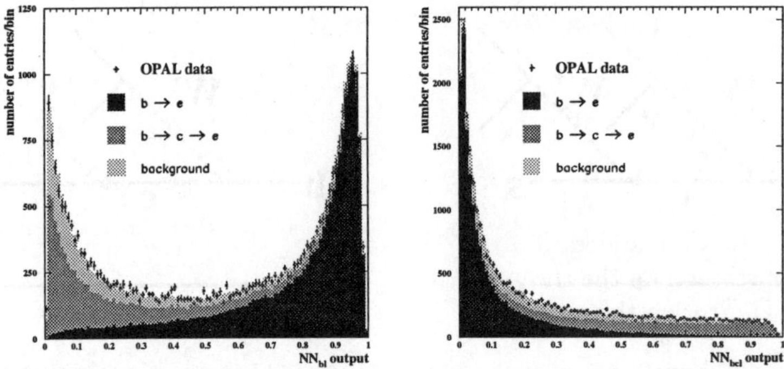

Figure 7: The outputs of two neural networks used by OPAL to distinguish $b \to \ell$ and $b \to c \to \ell$ events from all other events. The results are shown here for electrons only.

lected with a lepton opposite a b-tagged hemisphere. They also use charge information when a second lepton is found in the opposite hemisphere. The fit is done simultaneously to both samples. At OPAL [16], the total and transverse lepton momenta, jet charge and other jet shape variables have been combined into two neural networks trained respectively to distinguish $b \to \ell$ and $b \to c \to \ell$ events from all other events. The outputs of these two neural networks are shown in Figure 7. By not relying solely on the transverse lepton momentum, the model-dependence has been greatly reduced. The results for $BR(b \to X\ell)$ for these two analyses are:

$$OPAL = (10.83 \pm 0.10 \text{ (stat.)} \pm 0.20 \text{ (syst.)}^{+0.20}_{-0.13} \text{ (model))}\%$$

$$DELPHI = (10.65 \pm 0.11 \text{ (stat.)} \pm 0.23 \text{ (syst.)}^{+0.43}_{-0.27} \text{ (model))}\% \quad (\textit{prel.})$$

$$BR(b \to X\ell) = (10.82 \pm 0.23)\% \quad \text{(LEP EWWG averageFeb.99)}$$

where the LEP average was obtained in March 99 by the LEP Electroweak Working Group performing a fit to these results and all other electroweak parameters [17]. This number was updated for later summer conferences [14] since DELPHI presented an update and L3 showed a new preliminary result.

Starting from the inclusive measurement for $BR(b \to X\ell)$ given above, it is possible to extract the value of V_{cb}. One must first remove contributions from $b \to u$ transitions using the value given in Section 2.1, namely, $BR(b \to X_c\ell) = BR(b \to X\ell) - BR(b \to X_u\ell)$. Since $BR(b \to X_c\ell) = \tau_b \, \Gamma(b \to X_c\ell)$, one can extract V_{cb} from the theoretical expression derived for the semileptonic width and the average b hadron lifetime $\tau_b = (1.564 \pm 0.014)$ ps [10]. Using the calculation of [8] yields:

$$|V_{cb}| = 0.0419 \left(\frac{\mathrm{BR}(\mathrm{b} \to \mathrm{X}_c \ell)}{0.105}\right)^{1/2} \left(\frac{1.55}{\tau_b}\right)^{1/2} \times \left(1 - 0.012 \left(\frac{\mu_\pi^2 - 0.5\,\mathrm{GeV}/c^2}{0.1\,\mathrm{GeV}/c^2}\right)\right) \times$$
$$\left(1 \pm 0.015\,(pert.) \pm 0.010\,(m_b) \pm 0.012\,(1/m_Q^3)\right)$$

where the kinetic operator $\mu_\pi^2 = 0.5 \pm 0.1$ GeV. Following the discussion given in [3], the errors are added in quadrature, and then doubled to reflect the fact that the precision on the theoretical inputs is not well known. This gives a 5% theoretical error on the estimate of V_{cb}. From this, I obtain from the inclusive b $\to \ell$ LEP data:

$$|V_{cb}| = (42.0 \pm 0.5\,(\mathrm{exp.}) \pm 2.1\,(\mathrm{theory})) \times 10^{-3} \qquad (\text{LEP inclusive data}).$$

A similar calculation can be done for the $\Upsilon(4S)$ data. Using the same value for $\mathrm{BR}(\mathrm{b} \to \mathrm{X}_u \ell)$ (which is a good approximation) but a different lifetime corresponding to the B^0 and $B^\pm$ mixture found at lower energy, that is $\tau_B = 1.605 \pm 0.021$. With $\mathrm{BR}(\mathrm{b} \to \mathrm{X}\ell) = (10.45 \pm 0.21)\%$ [10], I obtain:

$$|V_{cb}| = (40.7 \pm 0.5\,(\mathrm{exp.}) \pm 2.0\,(\mathrm{theory})) \times 10^{-3} \qquad (\Upsilon(4S)\ \text{inclusive data}).$$

Assuming full correlation for the theoretical errors, my average for the LEP and $\Upsilon(4S)$ V_{cb} results from the inclusive branching fraction is:

$$|V_{cb}| = (41.1 \pm 0.4\,(\mathrm{exp.}) \pm 2.1\,(\mathrm{theory})) \times 10^{-3} \qquad (\text{all inclusive data}).$$

Note though that averaging these results is problematic given that the measured semileptonic branching fraction at the $\Upsilon(4S)$ resonance is smaller than $\mathrm{BR}(\mathrm{b} \to \mathrm{X}\ell)$ measured near the Z^0 resonance where contributions from b baryons are also included. From the observed difference in lifetimes, with $\tau_{\mathrm{b\ baryon}} \ll \tau_B$, the opposite should be seen.

3.2 Check on theoretical inputs

The CLEO collaboration has used its data to perform cross-checks on some of the theoretical inputs needed to determine V_{cb} from the inclusive semileptonic branching fraction. Two key players in the Operator Product Expansion (OPE) used in [8] are the kinetic operator μ_π^2 and the b quark mass m_b. These can be expressed in terms of two parameters: λ_1 and $\bar{\Lambda}$. Ideally, one would want to constrain λ_1 and $\bar{\Lambda}$ using experimental data and reduce the theoretical uncertainty on the OPE parameters (see the talk by R. Poling at ICHEP 98 for more details [18]). This can be done using the hadronic mass and the lepton energy moments.

The hadronic mass of the system in a b $\to$ c decay is given by:

$$M_X^2 \equiv M_B^2 + M_{\ell\nu}^2 - 2E_B E_{\ell\nu} + 2p_B p_{\ell\nu} \cos\theta_{\ell\nu,B}$$

where $\cos\theta_{\ell\nu,B}$ is the angle between the direction of the virtual W and the B meson, which cannot be evaluated precisely. But this last term can be

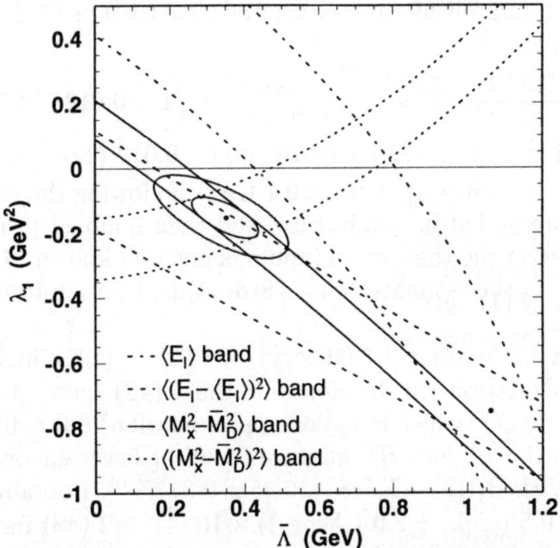

Figure 8: Constraints imposed by the hadronic mass and lepton energy moments on the λ_1 and $\bar\Lambda$ parameters of the Operator Product Expansion. Inconsistent results are obtained since there is no common overlap between the two sets of moments.

neglected since the B mesons are produced essentially at rest, implying that p_B is very small. By measuring M_X^2 and E_ℓ, and one can determine the four following moments:

$$\langle M_X^2 - \bar M_D^2 \rangle$$
$$\langle (M_X^2 - \bar M_D^2)^2 \rangle$$
$$\langle E_\ell \rangle \text{ and } \langle (E_\ell - \langle E_\ell \rangle))^2$$

which have been related to λ_1 and $\bar\Lambda$. By plotting these moments in the $\lambda_1 - \bar\Lambda$ plane, one obtains two sets of constraints on the two parameters λ_1 and $\bar\Lambda$. This is shown on Figure 8. Unfortunately, at present, inconsistent results are obtained from the hadronic mass moments and the lepton energy moments, reflecting either some experimental problems or a flaw in the theoretical predictions. More work is needed on this before final conclusions can be drawn.

3.3 V_{cb} from the exclusive method

The other method used to extract V_{cb} involves exclusive modes such as $B \to D^*\ell\nu$ and $B \to D\ell\nu$ decays, and is based on a form factor analysis. In the framework of Heavy Quark Effective Theory, the form factors can be expressed as functions of the square 4-velocity transfer: $(v_b - v_c)^2 = 2(1 - v_b \cdot v_c)$, where v_b and v_c represent the b and c quarks 4-velocity w.r.t. the light quark in the bound state. Defining:

$$w \equiv (v_b \cdot v_c) = (m_B^2 + m_{D*}^2 - q^2)/(2m_B m_{D*}),$$

one can express the differential decay rate $d\Gamma/dw$ as a function of w:

$$d\Gamma/dw = \mathcal{K}(w)\mathcal{F}^2(w)|V_{cb}|^2$$

where $\mathcal{K}(w)$ is the phase space factor, a known function of w which vanishes at the point of zero-recoil, i.e. when $w = 1$, and $\mathcal{F}(w)$ is the form factor. It can be written as $\mathcal{F}(1) \cdot g(w)$ where $g(w)$ is a function to be parametrised.

The parametrisation used by the LEP V_{cb} Working Group for the $B \to D^*\ell\nu$ channel is [19]

$$\mathcal{F}(w) = \mathcal{F}(1)\left[1 - 8\rho^2 z + (53\rho^2 - 15)z^2 - (231\rho^2 - 91)z^3\right]$$
$$\text{where } z = (\sqrt{w+1} - \sqrt{2})/(\sqrt{w+1} + \sqrt{2}).$$

Experimentally, one needs to evaluate the differential decay rate $d\Gamma/dw$ as a function of w, then extrapolate to the point where $w = 1$ to fit for $\mathcal{F}(1)|V_{cb}|$. This is shown in the left plot of Figure 9 for the DELPHI $B \to D^*\ell\nu$ data. The results obtained at LEP are shown in the right plot of Figure 9. Combining results from DELPHI, ALEPH and OPAL for $B \to D^*\ell\nu$ decays, the preliminary LEP average presented at the summer 1999 conferences is [13]:

$$\mathcal{F}_{D*}(1)|V_{cb}| = (33.8 \pm 0.9 \text{ (stat.)} \pm 1.9 \text{ (syst.)}) \times 10^{-3}$$
$$\rho^2 = 1.00 \pm 0.09 \pm 0.14 \text{ (linear fit)},$$

where ρ was obtained by neglecting terms of $\mathcal{O}(z^2)$. This is not yet a world average since the following results for $\mathcal{F}_{D*}(1)|V_{cb}|$:

$$CLEO = (35.1 \pm 2.0 \text{ (stat)} \pm 2.0 \text{ (syst)}) \times 10^{-3}$$
$$ARGUS = (39.2 \pm 3.9 \text{ (stat)} \pm 2.8 \text{ (syst)}) \times 10^{-3}$$

still have to be included, which will require careful analysis of the various contributions to the uncertainties. Using $\mathcal{F}_{D*}(1) = 0.88 \pm 0.05$ as derived in [3], the LEP V_{cb} Working Group obtains:

$$|V_{cb}| = (38.4 \pm 2.5 \text{ (exp.)} \pm 2.2 \text{ (theory)}) \times 10^{-3} \text{ (from } B \to D^*\ell\nu \text{ at LEP)}.$$

Similar results can also be derived from another exclusive channel using $B \to D\ell\nu$ decays. A recent CLEO result [20] can be averaged with an ALEPH result [21] taking into account correlations, which gives $\mathcal{F}_D(1)|V_{cb}|$:

$$CLEO = (41.6 \pm 4.7 \pm 3.7) \times 10^{-3}$$
$$ALEPH = (31.1 \pm 9.9 \pm 8.6) \times 10^{-3}$$
$$\mathcal{F}_D(1)|V_{cb}| = (39.9 \pm 5.5) \times 10^{-3} \text{ (my average)}.$$

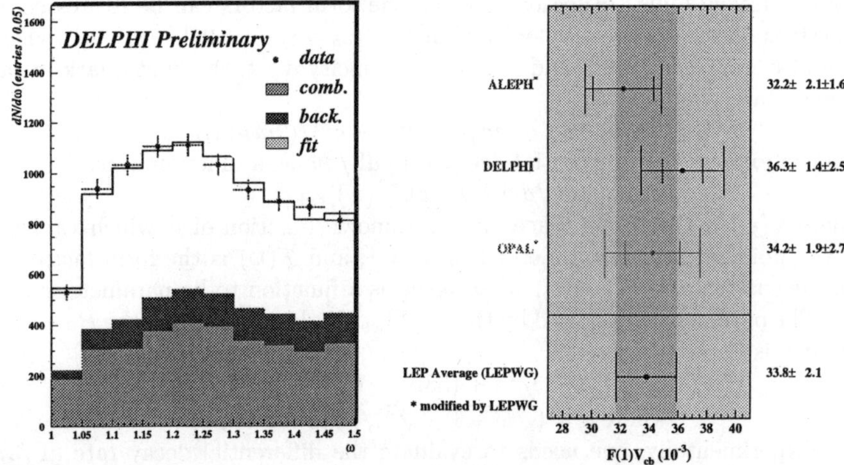

Figure 9: (Left) Fit to the DELPHI data for the number of candidate $B \rightarrow D^*\ell\nu$ events as a function of w. The value of $\mathcal{F}_1(1)|V_{cb}|$ is extracted from the extrapolation to the point of zero-recoil $w = 1$. (Right) LEP measurements of $\mathcal{F}_1(1)|V_{cb}|$ from $B \rightarrow D^*\ell\nu$ decays.

With $\mathcal{F}_D(1) = 0.98 \pm 0.07\,^{22}$, I obtain:

$$|V_{cb}| = (40.7 \pm 5.6\ (\text{exp.}) \pm 2.9\ (\text{theory})) \times 10^{-3} \quad (B \rightarrow D\ell\nu \text{ channel})$$

where the experimental and form factors uncertainties are shown separately.

3.4 Summary of V_{cb} measurements

Assuming the errors to be uncorrelated between the inclusive and exclusive method[a] gives the result shown in Table 1. Note that at this point, results from the $\Upsilon(4S)$ (CLEO and ARGUS) have not been included. Although the result given in Table 1 is not an official average, it gives a good indication of the precision that has already been achieved on the V_{cb} measurement.

4 Constraints on the unitarity triangle

The four parameters of the Wolfenstein parametrisation are λ, A, ρ, η where λ is the Cabibbo angle and $A = |V_{cb}|^2/\lambda^2$. Information on ρ and η can be obtained from Δm_d, Δm_s, $|V_{ub}|/|V_{cb}|$ and ϵ_K, as can be seen in [1] and [2]. Their values, as they were known in February 1999, are shown in Table 2. These

---

[a] A more careful treatment of the correlations can be found in [13] where V_{cb} results from $B \rightarrow D^*\ell\nu$ and the inclusive methods have been combined but for the LEP data only.

Table 1: Results for V_{cb} from various methods. The results from CLEO and ARGUS are missing from the average for the $B \rightarrow D^*\ell\nu$ exclusive channel.

method	$V_{cb}(\times 10^{-3})$
BR(B $\rightarrow$ X$_c\ell$)	41.1 ± 0.4 (exp.) ± 2.1 (theory)
$B \rightarrow D^*\ell\nu$ (LEP)	38.4 ± 2.5 (exp.) ± 2.2 (theory)
$B \rightarrow D\ell\nu$	40.7 ± 5.6 (exp.) ± 2.9 (theory)
V_{cb} (my average)	40.3 ± 1.7

values have not changed by much, since we now have:

$$|V_{ub}| = (3.53 \pm 0.59) \times 10^{-3}$$
$$|V_{cb}| = (40.3 \pm 1.7) \times 10^{-3}$$
$$|V_{ub}|/|V_{cb}| = 0.088 \pm 0.015$$

where the errors on the ratio have been calculated taking into account correlations between the theoretical inputs for V_{ub} and V_{cb}. Hence, the same conclusions still hold.

The results of the fit to all these constraints yield values for ρ and η which can be used to construct the unitarity triangle in the complex plane. A non-zero value for η means that the CKM matrix would have a complex phase, which is needed within the Standard Model to allow for CP-violation. The values obtained from the fit are:

$$\rho = 0.160^{+0.094}_{-0.070}$$
$$\eta = 0.381^{+0.061}_{-0.058}$$

under the assumption of a complex CKM matrix. The result indicates that the CKM matrix would indeed be complex. Alternatively, one could assume the CKM matrix to be real[23]; the fit still yields a consistent result but for ρ only, namely $\rho = 0.321^{+0.053}_{-0.056}$.

The results of the fit to ρ and η can be seen in Figure 10, with the direct measurement of one side of the triangle ($|V_{ub}|/|V_{cb}|$) and the constraints from the other parameters. These **indirect** constraints are already as stringent as what will be reached by the BaBar Collaboration alone using their full data sample, as reported in the BaBar book. Of course, only **direct** measurements of all angles at B factories and future hadron colliders will provide the final answer on the complex phase η of the CKM matrix and CP violation.

Table 2: Inputs used by Salvatore Mele to produce indirect constraints on the unitarity triangle as of February 1999.

	February 99
λ	0.2196 ± 0.0023
$\|V_{cb}\|$	$(39.5 \pm 1.7) \times 10^{-3}$
A	0.819 ± 0.035
$\|\epsilon_K\|$	$(2.280 \pm 0.019) \times 10^{-3}$
Δm_d	(0.471 ± 0.016) ps^{-1}
Δm_s	> 12.4 ps^{-1} (95% C.L.)
$\|V_{ub}\|/\|V_{cb}\|$	0.093 ± 0.016

Acknowledgements

I wish to thank the organisers of the Physics in Collision 99 conference for their kind invitation to participate in a very interesting meeting. I also want to thank my colleagues from the LEP Heavy Flavour Steering Group as well as the V_{cb} and V_{ub} Working Groups for their essential input and many contributions to this talk. Special thanks to Richard Hawkings for his careful reading of this document and his many constructive suggestions.

References

1. S. Mele, *Phys. Rev.* D **59** 113011 (1999).
2. Constraints on the parameters of the CKM matrix by end 1998, F. Parodi, P. Roudeau, A. Stocchi, *hep-ex/9903063*.
3. Informal Workshop on the derivation of V_{cb} and V_{ub}: Experimental Status and Theory Uncertainties, accessible form http://www.cern.ch/LEPHFS.
4. LEP V_{ub} Working Group, Summer 1999 averages available from http://home.cern.ch/ battagl/vub/vub.html.
5. DELPHI Collaboration, contributed paper to ICHEP-99, #4.521 DELPHI 99-110 CONF 297
6. M. Acciarri *et al.*, L3 Collaboration, *Phys. Lett.* B **436** 174 (1998)
7. R. Barate *et al.*, ALEPH Collaboration, *Eur. Phys. J.* C **6** 555 (1999).
8. I. Bigi, M. Shifman, N. Uraltsev, *Annu. Rev. Part. Sci.* **47** 591 (1997).
9. N. Uraltsev, *hep-ph/990552*. Update of the calculations presented in [8].
10. Review of Particle Physics, C. Caso *et al.* (Particle Data Group), *Eur. Phys. J.* C **3** 1998 (1).
11. B.H. Behrens *et al.*, CLEO Collaboration, *hep-ex/9905056*.
12. J. Alexander *et al.*, CLEO Collaboration, *Phys. Rev. Lett.* **77** 5000

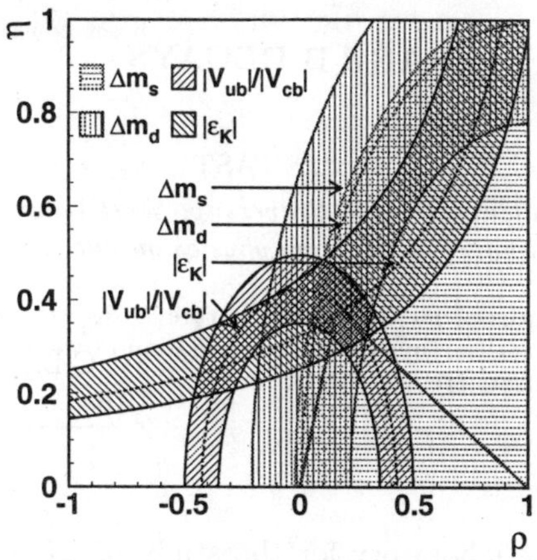

Figure 10: Various indirect constraints on the unitarity triangle. A non-zero value for η is needed within the Standard Model to allow for CP-violation.

(1996).

13. LEP V_{cb} averages for the summer 99 conferences, available from http://www.cern.ch/LEPVCB/.

14. Semileptonic b branching fraction at LEP, P. Gagnon, talk given in Parallel Session #5 at the ICHEP 99, Tampere, Finland.
See http://neutrino.pc.helsinki.fi/hep99/transparencies/session-05/.

15. DELPHI Collaboration, DELPHI 98-122 CONF 183, 22 June 1998, ICHEP'98 contributed paper #129.

16. G. Abbiendi *et al.*, OPAL Collaboration, CERN-EP/99-078, submitted to Eur. Phys. J. C.

17. Electroweak Heavy Flavour Results presented at the 1999 winter conferences, LEP Electroweak Working Group, LEPHF 99-01
http://www.cern.ch/LEPEWWG/heavy.

18. Selected results on $b \to c\ell\nu$ from CLEO, R. Poling, talk presented at the ICHEP 98 meeting, Vancouver, Canada. Published in the conference proceedings, World Scientific.

19. I. Caprini, L. Lellouch, M. Neubert, *hep-ph/9712417*.

20. J. Bartelt *et al.*, CLEO Collaboration, *Phys. Rev. Lett.* **82** 3746 (1999).

21. D. Buskulic *et al.*, Aleph Collaboration, *Phys. Lett.* B **395** 373 (1997).

22. I. Caprini, M. Neubert, *Phys. Lett.* B **380** 376 (1996).

23. H. Georgi and S. Glashow, *Phys. Lett.* B **451** 372 (1999).

RARE B DECAYS

JAMES FAST

1396 Physics Building, Purdue University, West Lafayette, IN 47907
E-mail: jfast@purpcj.physics.purdue.edu

The experimental status of measurements in rare B decays are presented. The implications of these measurements on determination of the Cabibbo-Kobayashi-Maskawa matrix elements are also discussed.

1 Introduction

Rare decays provide a laboratory for the study of CP violation within the standard model, as well as a window into physics beyond the standard model. Within the framework of the standard model the CKM matrix appears in the weak charged current,

$$
J_\mu = (\bar{u}\ \bar{c}\ \bar{t})_L\ \gamma_\mu
\begin{pmatrix}
V_{ud} & V_{us} & V_{ub} \\
V_{cd} & V_{cs} & V_{cb} \\
V_{td} & V_{ts} & V_{tb}
\end{pmatrix}
\begin{pmatrix}
d \\
s \\
b
\end{pmatrix}_L ,
\tag{1}
$$

rotating the flavor eigenstate basis into the weak eigenstate basis. The unitarity of this matrix leads to the relation $\Sigma V_{ik} V_{jk}^* = 0$ for $i \neq j$ which can be graphically expressed as a triangle, Figure 1, the three sides of which are related to the magnitudes of the CKM elements while the angles are related to the phases of the CKM elements. Measuring all three sides and all three angles of the unitarity triangle allows for a consistency check of the standard model; any deviations from a closed triangle indicate non-standard model processes. Extensions to the standard model, such as SUSY, typically transform the unitarity triangle into a quadrilateral.

The study of rare B decays spans a large range of physics processes; hadronic decays ($b \to u$ and $b \to s(d)$), radiative decays ($b \to s(d)\gamma$), charmless semi-leptonic decays ($b \to ul\nu$), leptonic decays ($b \to sl^+l^-$, $b \to l^+l^-$, $B \to l\nu$) and decays forbidden in the standard model. The first measurement of a rare B decay process came from the CLEO collaboration in 1993 [1] with an analysis of the decays [a] $B^0 \to K^+\pi^-/\pi^+\pi^-$, where the two modes were not separable with the available statistics at that time, and then in 1995 with the discovery [2] of the radiative penguin decay $b \to s\gamma$. Subsequently the LEP collaborations have published results on radiative decays and two-body charmless decays confirming CLEO's initial observations. In addition,

---

[a] charge conjugate states are implied throughout this article

a large effort has been made by CLEO, the LEP collaborations, D∅ and CDF to search for a broad range of rare B decays. This article focuses on the most recent results in charmless hadronic decays of B mesons.

2 CKM measurements and CP violation

The ultimate goal of rare B decay measurements is the extraction of the CKM matrix elements and direct measurement of CP violation in the B sector. Figure 1 shows the unitarity triangle with physics processes which can be used to extract the CKM elements and angles (phases).

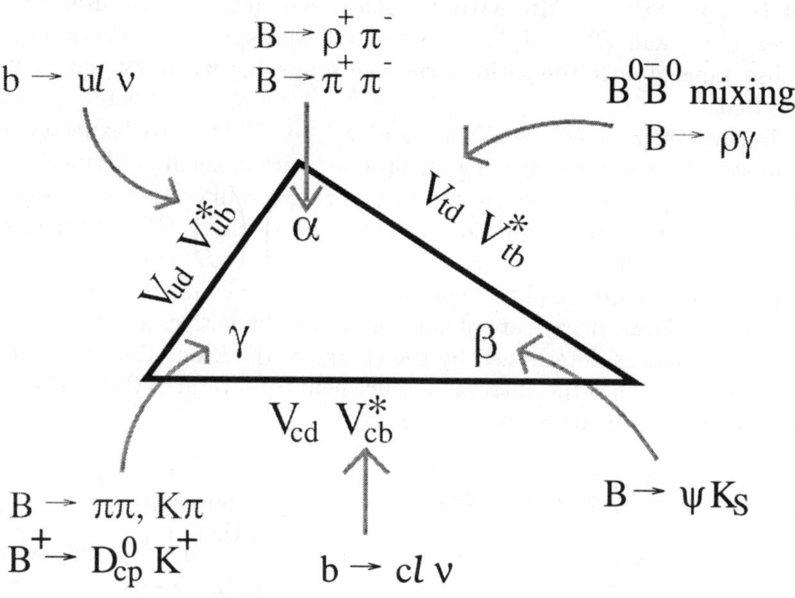

Figure 1. Unitarity triangle and physics processes which can be used to extract the various elements and phases.

Decays such as $B^0 \rightarrow \pi^+\pi^-$ and $B^0 \rightarrow \pi^\pm\rho^\mp$ are expected to be dominated by the $b \rightarrow u$ spectator transition, and measurements of their branching fractions could be used to extract a value for $|V_{ub}|$, although the $b \rightarrow u l\nu$ semileptonic decay measurements will likely continue to provide the most accurate determination of $|V_{ub}|$. These decays can also be used to measure CP violation in the B sector at both asymmetric B factories [3] and hadron colliders [4]. Since the $\pi^+\pi^-$ final state is a CP eigenstate, CP violation can arise from interference between the amplitude for direct decay and the amplitude for the process in which the B^0 first mixes into a $\bar{B}^0$ and then decays. Measurement of the time evolution of the rate asymmetry leads to a measurement of $\sin 2\alpha$. If the $B^0 \to \pi^+\pi^-$ decay has a non-negligible contribution from the $b \to dg$ penguin, interference between the spectator and penguin contributions will contaminate the measurement of CP violation via mixing [5], an effect known as "penguin pollution." If this is the case, the penguin and spectator effects can be disentangled by also measuring the isospin-related decays $B^0 \to \pi^0\pi^0$ and $B^\pm \to \pi^\pm\pi^0$ [6]. Alternatively, SU(3) symmetry can be used to relate $B^0 \to \pi^+\pi^-$ and $B^0 \to K^+\pi^-$ [7]. Penguin and spectator effects may then be disentangled once the ratio of the two branching fractions and $\sin 2\beta$ are measured.

Decays such as $B^0 \to K^+\pi^-$ and $B^0 \to K^{*+}\pi^-$ are expected to be dominated by the $b \to sg$ penguin process, with a small contribution from a Cabibbo-suppressed $b \to u$ spectator process. Interference between the penguin and spectator amplitudes can give rise to direct CP violation, which will manifest itself as a rate asymmetry for decays of B^0 and $\bar{B}^0$ mesons, but the presence of hadronic phases complicates the extraction of the CP violation parameters. These decays are of particular interest as they are "self-tagging", i.e. the B flavor is determined by the charge of the kaon. Recent theoretical work [8] indicates that the current measurements of several of these self-tagging decays can be used to constrain $\sin \gamma$.

3   Charmless hadronic decays

The first measurement of charmless hadronic decays came in 1993 with the measurement of the combined branching ratio for $B^0 \to K^+\pi^-/\pi^+\pi^-$ by the CLEO collaboration [1]. At that time there were insufficient statistics to separate the two decay modes using the available particle identification. There are now over 50 candidates in this event sample using about 50% of the available data from CLEO. Signals have also been observed in the associated decays with either a neutral kaon or pion.

For some time the dominant decay diagram appeared to be the $b \to sg$ penguin, with no statistically significant evidence for exclusive $b \to u$ decays. This is no longer the case. CLEO now has measurements of two decays dominated by the $b \to u$ transition, $B^+ \to \rho^0\pi^+$ and $B^0 \to \rho^\pm\pi^\mp$; see Figures 2 and 3.

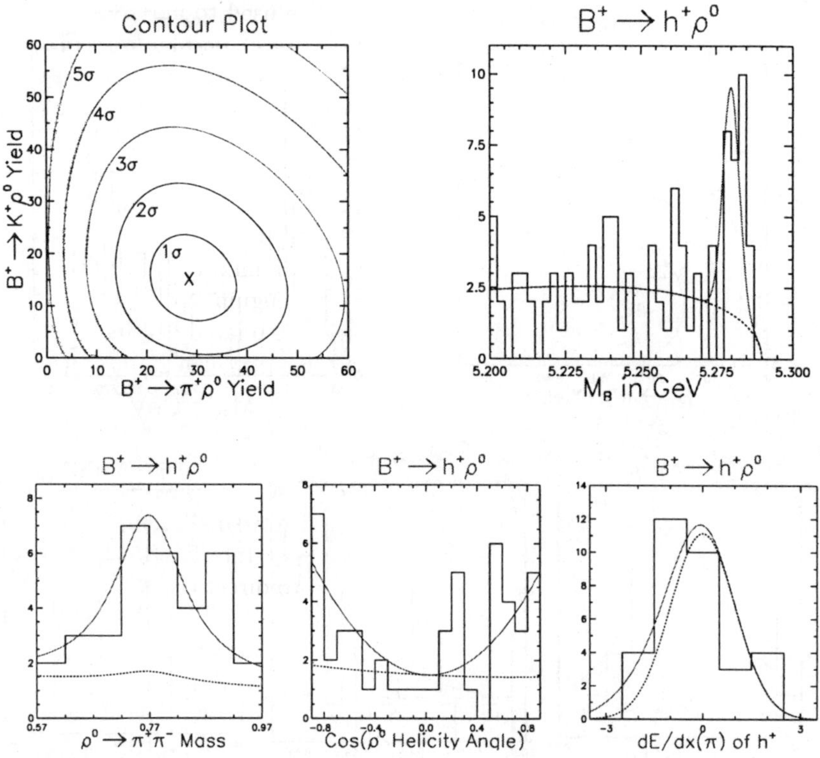

Figure 2. Likelihood fit results and projection plots for CLEO analysis of $B^{\pm} \to \rho^0 \pi^{\pm}$.

We summarize in Tables 1-3 the most recent branching ratio measurements and upper limits on charmless hadronic decays from CLEO.

To summarize, there are now measurements in all three charge states of the decay $B \to K\pi$ with about 25% accuracy. the decays to the $\pi\pi$ final state are not observed at a statistically significant level. This situation implies that the measurement of the CKM angle α at the B factories will be hampered both by low rate and theoretical complications from the possibly large $b \to d$ contribution. The relatively large observed rate for the decay $B \to \rho\pi$, on the other hand, makes this decay a very attractive alternative for determination of α. The measurements of the charged B decays to $\pi\pi$ and $K\pi$ can be used to constrain the CKM angle γ already. Using the method proposed by Neubert

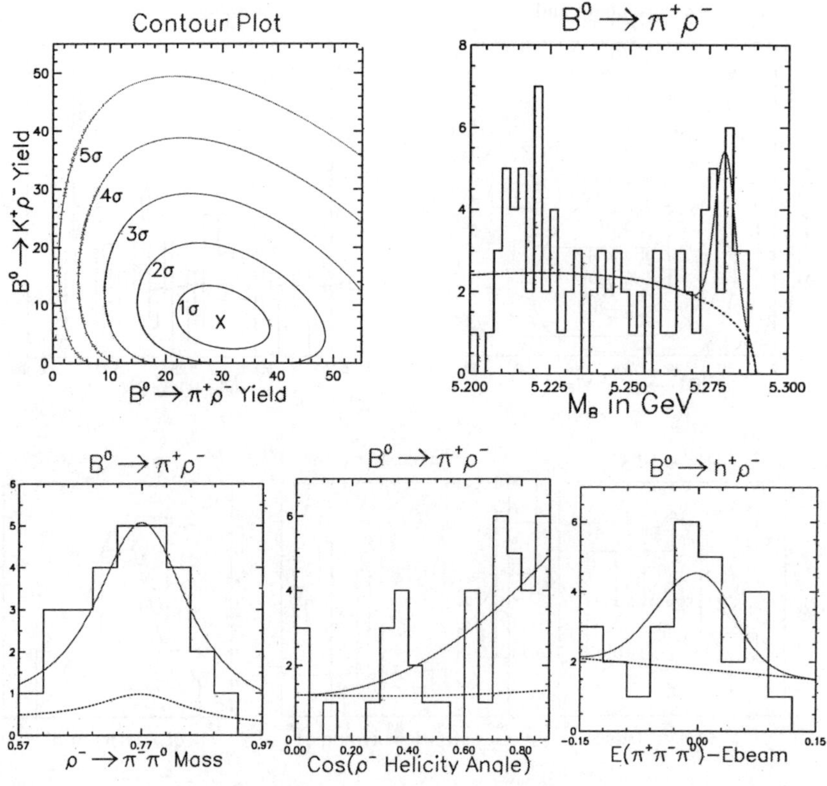

Figure 3. Likelihood fit results and projection plots for CLEO analysis of $B^0 \to \rho^\pm \pi^\mp$.

and Rosner [8] the CLEO data *excludes* the region $|\gamma| \geq 70°$ at 90%CL.

4   Electroweak penguins and CP violation studies

The electroweak penguin decay $b \to s\gamma$ was first observed by CLEO in 1993 [2]. The Standard Model prediction for this decay rate is $(3.3 \pm 0.3) \times 10^{-4}$. An observed rate inconsistent with this value would point to new physics in the loop diagram, possible charged Higgs, SUSY or anomalous $WW\gamma$ couplings. The CLEO result for this branching ratio, $(3.15 \pm 0.35 \pm 0.32 \pm 0.26) \times 10^{-4}$, is in good agreement with the standard model prediction. CLEO has extended this analysis to a search for CP asymmetry in the decay. The method of

Table 1. Measurements and/or upper limits on $K\pi$ and $\pi\pi$ modes from CLEO

Mode	$\mathcal{B}$ $(\times 10^{-5})$	Theory $\mathcal{B}$ $\times 10^{-5}$)
$\pi^{\pm}\pi^{\mp}$	< 0.84	0.8-2.6
$\pi^{\pm}\pi^{0}$	< 1.6	0.4-2.0
$\pi^{0}\pi^{0}$	< 0.9 †	
$K^{\pm}\pi^{\mp}$	$1.4 \pm 0.3 \pm 0.2$	0.7-2.4
$K^{\pm}\pi^{0}$	$1.5 \pm 0.4 \pm 0.3$	0.3-1.3
$K^{0}\pi^{\pm}$	$1.4 \pm 0.5 \pm 0.2$	0.8-1.5
$K^{0}\pi^{0}$	< 4.1	
$K^{\pm}K^{\mp}$	< 0.23	-
$K^{\pm}\overline{K}^{0}$	< 0.93	0.07-0.13
$K^{0}\overline{K}^{0}$	< 1.7	-

Table 2. Upper limits on $K\phi$ and $\pi\phi$ modes from CLEO

Mode	$\mathcal{B}$ $(\times 10^{-5})$
$K^{\pm}\phi$	< 0.59
$K^{0}\phi$	< 2.84
$\pi^{\pm}\phi$	< 0.40
$\pi^{0}\phi$	< 0.54

Table 3. Recent CLEO results for $\rho\pi$ and $K^{*}\pi$

Mode	$\mathcal{B}$ $(\times 10^{-5})$
$\rho^{0}\pi^{\pm}$	$1.5 \pm 0.5 \pm 0.4$
$\rho^{\pm}\pi^{\mp}$	$3.5^{+1.1}_{-1.0} \pm 0.5$
$K^{*\pm}\pi^{\mp}$	$2.2^{+0.8+0.4}_{-0.6-0.5}$

psuedo-reconstruction is used to obtain a sample of 65 events where the flavor of the strange quark is tagged through the kaon charge in the final state. The exact final state, which includes up to 4 pions, is not required to be identified correctly, hence the "pseudo' nature of the reconstruction technique. Once this sample is selected an observed asymmetry is calculated,

$$A_{meas} = \frac{N_{K^-} - N_{K^+}}{N_{K^-} + N_{K^+}} 0.13 \pm 0.11.$$

Misreconstruction is studied in Monte Carlo and a correction factor applied to get the true CP asymmetry,

$$A_{CP} = 0.16 \pm 0.14 \pm 0.05.$$

This result has been obtained with about one third of the available CLEO data so one can expect a significant improvement in the accuracy of this result in the near future.

References

1. M. Battle *et al.*, Phys. Rev. Lett. **71**, 3922 (1993).
2. M. S. Alam *et al.*, Phys. Rev. Lett. **74**, 2885 (1995).
3. M. T. Cheng *et al.* (BELLE Collaboration) KEK report 95-1 (1995); D. Boutigny *et al.* (BaBar Collaboration) SLAC Report SLAC-R-95-457 (1995); K. Lingel *et al.* (CESR-B Physics Working Group), Cornell Report CLNS 91-1043 (1991).
4. See for example S. Erhan in *The Proceedings of the Workshop on Beauty '93* (ed. P. E. Schlein), Melink, Czech Republic, Nucl. Instr. and Meth. A **333**, 213 (1993), and references therein.
5. M. Gronau, Phys. Rev. Lett. **63**, 1451 (1989).
6. M. Gronau and D. London, Phys. Rev. Lett. **65**, 3381 (1990).
7. M. Gronau, J. L. Rosner and D. London, Phys. Rev. Lett. **73**, 21 (1993).
8. M. Neubert and J. L. Rosner, Phys. Rev. Lett. bf 81, 5076 (1998). R. Fleischer and T. Mannel, Phys. Rev. D **57**, 2752 (1998).

The Sensitivity Frontier.

R. S. Tschirhart

Fermi National Accelerator Laboratory, Batavia,
Il 60510, USA
E-mail: tsch@fnal.gov

The status and prospects of rare kaon decay frontier is reviewed. Particular emphasis is given to the processes $K_L \to \pi^0 \nu \bar{\nu}$ and $K^+ \to \pi^+ \nu \bar{\nu}$, which hold the promise of quantitatively testing the structure of CP violation in the Standard Model.

1   Introduction.

The field of rare kaon decay physics was born some 35 years ago now with the discovery of the CP violating process $K_L \to \pi\pi$ at a branching fraction of $\sim 2 \times 10^{-3}$. Since then the evolution of the rare decay frontier has advanced by nine orders of magnitude culminating recently with the rarest particle decay every measured [1]: $B(K_L \to e^+ e^-) = (9 \pm 4) \times 10^{-12}$. This dramatic march in sensitivity has been fueled by ever increasing kaon beam intensity and detector performance. The rate capability [3] and precision [4] of the current round of detector systems are at or beyond state of the art.

Every step in increased sensitivity has yielded insights ranging from the GIM mechanism, manifest in the very low branching fraction [5] of $K_L \to \mu^+ \mu^-$, to the recent observation of a large T-odd asymmetry [6] in $K_L \to \pi^+ \pi^- e^+ e^-$ decays. The sensitivity frontier currently at the 10^{-12} level is defined by the search for the lepton flavor violating process [2] $K_L \to \mu^\pm e^\mp$. Searches for this and other [7] lepton flavor violating processes limit the mass of horizontal gauge bosons to more than about [5] $220 \, TeV/c^2$ assuming weak (G_F) coupling. This is a classic example of how the rare decay frontier translates into reach beyond the current energy frontier.

The rare decay frontier is also highlighted in the literature today [8] [9] [10] by the enticing possibility of measuring the ultra-rare decays $K^+ \to \pi^+ \nu \bar{\nu}$ and $K_L \to \pi^0 \nu \bar{\nu}$. These processes are expected to occur with branching fractions of $(8 \pm 3) \times 10^{-11}$ and $(3 \pm 1) \times 10^{-11}$, which depend directly on the CP violating parameters [11] ρ and η of the Standard Model. The ratio of these processes to the far more common Ke3 $(K \to \pi e \nu)$ process can be calculated with precision in the Standard Model. The high level of interest today in the $K \to \pi \nu \bar{\nu}$ processes is driven by their theoretical cleanliness. Together with measurements in the B-meson system of $\sin(2\beta)$, and possibly x_s/x_d from B_s mixing, these four measurements compose an unimpeachable suite of predictions with which to over constrain the (ρ, η) parameterization of the Standard Model.

Within the last year there has been a rich harvest of rare kaon decay results from experiments at Brookhaven, Fermilab and CERN. These results together with recent precision measurements of $Re(\epsilon'/\epsilon)$ (see reports in these proceedings) were the subject of the week-long Kaon99 conference in June of 1999. The proceedings [12] of that conference are an excellent reference for a detailed discussion of the the state of the rare decay field. This report will focus on the physics case and prospects for $K \to \pi\nu\bar{\nu}$ measurements.

2 Physics reach on the $\bar{\rho} - \bar{\eta}$ plane of $K \to \pi\nu\bar{\nu}$

The physics of the $K \to \pi\nu\bar{\nu}$ decay has been well and thoroughly discussed in several recent papers [8] [9] [10]. Figure 1, which is reproduced from reference 9, shows the dependence of the possible measurements on the apex coordinates $(\bar{\rho}, \bar{\eta})$ of the unitarity triangle, where $\bar{\rho}$ and $\bar{\eta}$ are the modified Wolfenstein parameters [11].

The dependence of the $K_L \to \pi^0\nu\bar{\nu}$ branching fraction on CKM parameters is shown below in equation 1. The branching ratio of $K_L \to \pi^0\nu\bar{\nu}$ is a direct measurement of the $\bar{\eta}$ parameter, and is essentially free of theoretical uncertainties. The function $X^2(x_t)$ is a well determined slowly varying function of the top quark mass.

$$Br[K_L \to \pi^0\nu\bar{\nu}\,] = 1.80 \times 10^{-10}[A^4 X^2(x_t)\bar{\eta}^2] \tag{1}$$

A measurement of the $K_L \to \pi^0\nu\bar{\nu}$ branching fraction with a precision of $\pm 10\%$ directly corresponds to a measurement of $\bar{\eta}$ with $\pm 5\%$ uncertainty. The connection of the $K^+ \to \pi^+\nu\bar{\nu}$ branching fraction to CKM parameters is less direct, but the attendant theoretical uncertainties are under control at the $\pm 5\%$ level.

The branching ratio of $K^+ \to \pi^+\nu\bar{\nu}$ is proportional to the area of the labeled circle in figure 1. This circle is described with central parameter values by:

$$Br[K^+ \to \pi^+\nu\bar{\nu}\,] = [0.44 \pm 0.1] \times 10^{-10}[(1.4 - \bar{\rho})^2 + \bar{\eta}^2] \tag{2}$$

The shift of this circle's origin by $+0.4$ from point B is the due to the contribution of the charmed quark in the matrix element.

The precision with which the magnitude of V_{td} can be determined from $K^+ \to \pi^+\nu\bar{\nu}$ is one half the fractional uncertainty in the branching ratio measurement combined in quadrature with the uncertainties in the other parameters including V_{cb} and the mass of the charmed quark. Buras and col-

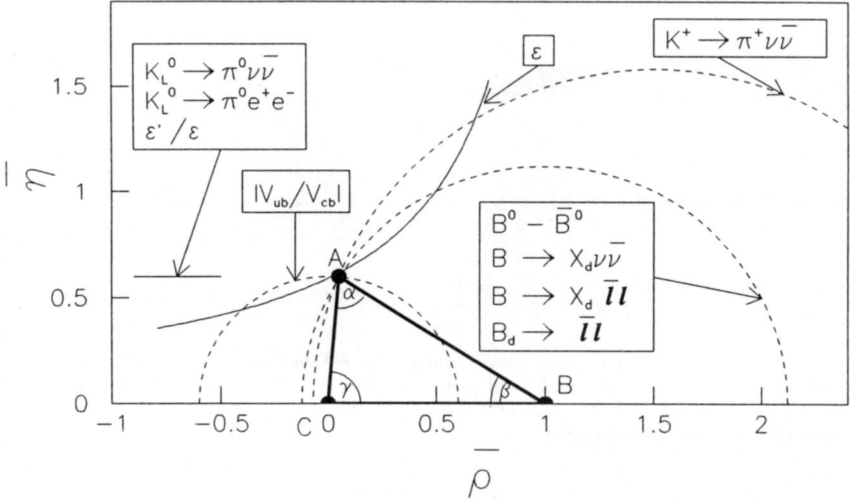

Figure 1: The ideal Unitarity Triangle. For illustrative reasons the value of $\bar{\eta}$ has been chosen to be higher than the fitted central value (reproduced from reference 9).

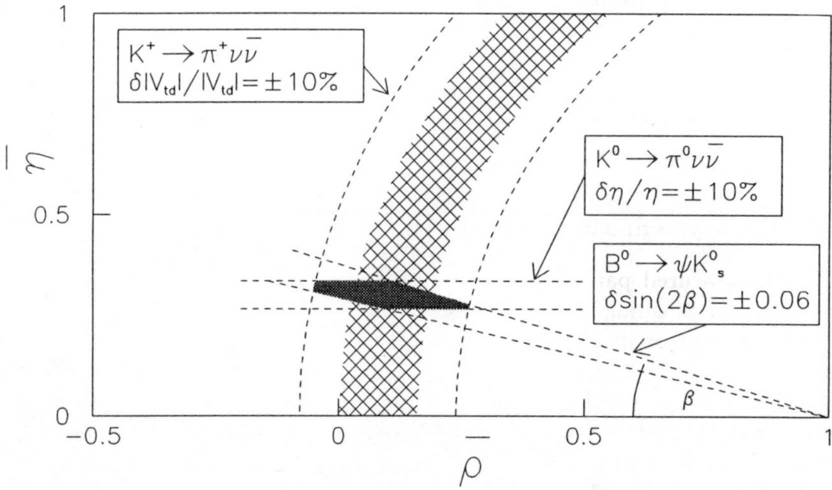

Figure 2: A possible outcome from measurements of $K^+ \to \pi^+ \nu \bar{\nu}$, $K_L \to \pi^0 \nu \bar{\nu}$ and $B^0 \to \psi K^0_s$. The hashed annulus and band in $\bar{\eta}$ indicate the statistical precision of a 100 event $K \to \pi \nu \bar{\nu}$ experiments. In the case of $K_L \to \pi^0 \nu \bar{\nu}$, the total band indicates the combined statistical and theoretical errors.

leagues[8][9][10] has projected these uncertainties and concludes that a 10% measurement of the $K^+ \to \pi^+ \nu \bar{\nu}$ branching ratio will yield a 10% determination of V_{td} . This error has a somewhat larger contribution from the uncertainties in the other parameters than from the statistics of the branching ratio measurement.

The measurement of the branching ratio of $K^+ \to \pi^+ \nu \bar{\nu}$ to the 10% precision level (100 events observed) will yield essentially all the available information which the $K^+ \to \pi^+ \nu \bar{\nu}$ decay mode can provide in determining the parameters of the CKM matrix.

Figure 2 shows a possible outcome on the $\bar{\rho} - \bar{\eta}$ plane of pre-LHCB meson and kaon experiments. For the case of $K^+ \to \pi^+ \nu \bar{\nu}$ the 5% statistical error is required to achieve a total error of 10% including theoretical uncertainties as estimated [8] [9] [10] by Buras are shown. Ten percent measurements of CKM parameters in the two rare kaon decays match well the 10% determination of $\sin(2\beta)$ anticipated from the present generation of upcoming B meson experiments at the B-factories and colliders. Note that $K^+ \to \pi^+ \nu \bar{\nu}$ and $\sin(2\beta)$ are essentially orthogonal measurements on the $\bar{\rho} - \bar{\eta}$ plane for the solution assumed [0.04, 0.32] in the figure.

3 Related B-system Measurements and Theory

With two unmeasured parameters in the CKM matrix, parameterized here as $\bar{\rho}$ and $\bar{\eta}$, any meaningful test of the source of CP violation requires at least three different quantities to be independently measured. The spectrum of possible measurements include the ultra-rare charged and neutral decays, $K^+ \to \pi^+ \nu \bar{\nu}$ and $K_L \to \pi^0 \nu \bar{\nu}$ in the kaon system, and several observables in the B meson system including mixing, hadronic, semi-leptonic, and leptonic decay modes. Each of these measurements is difficult; more than a few may turn out to be impossible.

The goal of this whole experimental program is a critical test of the source of CP violation in nature. This requires that the measurements must have sufficient integrity to compel us to abandon the Standard Model as the sole source of CP violation should that be the outcome. Each measurement has both experimental and theoretical liabilities. Experimental problems can be overcome with a different or improved experimental technique to yield a useful measurement. Theoretical limitations in the relationship between the experimental observables and the physics parameters of interest compromise whole classes of measurements until those limitations can be overcome.

The ultra-rare modes $K^+ \to \pi^+ \nu \bar{\nu}$ and $K_L \to \pi^0 \nu \bar{\nu}$ are identified by several authors as modes where the limitations of theory are contained and quantified. In the case of $K_L \to \pi^0 \nu \bar{\nu}$ the theoretical uncertainties are negligible, and in $K^+ \to \pi^+ \nu \bar{\nu}$ the theoretical uncertainty of the charm quark contribution is small and under control. The great attractiveness of these kaon decays is that, while the measurements are extremely challenging experimentally, there are certainly meaningful physics results when the branching ratios are reliably measured. As illustrated in figure 2; the two measurements in the kaon system can, by themselves, cleanly determine the two unmeasured CKM parameters. They cannot test the Standard Model's contribution to CP violation without a third, independent, measurement.

In the B meson system this story is much longer and more complicated. The same authors have studied the theoretical limitations of the B sector measurements [8] [9] [10]. A summary of their conclusions is as follows:

1. The present uncertainty the magnitude of V_{td} measured with the B_d mixing parameter x_d is already dominated by the uncertainties in the lattice calculations of F_{B_d} and B_{B_d}. This is a theoretical limitation common to most of the hadronic B decays.

2. The B_s/B_d ratios of these observables is significantly less sensitive to lattice calculation uncertainties. These require matching measurements from hadronic B production experiments with good enough tagging to distinguish B_s from B_d events to complement the measurements in the B_d decay expected from the B factories.

3. Among the conventional measurements of the unitarity triangle angles α, β and γ expected from the B factories only β angle as measured with $B^0 \to \psi K_s^0$ is free from significant theoretical uncertainties.

4. α measured with $B^0 \to \pi^+\pi^-$ suffers from significant "penguin pollution" which must be overcome before this measurement can be useful for CKM parameter determinations. The isospin analysis required to control this problem requires measurements of modes which are probably too low in branching ratio to be well measured with the available luminosity.

5. The B sector analogies of the $K \to \pi\nu\bar{\nu}$ decays; $B^0 \to X_d\nu\bar{\nu}$, $B^0 \to X_s\nu\bar{\nu}$, $B_d^0 \to \mu^+\mu^-$ and $B_s^0 \to \mu^+\mu^-$ are likewise theoretically clean. The Standard Model branching ratios for these modes are believed to be beyond the sensitivity of all proposed experiments.

6. x_s/x_d may make a useful contribution to constraining the apex of the unitarity triangle provided a measurement and not just a lower limit can be achieved.

The B sector experiments have more than ample experimental insurance; two experiments are now being commissioned (Babar, Belle), and four experiments are under construction: Babar, Belle, CDF, CLEO-III, Dzero, and HERA-B. In addition, there is a vigorous ongoing R&D effort for the BTeV experiment at Fermilab. The liabilities of these experiments appear to be mainly theoretical; only the measurement of $\sin(2\beta)$ is known to be unambiguous.

4 Status and prospects for precision measurements of $K \to \pi\nu\bar{\nu}$

In the kaon sector the claim of our theoretical colleagues is that both ultra-rare modes are well understood. The challenge in the kaon measurements is all experimental; and formidable. Presently there is the BNL $K^+ \to \pi^+\nu\bar{\nu}$ experiment

E787 which has published observation of one clean event [13] which corresponds to a branching fraction of $(4^{+9.7}_{-3.5}) \times 10^{-10}$. The theoretical expectation [8] for this process is $(8 \pm 3) \times 10^{-11}$. A central value of 4×10^{-10} is inconsistent with this prediction, and would be a signal of physics outside the Standard Model. Recently at the Kaon99 conference the BNL E787 collaboration reported on a preliminary analysis of about $\times 2.5$ the published data where no new events were found. Although the final result based on this data set is still in preparation, it is safe to say that the branching fraction including this new data will be more in line with the Standard Model expectation.

A series of direct searches for $K_L \to \pi^0 \nu \bar{\nu}$ decay have been carried out by the Fermilab E799 and KTeV collaborations, culminating in an upper limit of $5.9 \times 10^{-7}, (90\% \ C.L.)$ [14] which has been submitted for publication. The improvement in sensitivity has been impressive, but still far from the expected branching fraction of $(3 \pm 1) \times 10^{-11}$. The best limit to date required that the π^0 undergo Daltiz decay $(\pi^0 \to e^+ e^- \gamma)$ to better constrain the kinematics and control backgrounds. Unfortunately this requirement reduces the sensitivity of the search by $\times \ 70$ at the outset. The KTeV collaboration has also reported [15] a limit on $K_L \to \pi^0 \nu \bar{\nu}$ of $1.6 \times 10^{-6}, (90\% \ C.L.)$ based on a brief test run using the dominant $\pi^0 \to \gamma\gamma$ channel. This test run identified the leading backgrounds, and provides crucial information for the design of future experiments aiming for the expected signal range.

Two proposals to actually measure $K_L \to \pi^0 \nu \bar{\nu}$ at BNL [16] and KEK [17] have been scientifically approved by the respective laboratories, and these collaborations are now in the technical design phase. There are two additional proposals at Fermilab; a proposal for a $K_L \to \pi^0 \nu \bar{\nu}$ experiment (KAMI [18], Kaons At the Main Injector), and a proposal for a $K^+ \to \pi^+ \nu \bar{\nu}$ experiment (CKM [19], Charged Kaons at the Main injector). Both experiments are motivated by the very high flux of protons available from Fermilab Main Injector. The KAMI experiment proposes to evolve from the existing KTeV apparatus and beamline, whereas the CKM proposal is a new experiment and beamline. The Fermilab proposals have received strong encouragement from the laboratory management to develop the initial proposals into full technical reports to be reviewed in the spring of 2001.

In parallel with these design efforts for future $K \to \pi \nu \bar{\nu}$ experiments, The KTeV rare kaon decay program continues to run and expects to triple the sensitivity of the 1997 run, and the Brookhaven E787 experiment has been approved for two more years of running as experiment [20] E949. The E949 collaboration is in the process now of upgrading their detector and expects to commence running in 2001. These upgrades and the additional running time will push the charged-K frontier to the 1×10^{-11} single event sensitivity, which will provide of sample of ~ 8 $K^+ \to \pi^+ \nu\bar{\nu}$ events at the Standard Model ranching fraction.

5 Summary

Figure 2 displays a test of the source of CP violation in nature using only measurements known to be free of theoretical problems. If these three measurements can be successfully accomplished and the three regions of the $\bar{\rho}$ - $\bar{\eta}$ plane do not mutually overlap consistent with one point, the hypothesis that the Standard Model is the sole source of CP violation is untenable. The only alternatives will be to attack the veracity of the measurements or to invoke new physics. Other measurements in both the B and kaon sectors can, and hopefully will, make comparable contributions. These will be extremely valuable in confirming the Standard Model CP hypothesis. Any of those measurements which can be called into question over issues of the reliability of lattice calculations, penguin contributions or other theoretical weaknesses will not have the integrity to falsify the hypothesis of the Standard Model as the sole source of CP violation.

In summary, there is compelling case now for precision measurements (100 event experiments) of the $K \to \pi\nu\bar{\nu}$ decay modes. These two measurements together with $\sin(2\beta)$ and possibly x_s/x_d form a set of unimpeachable measurements with which to overconstrain and test the Standard Model source of CP violation. Vigorous efforts are underway now to design, and hopefully build experiments that will reach the goal of precision measurements of $K \to \pi\nu\bar{\nu}$ decay.

References

1. D. Ambrose, et.al., Phys.Rev.Lett. **81**, 4309-4312 (1998).
2. "A new limit on the Lepton Number Violating process $K_L \to \mu^\pm e^\mp$": D. Ambrose, et.al., Phys.Rev.Lett. **81**, 5734-5737 (1998). A review of lepton number violating processes and other rare kaon decays can be found in the 1999 Lepton-Photon proceedings of Bill Molzon.
3. The BNL E871 experiment at Brookhaven is an example of a very high rate kaon decay experiment. The flux taken by this experiment allows measurement of the rarest hadron decay observed to date: $K_L \to e^+ e^-$.
4. The KTeV experiment at Fermilab is an example of a moderate rate high precision kaon decay experiment. The experiment features a pure CsI calorimeter, that is the highest resolution calorimeter today in High Energy Physics. The calorimeter enables the efficient and precise reconstruction of kaon decays with multiple photons.

5. <u>Review of Particle Properties</u>, C. Caso et. al., European Physical Journal C bf 3 1, (1998).

6. "Observation of CP violation in $KL \to \pi^+\pi^- e^+ e^-$". Submitted to Physical Review Letters, e-Print Archive: **hep-ex/9908020**.

7. BNL experiment E865 searches for the Lepton Flavor Violating process: $K^+ \to \pi^+ \mu^+ e^-$.

8. "The rare decays $K \to \pi\nu\bar\nu$ $B \to X\nu\bar\nu$ and $B \to l\bar{l}$, an update." A. Buras et. al., Nucl.Phys. **B548**, 309-327 (1999). e-Print Archive: **hep-ph/9901288**

9. "CKM MATRIX: PRESENT AND FUTURE." A. Buras et. al., e-Print Archive: **hep-ph/9711217**

10. "$K \to \pi\nu\bar\nu$ AND HIGH PRECISION DETERMINATIONS OF THE CKM MATRIX": A. Buras et. al., Phys.Rev. **D54** 6782-6789, (1996) e-Print Archive: **hep-ph/9607447**

11. $\bar\rho = \rho[1 - \lambda^2/2]$, $\bar\eta = \eta[1 - \lambda^2/2]$ with ρ and η the usual Wolfenstein parameters. L. Wolfenstein, Phys. Rev. Lett. **51**, 1945 (1983).

12. Proceedings of the Chicago Conference on Kaon Physics, the University of Chicago, June 21-26, 1999, to be published by University of Chicago Press, J. Rosner and B. Winstein editors.

13. S. Adler, et.al., Phys. Rev. Lett. **79**, 2204 (1997).

14. "Search for the Decay $K_L \to \pi^0\nu\bar\nu$ using $\pi^0 \to e^+e^-\gamma$". Submitted to Phys. Rev. Lett., July 8th 1999.

15. "Search for the Decay $K_L \to \pi^0\nu\bar\nu$ ". J. Adams, et.al., Phys. Lett. **447**, 240 (1999).

16. Details of the BNL E926 experiment can be found on the following website: http://sitka.triumf.ca/e926/

17. T. Inagaki *et al.*, "Measurement of the $K_L \to \pi^0\nu\bar\nu$ decay," KEK proposal, (June 1996).

18. Details of the KAMI experiment can be found on the following website: http://ppd.fnal.gov/experiments/ktev/kami/kami.html

19. Details of the CKM experiment can be found on the following website: http://www.fnal.gov/projects/ckm/Welcome.html

20. Details of the E949 experiment can be found on the following website: http://www.phy.bnl.gov/e949/

STUDIES OF THE CP, T AND CPT SYMMETRIES IN THE NEUTRAL KAON SYSTEM

R. LE GAC

Centre de Physique des Particules de Marseille
CNRS/IN2P3–Université de la Méditerranée
163 avenue de Luminy, Case 907,
F–13288 MARSEILLE, France
E-mail: legac@in2p3.fr

In this contribution, we review the experimental results on the studies of the CP, T and CPT symmetries in the neutral kaon system.

1 Introduction

The discrete symmetries C, P and T and their combinations CP and CPT have been thoroughly studied in the neutral kaon system. The CP violation discovered in the $K_L \to \pi^+\pi^-$ decays by J.H. Christenson, J.W. Cronin, V. Fitch and R. Turlay in 1964 [1], is now well established. But the origin of the CP violation is still not understood. The superweak model reproduces CP violation introducing a new interaction [2]. The Standard Model can accommodates CP violation via a phase in the Cabbibo–Kobayashi–Maskawa matrix describing the quark mixing [3]. The Standard Model also predicts a CP violation in the decay amplitudes (direct CP violation).

The experimental situation was far from being complete when the latest generation of experiments started. Expected CP violation in the K_S, and direct observation of the time–reversal violation were not reported. There was no clear evidence of direct CP violation.

Another strong motivation for studying these symmetries is related to the CPT theorem. Demonstrated in the middle of the 50's [4], it tells us that local field theories, Lorentz invariant, obeying to the Bose–Einstein and Fermi–Dirac spin–statistics are invariant under the CPT combined operator. As a consequence mass and lifetime, are equal for a particle and its antiparticle. A failure in these equalities would reveal new physics. Violation of the CPT symmetry is expected at the Planck scale when quantum gravitational effects are taken into account [5]. Other sources of CPT violation appear in string theory describing extended objects which violate assumptions at the origin of the CPT theorem [6].

Therefore, the measurement of *Direct CP violation*, *Time reversal violation* and the direct test of *CPT invariance*, are desirable to make progress. At present, the neutral kaon system is still the most promising place to study these symmetries.

In this contribution, we give an outlook of the neutral kaon phenomenology in the general framework where the CP, T and CPT symmetries can be violated. We present experimental methods employed to study these symmetries and we review the results.

2 Outlook of the neutral kaon phenomenology

A detailed description of the neutral kaon phenomenology can be found in many textbooks [7]. Here, we only define the phenomenological parameters describing a possible violation of CP, T and CPT and we relate them to experimental quantities. We follow the time evolution of a neutral kaon. We divide it in two steps: the mixing and the decay.

2.1 The mixing

The neutral kaon particle is peculiar because it is the lightest particle containing a strange quark. Its creation is governed by the strong interaction but it cannot decay via strong processes since strong interaction conserves strangeness. The decay can only involve weak processes, violating strangeness conservation. As a consequence, an initial K^0 (resp. $\overline{K}^0$) evolves into a superposition of K^0 and $\overline{K}^0$. The time evolution of the system is described by the Wigner–Weisskopf formalism [8] using an effective Hamiltonian; $\Lambda = M - (i/2)\Gamma$. Four states are introduced:

1. The K^0 and $\overline{K}^0$ states are the eigenstates of the strong interaction. They have a well defined strangeness. Their mass and decay width are equal.

2. The K_S and K_L states are the eigenstates of the effective Hamiltonian with the eigenvalues $\lambda_{S,L} = m_{S,L} - (i/2)\Gamma_{S,L}$:

$$|K_S\rangle = p_S|K^0\rangle + q_S|\overline{K}^0\rangle$$

$$|K_L\rangle = p_L|K^0\rangle + q_L|\overline{K}^0\rangle$$

They have a well defined mass, m_S and m_L, and decay width, Γ_S and Γ_L, but the strangeness is undefined.

The nice behavior of the neutral kaon system appears when the probability to be in the K^0 state at a time τ is expressed in the $(K^0, \overline{K}^0)$ basis:

$$|K^0_{t=\tau}\rangle = \left\{ \left[\frac{p_S}{q_S} e^{-i\lambda_S\tau} - \frac{p_L}{q_L} e^{-i\lambda_L\tau} \right] |K^0_{t=0}\rangle + \left[e^{-i\lambda_S\tau} - e^{-i\lambda_L\tau} \right] |\overline{K}^0_{t=0}\rangle \right\}$$

It is governed by the $\Delta S = 0$ and $\Delta S = 2$ transitions. In the $\Delta S = 0$ transition, the strangeness of the initial neutral kaon is preserved while in the $\Delta S = 2$ transition, the initial $\overline{K}^0$ state oscillates into the K^0 state at a time τ. These two amplitudes are sensitive to the T and CPT symmetries. If the probability of oscillation, $P(K^0_{t=0} \to \overline{K}^0_{t=\tau})$ is different from $P(\overline{K}^0_{t=0} \to K^0_{t=\tau})$, the time–reversal symmetry is violated. T violation in the mixing is described by the complex parameter

$$\varepsilon = \frac{\Lambda_{\overline{K}^0, K^0} - \Lambda_{K^0, \overline{K}^0}}{2(\lambda_S - \lambda_L)},$$

where $\Lambda_{i,j}$ are the off-diagonal elements of the effective Hamiltonian. In the same way, if the $\Delta S = 0$ amplitudes are different between initial K^0 and $\overline{K}^0$ state, the CPT symmetry is violated in the mixing. It is parametrized by

$$\delta = \frac{\Lambda_{\overline{K}^0, \overline{K}^0} - \Lambda_{K^0, K^0}}{2(\lambda_S - \lambda_L)},$$

where δ is complex.

At this stage, it is interesting to express the K_S and K_L states in the K_1 and K_2 basis where the K_1 and K_2 are the CP eigenstates with the CP eigenvalues $+1$ and -1:

$$K_S = K_1 + \varepsilon_S K_2$$
$$K_L = K_2 + \varepsilon_L K_1$$

where $\varepsilon_S = \varepsilon + \delta$ and $\varepsilon_L = \varepsilon - \delta$. The latter expression shows that the K_S and the K_L are not anymore equal to the CP eigenstate K_1 and K_2. The K_L (resp. K_S) contains a small admixture of K_1 (resp. K_2). This contamination is at the origin of the CP violation measured in K_L decays into the CP eigenstate $\pi^+\pi^-$ (CP=+1).

2.2 *The decay*

At the end of the mixing phase a neutral kaon decays either into an hadronic or into a semileptonic final state. The decay amplitudes are sensitive to CP and CPT symmetries. The decay amplitudes obey to the following equations when the CP and CPT symmetries hold[1]:

CP $\quad \langle 2\pi|H_W|K^0\rangle = \pm\langle 2\pi|H_W|\overline{K}^0\rangle \quad\quad \langle \pi^- l^+ \nu_l|H_W|K^0\rangle = \pm\langle \pi^+ l^- \overline{\nu}_l|H_W|\overline{K}^0\rangle$

CPT $\quad \langle 2\pi|H_W|K^0\rangle = \pm\langle 2\pi|H_W|\overline{K}^0\rangle^* \quad\quad \langle \pi^- l^+ \nu_l|H_W|K^0\rangle = \pm\langle \pi^+ l^- \overline{\nu}_l|H_W|\overline{K}^0\rangle^*$

These equalities are not satisfied anymore when CP (resp. CPT) symmetry is violated in the decay amplitudes. To allow such effects, we use the parameterization of

---

1. Phase convention: $CP|K^0\rangle = |\overline{K}^0\rangle$, $CP|\overline{K}^0\rangle = |K^0\rangle$ and $CPT|K^0\rangle = |\overline{K}^0\rangle$, $CPT|\overline{K}^0\rangle = |K^0\rangle$.

the decay amplitudes given in Table 1 [9]. For the 2π final state, the index I is related to the isopin of the $\pi\pi$ state and δ_I is a strong phase shift due to final state interaction. $Re(A_I)$, $Re(a)$, $Re(c)$ are CP, T and CPT conserving. $Im(A_I)$, $Im(a)$, $Im(c)$ violate CP while the B, b and d terms violate CPT. In addition, the c and d terms violate the $\Delta S=\Delta Q$ rule.

Table 1: Parameterization of the neutral decay amplitudes allowing a violation of CP and CPT [9].

	K^0	$\overline{K}^0$
$\lvert\pi\pi, I\rangle$	$(A_I + B_I)e^{i\delta_I}$	$(A^*_I - B^*_I)e^{i\delta_I}$
$\lvert\pi^+ l^- \nu_l\rangle$	$a + b$	$c^* - d^*\ (\Delta S \neq \Delta Q)$
$\lvert\pi^- l^+ \nu_l\rangle$	$c + d\ (\Delta S \neq \Delta Q)$	$a^* - b^*$

Four parameters are introduced in the phenomenology to take into account a possible violation of CP and CPT in the decay amplitudes. The first one, ε', is well known. It describes the CP violation in the K^0, $\overline{K}^0 \to 2\pi$ decay amplitudes, namely the direct CP violation [9, 10]:

$$\varepsilon' = \frac{1}{\sqrt{2}}\frac{ReA_2}{ReA_0}\left\{\left(\frac{ImA_2}{ReA_2} - \frac{ImA_0}{ReA_0}\right) - i\left(\frac{ReB_2}{ReA_2} - \frac{ReB_0}{ReA_0}\right)\right\}\exp\left\{i\left(\delta_2 - \delta_0 + \frac{\pi}{2}\right)\right\}.$$

It is also sensitive to CPT violation in the decay amplitudes, in this general framework. The tree other parameters express the CPT violation in the decay amplitudes: ReB_I/ReA_I where $I = 0, 2$, describe the CPT violation in neutral kaon decays into 2π while $Re(y) = -Re(b)/Re(a)$ describes the CPT violation in the semileptonic decay amplitudes, when the $\Delta S=\Delta Q$ rule holds.

In addition, two parameters are introduced to take into account a possible violation of the $\Delta S=\Delta Q$ rule:

$$x = \frac{\langle\pi^- e^+ \nu | H_W | \overline{K}^0\rangle}{\langle\pi^- e^+ \nu | H_W | K^0\rangle} = \frac{c^* - d^*}{a + b}$$

$$\bar{x} = \frac{\langle\pi^+ e^- \bar{\nu} | H_W | K^0\rangle}{\langle\pi^+ e^- \bar{\nu} | H_W | \overline{K}^0\rangle} = \frac{c^* + d^*}{a - b}$$

They are equal to the ratio between the amplitude violating the $\Delta S=\Delta Q$ rule and the amplitude satisfying the $\Delta S=\Delta Q$ rule. They can be combined to form:

$$x_+ = \frac{(x + \bar{x})}{2} = \frac{c^*}{a}$$

$$x_- = \frac{(x - \bar{x})}{2} = \left(\frac{d^*}{b}\right)y$$

where x_+ and x_- describe a violation of the $\Delta S = \Delta Q$ rule in CPT-conserving and CPT-violating amplitudes respectively.

2.3 The observables

Several experimental quantities have been defined to directly measure the violation of CP, T and CPT symmetries in the hadronic and semileptonic final states. Here, we remind the definition of the main observables and we give their expressions as a function of the phenomenological parameters.

η_{+-} and η_{00} measure CP violation effects in the 2π final states [10]:

$$\eta_{+-} = \frac{A(K_L \to \pi^+\pi^-)}{A(K_S \to \pi^+\pi^-)} = |\eta_{+-}|e^{i\varphi_{+-}} = \left[\varepsilon + i\frac{\mathrm{Im}A_0}{\mathrm{Re}A_0}\right] + \left[\frac{\mathrm{Re}B_0}{\mathrm{Re}A_0} - \delta\right] + \varepsilon'$$

$$\eta_{00} = \frac{A(K_L \to \pi^0\pi^0)}{A(K_S \to \pi^0\pi^0)} = |\eta_{00}|e^{i\varphi_{00}} = \left[\varepsilon + i\frac{\mathrm{Im}A_0}{\mathrm{Re}A_0}\right] + \left[\frac{\mathrm{Re}B_0}{\mathrm{Re}A_0} - \delta\right] - 2\varepsilon'$$

They are equal to the ratio between the CP violating K_L decay amplitude and the CP conserving K_S decay amplitude, in the $\pi^+\pi^-$ and $\pi^0\pi^0$ final states. They are sensitive to three effects. The first one violates T in the mixing and CP in the decay amplitudes. The second one violates CPT in the mixing and in the decay amplitudes. The third component violates both CP and CPT in the decay amplitudes.

The charge asymmetry in the semileptonic final state,

$$\delta_l = \frac{R(K_L \to \pi^- l^+ \nu_l) - R(K_L \to \pi^+ l^- \nu_l)}{R(K_L \to \pi^- l^+ \nu_l) + R(K_L \to \pi^+ l^- \nu_l)} = 2\mathrm{Re}(\varepsilon - \delta) - 2\mathrm{Re}(y) - 2\mathrm{Re}(x_-),$$

is sensitive to a possible violation of T and CPT in the mixing, a possible violation of CPT in the decay amplitudes as well as a possible violation of the $\Delta S = \Delta Q$ rule.

The Kabir asymmetry [11] measures directly the time–reversal non-invariance:

$$A_T = \frac{R(\overline{K}^0_{t=0} \to K^0_{t=\tau}) - R(K^0_{t=0} \to \overline{K}^0_{t=\tau})}{R(\overline{K}^0_{t=0} \to K^0_{t=\tau}) + R(K^0_{t=0} \to \overline{K}^0_{t=\tau})},$$

since it compares the probability of oscillation between K^0 and $\overline{K}^0$ states. The time dependent decay rate asymmetry depends on the time–reversal violation but also on a possible violation of CPT in the decay amplitudes and on a possible violation of the

$\Delta S = \Delta Q$ rule when the strangeness of the neutral kaon at the decay time is determined through semileptonic final states [12]:

$$A_T = \frac{R(\overline{K}^0_{t=0} \to \pi^- e^+ \nu_e)(\tau) - R(K^0_{t=0} \to \pi^+ e^- \overline{\nu}_e)(\tau)}{R(\overline{K}^0_{t=0} \to \pi^- e^+ \nu_e)(\tau) + R(K^0_{t=0} \to \pi^+ e^- \overline{\nu}_e)(\tau)} \to 4\mathrm{Re}(\varepsilon - y - x_-),\ \tau \gg \tau_S$$

Other time dependent decay rate asymmetries, $A_{\Delta m}$ [13], A_δ [14] are built to measure Δm, the difference of mass between the K_L and K_S, and to directly test the CPT invariance.

3 The experimental methods

The methods used to study CP, T and CPT symmetries in the neutral kaon system evolve with time. All of them are employed in the latest generation of experiments: CPLEAR and NA48 at CERN, KLOE at Frascati and KTeV at Fermilab.

3.1 KTeV and NA48

The method exploited by KTeV and NA48 is very similar to the technic of the pioneer experiment [1]. They use simultaneous K_S and K_L beams and detect their decay products. Their detectors are shown on Figure 1, but they are described in detail elsewhere in these proceedings [15].

The method is based on the measurement of the K_S and K_L decay rates in the $\pi^+\pi^-$ and $\pi^0\pi^0$ final states. The ratio of decay rates $R(K_L \to \pi^+\pi^-)/R(K_S \to \pi^+\pi^-)$ determines $|\eta_{+-}|^2$ while the double ratio between the $\pi^+\pi^-$ and $\pi^0\pi^0$ final states is sensitive to the direct CP violation:

$$\mathrm{Re}\left(\frac{\varepsilon'}{\varepsilon}\right) \approx \frac{1}{6}\left\{\frac{R(K_L \to \pi^+\pi^-)/R(K_S \to \pi^+\pi^-)}{R(K_L \to \pi^0\pi^0)/R(K_S \to \pi^0\pi^0)} - 1\right\}.$$

They also study rare decays of the $K_L \to \pi^+\pi^- e^+ e^-$, $\pi^0 e^+ e^-$, $\pi^0 \mu^+ \mu^-$, ... to constraint the Cabbibo–Kobayashi–Maskawa matrix, describing the quark mixing in the Standard Model. In addition, KTeV determines the mass difference between K_L and K_S states, the lifetime of the K_S and the phase difference $\varphi_{+-} - \varphi_{00}$ taking benefit of its regenerator.

Currently, KTeV and NA48 take data. They will reach an accuracy on $\mathrm{Re}(\varepsilon'/\varepsilon)$ at a level of a few 10^{-4}.

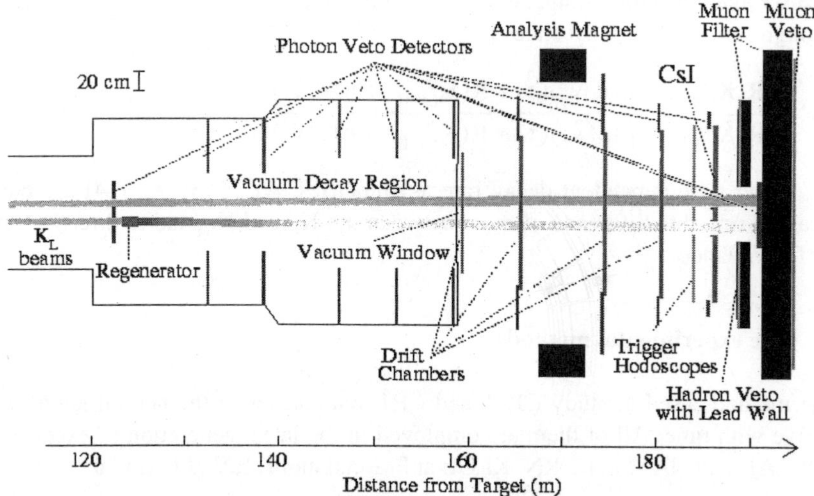

Figure 1: The KTeV experiment at Fermilab

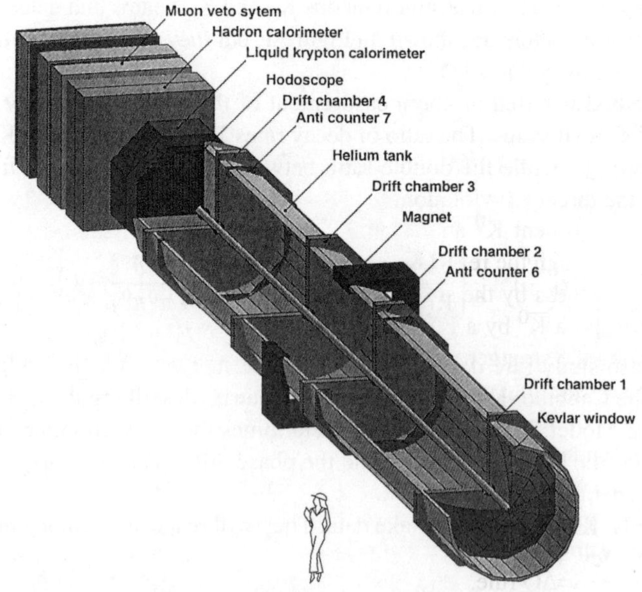

Figure 2: The NA48 experiment at CERN

3.2   CPLEAR

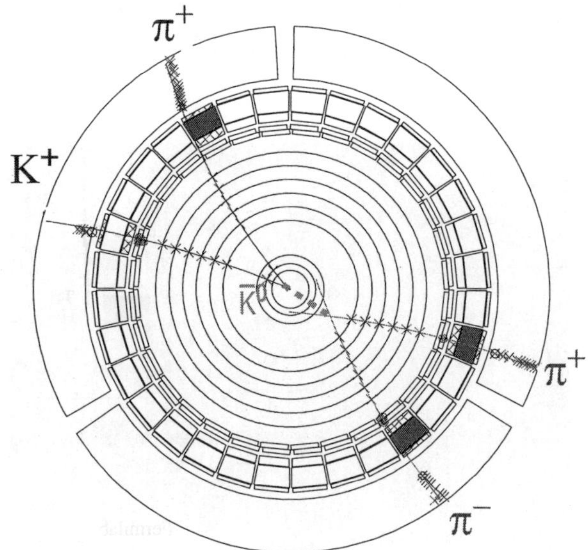

Figure 3: A CPLEAR event where a $\overline{K}^0$ decays into $\pi^+\pi^-$

The CPLEAR collaboration developed a new experimental approach to measure CP violating effects from interference between K_S and K_L amplitudes with reduced systematic errors [16]. A CPLEAR event is shown on Figure 3 and the detector is described in [17].

In that experiment K^0 and $\overline{K}^0$ states are produced in about 4×10^{-3} of the $\bar{p}p$ annihilations through the reactions $\bar{p}p \rightarrow K^-\pi^+K^0$ and $\bar{p}p \rightarrow K^+\pi^-\overline{K}^0$. The conservation of strangeness by the strong interaction implies that a K^0 is always accompanied by a K^- and a $\overline{K}^0$ by a K^+. Thus, the identification of the charged kaon and its charge sign tag the strangeness of the neutral kaon at the production time. When the neutral kaon decays into a semileptonic final state, the sign of the lepton charge tags the strangeness of the neutral kaon at the decay time.

Time dependent decay rate asymmetries between initial K^0 and $\overline{K}^0$ state are built using these tagging procedures. This method allows to measure CP violation effects in the 2π and 3π final states, to directly test T and CPT invariance in the semileptonic decay, to measure the mass difference between K_L and K_S independently of φ_{+-}, and to test the $\Delta S=\Delta Q$ rule.

The systematic errors are small since the detector efficiencies, the geometrical acceptance and the background contamination are much the same for K^0 and $\overline{K}^0$, thus canceling in the measured asymmetries.

Data taking was completed at the end of 1996. Most of the final results were published between 1997 and 1999 improving the current knowledge on the CP, T and CPT symmetries in the neutral kaon system.

3.3 KLOE

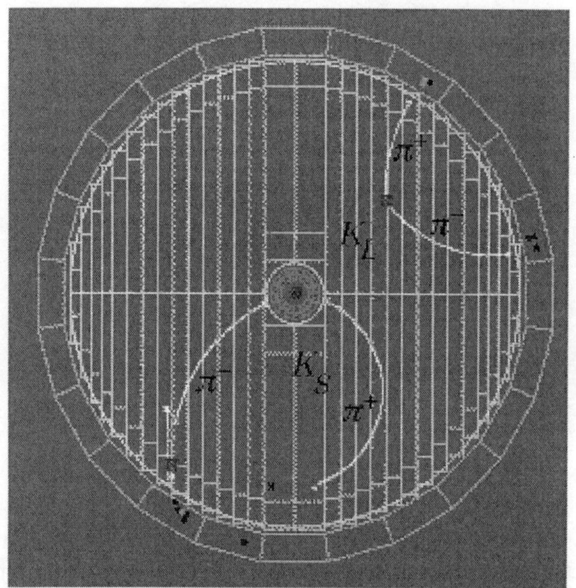

Figure 4: The first CP violating event recorded by KLOE in April 99

The KLOE detector is installed on the ϕ factory DAΦNE at Frascati. They use the ϕ decays into the coherent $K_S K_L$ final state and reconstruct the decay products of the neutral kaons. They measure time dependent decay rate asymmetries between $K_S \to f_1$ and $K_L \to f_2$. By choosing the final states f_1 and f_2 adequately, they measure CP violation effects in the 2π and 3π final states ($f_1 = 2\pi, 3\pi$ and $f_2 = \pi l \nu$), direct CP violation in the 2π final states ($f_1 = \pi^+\pi^-$ and $f_2 = \pi^0\pi^0$), test the T and CPT invariance [18].

Data taking started in Spring 1999. Figure 4 shows the first CP violating event recorded in April 99 where a K_S and a K_L decay into $\pi^+\pi^-$.

4 Overview of the results

The results on the study of the CP, T and CPT symmetries are summarized in the following subsections.

The recent results on the direct CP violation, reported by the KTeV ($\mathrm{Re}(\varepsilon'/\varepsilon) = (28\pm4.1)\times10^{-4}$) and by the NA48 collaboration (($18.5\pm7.3)\times10^{-4}$) confirm the earlier CERN results. For a detailed description, we refer the reader to reference [15]. Direct CP violation is now observed without doubt.

4.1 CP violation

The most accurate determination of the CP violation in the $\pi^+\pi^-$ final state was obtained by CPLEAR. They reconstructed 70×10^6 of $K^0,\overline{K}^0\to\pi^+\pi^-$ with a decay time above 1 τ_S. The background contamination is very small except at large decay time. It comes mainly from the neutral kaons decay into semileptonic final states.

Corrections were applied to the data to take into account the difference in tagging efficiencies between K^0 and $\overline{K}^0$. They also corrected regeneration effects appearing when a neutral kaon crosses the detector.

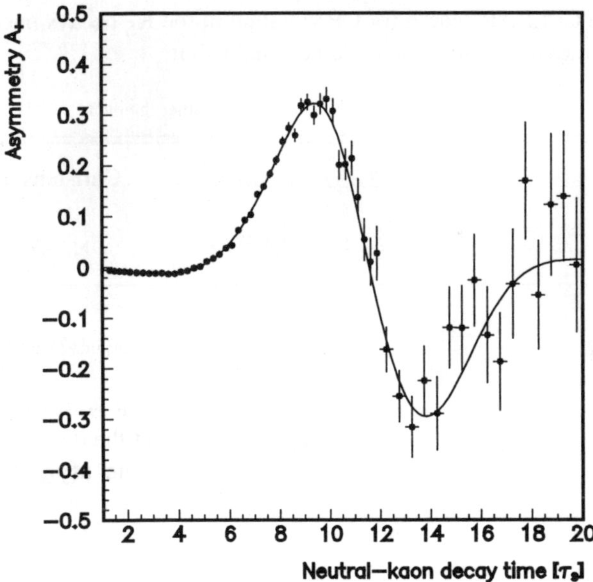

Figure 5: The time–dependent asymmetry A_{+-} as a function of the neutral kaon decay time [19].

Time dependent decay rate asymmetry between initial K^0 and $\overline{K}^0$ isolates the interference term between K_S and K_L amplitudes. It depends on $\left|\eta_{+-}\right|$, φ_{+-} and Δm:

$$A_{+-}^{\exp} = \frac{\overline{N}(\tau)-\alpha N(\tau)}{\overline{N}(\tau)+\alpha N(\tau)} = -\frac{2\left|\eta_{+-}\right|\cos(\Delta m\tau-\varphi_{+-})\exp\left(\frac{\Gamma_S-\Gamma_L}{2}\right)}{1+\left|\eta_{+-}\right|^2\exp((\Gamma_S-\Gamma_L)\tau)}.$$

The $K^0/\overline{K}^0$ normalization factor $\alpha = [1 + 4\mathrm{Re}(\varepsilon_L)] \times \xi$, corrects for the slight difference in the two decay rates due to the parameter ε_L, as well as for the tagging efficiency ξ of K^0 relative to $\overline{K}^0$.

The experimental asymmetry is shown on Figure 5 together with the result of the fit. With Δm and the decay widths fixed to their world average values, they obtained [19]:

$$|\eta_{+-}| = (2.264 \pm 0.023_{\mathrm{stat}} \pm 0.026_{\mathrm{syst}} \pm 0.007_{\tau_s}) \times 10^{-3}$$

$$\varphi_{+-} = (43.19 \pm 0.53_{\mathrm{stat}} \pm 0.28_{\mathrm{syst}} \pm 0.42_{\Delta m})^\circ$$

Systematic uncertainties are dominated by uncertainties in the correction of the regeneration effects. The results have approximately the same error as the weighted average of all previous experiments.

Results on CP violation in other decay modes are summarized in Table 2. CP violation was well established in K_L decays into $\pi^0\pi^0$ and $\pi^+\pi^-\gamma$ final states since a long time. The CPLEAR search for CP violation in the K_S decays into 3π improved previous limit measurements, but could not establish it.

Table 2: CP violation in decay modes other than $\pi^+\pi^-$

$K_L \rightarrow \pi^0\pi^0$	$|\eta_{00}| = (2.33 \pm 0.18) \times 10^{-3}$	Christenson 79 [20]
	$\varphi_{00} = (44.5 \pm 2.5)^\circ$	NA31 90 [21]
$K_L \rightarrow \pi^+\pi^-\gamma$	$|\eta_{+-\gamma}| = (2.359 \pm 0.074) \times 10^{-3}$	E773 95 [22]
	$\varphi_{+-\gamma} = (43.8 \pm 4.0)^\circ$	
$K_S \rightarrow \pi^+\pi^-\pi^0$	$\mathrm{Re}(\eta_{+-0}) = (-2 \pm 8) \times 10^{-3}$	CPLEAR 97 [23]
	$\mathrm{Im}(\eta_{+-0}) = (-2 \pm 9) \times 10^{-3}$	
$K_S \rightarrow \pi^0\pi^0\pi^0$	$\mathrm{Re}(\eta_{000}) = 0.18 \pm 0.15$	CPLEAR 98 [24]
	$\mathrm{Im}(\eta_{000}) = 0.15 \pm 0.20$	

4.2 *Time–reversal symmetry non–conservation*

In the domain of the time–reversal non–conservation, CPLEAR made a breakthrough with the first direct observation of the time–reversal symmetry violation [12]. The KTeV collaboration reported on a T–odd effects in the $K_L \rightarrow \pi^+\pi^-e^+e^-$

[25]. It might be interpreted as an indirect test of a time-reversal violation assuming unitarity, CPT invariance and neglecting final state interactions [26].

To study the time–reversal symmetry, CPLEAR used semileptonic data where the strangeness of the neutral kaon is tagged at the production as well as at the decay time. Their final data set contains 1.2×10^{6} events with a decay time above 1 τ_S.

Corrections were applied to the data to take into account difference in the tagging efficiency between K^0 and $\overline{K}^0$ at the production as well as at the decay time and to correct for regeneration effects.

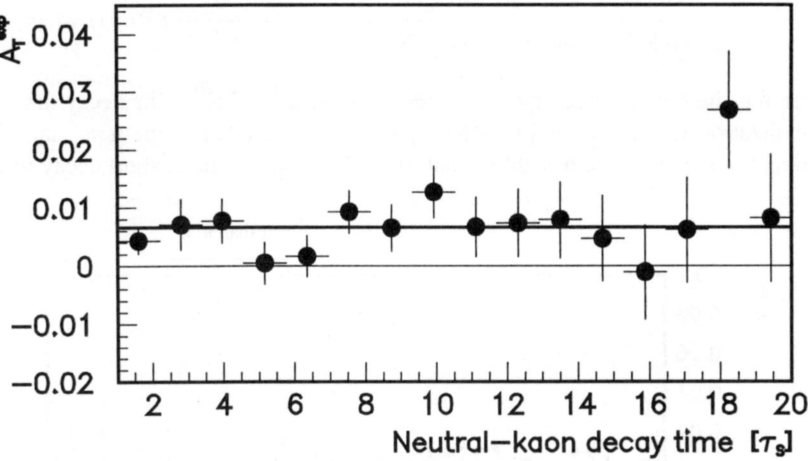

Figure 6: The experimental asymmetry A_T as a function of the neutral kaon decay time.

The experimental asymmetry is shown on Figure 6 together with the results of the fit. The data points scatter around a positive and constant offset, the average being

$$\langle A_T^{exp} \rangle = (6.6\pm1.3_{stat}\pm1.0_{syst})\times10^{-3} \;[12].$$

The systematic errors are dominated by uncertainties in the tagging correction at the decay vertex. Residual effects relate to a possible violation of CPT in the decay amplitude and to a possible violation of the $\Delta S=\Delta Q$ rule in the CPT-conserving and CPT-violating amplitudes, are negligible. They estimated the sum of these effects and they obtained: $Re(y + x_-) = (-2\pm3)\times10^{-4}$ [27]. This is the direct evidence of time–reversal violation.

332

4.3 Test of the CPT invariance

In the past, several studies tested indirectly the CPT invariance. Most of them neglect either a possible violation of CPT in the decay amplitudes or a possible violation of the $\Delta S = \Delta Q$ rule. The additional information provided by the semileptonic data recorded by CPLEAR allow a direct test of the CPT invariance without assumptions.

They built the experimental asymmetry [14]

$$A_\delta^{exp} = \frac{\eta N(\overline{K}^0_{t=\tau} \to e^+\pi^-\nu) - \alpha_{2\pi} N(K^0_{t=\tau} \to e^-\pi^+\overline{\nu})}{\eta N(\overline{K}^0_{t=\tau} \to e^+\pi^-\nu) + \alpha_{2\pi} N(K^0_{t=\tau} \to e^-\pi^{+|}\overline{\nu})}$$

$$+ \frac{\eta N(\overline{K}^0_{t=\tau} \to e^-\pi^+\overline{\nu}) - \alpha_{2\pi} N(K^0_{t=\tau} \to e^+\pi^-\nu)}{\eta N(\overline{K}^0_{t=\tau} \to e^-\pi^+\overline{\nu}) + \alpha_{2\pi} N(K^0_{t=\tau} \to e^+\pi^-\nu)} \to 8\mathrm{Re}(\delta) \text{ when } \tau \gg \tau_S$$

where η is the ratio of tagging efficiencies between K^0 and $\overline{K}^0$ at the decay time. The normalization factor $\alpha_{2\pi} = [1 + 4\mathrm{Re}(\varepsilon_L)] \times \xi$ is measured in the $\pi^+\pi^-$ data. This asymmetry depends on a possible violation of the $\Delta S = \Delta Q$ rule at short decay time. It

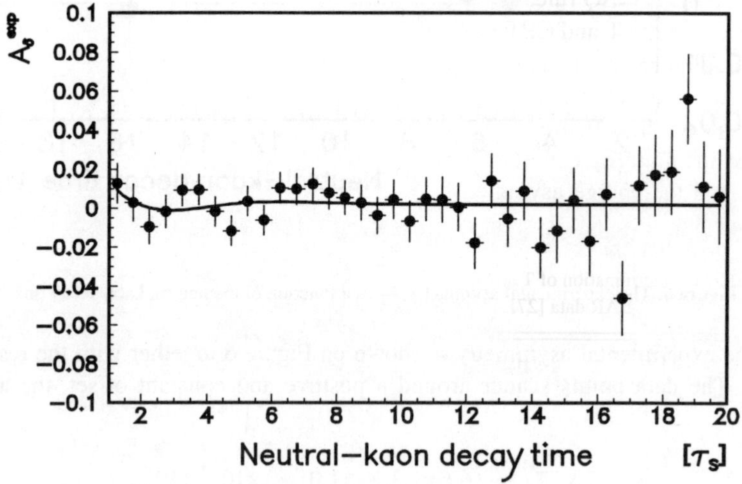

Figure 7: The asymmetry A_δ as a function of the neutral kaon decay time [14].

depends exclusively on a possible violation of CPT in the mixing, for long decay time.

The experimental asymmetry is shown on Figure 7 together with the results of the fit. The values obtained are given in Table 3 [14]. They are compatible with the CPT invariance and the validity of the $\Delta S=\Delta Q$ rule. The accuracy on $Re(\delta)$ reaches a level of 10^{-4} improving by 2 orders of magnitude the limit obtained by re-analysis of older measurements [28].

Table 3: Results of the direct test of the CPT invariance by CPLEAR [14].

$Re(\delta)$	$(3.0\pm3.3_{stat}\pm0.6_{syst})\times10^{-4}$
$Im(\delta)$	$(-1.5\pm2.3_{stat}\pm0.3_{syst})\times10^{-2}$
$Re(x_-)$	$(0.2\pm1.3_{stat}\pm0.3_{syst})\times10^{-2}$
$Im(x_+)$	$(1.2\pm2.2_{stat}\pm0.3_{syst})\times10^{-2}$

In the CPLEAR measurement the determination of the CPT violation in the mixing does not rely on unitarity conditions, CPT invariance in the decay amplitude and validity of the $\Delta S=\Delta Q$ rule.

To access the T and CPT violation in the mixing, CPT violation in semileptonic decay amplitudes and to a possible violation of the $\Delta S=\Delta Q$ rule, the two parts of the asymmetry A_δ, given above, are fitted simultaneously [27]. To disentangle these different sources of violation, constraints in the fitting procedure were added. They imposed the unitarity condition through the Bell–Steinberger relation [29] and they constrained the charge asymmetry, δ_l, in the K_L semileptonic decays to the world average value.

Table 4: Determination of T and CPT violation parameters using the unitarity conditions and CPLEAR data [27]. Error include statistic and systematics uncertainties

T	mixing	$Re(\varepsilon)$	$(164.9\pm2.5)\times10^{-5}$
CPT	mixing	$Re(\delta)$	$(2.4\pm2.8)\times10^{-4}$
		$Im(\delta)$	$(2.4\pm5.0)\times10^{-5}$
CPT	decay amplitude	$Re(y)$	$(0.3\pm3.1)\times10^{-3}$
$\Delta S=\Delta Q$		$Re(x_-)$	$(-0.5\pm3.0)\times10^{-3}$
		$Im(x_+)$	$(-2.0\pm2.7)\times10^{-3}$

Their results are given in Table 4 [27]. Time–reversal violation is the dominant effects while others are compatible with CPT invariance and validity of the $\Delta S=\Delta Q$ rule. The accuracy on $\mathrm{Im}(\delta)$ is improved by three orders of magnitude. For the first time the CPT invariance in the semileptonic decay amplitudes is tested. The accuracy reached is at a level of 10^{-3}. The sensibility on a possible violation of the $\Delta S=\Delta Q$ rule is at level of 10^{-3}. The precision on $\mathrm{Re}(\delta)$ is similar to the previous CPLEAR determination obtained without constraints.

5   Conclusions

CP violation is well established in the neutral kaon decays in the 2π final states using different experimental approaches. However the observation of the CP violation in the 3π final states is still missing.

For the first time, the time–reversal violation was measured directly by the CPLEAR collaboration. Its magnitude is equal to the magnitude of the CP violation observed in the 2π final states.

There is no evidence for the violation of the CPT invariance. The limit of a possible CPT violation in the mixing is below the T violation effect by at least one order magnitude. In addition there is no evidence of CPT violation in the decay amplitudes.

The values of $\mathrm{Re}(\delta)$ and $\mathrm{Im}(\delta)$ given in Table 4 and are combined to determine the difference of mass and decay width between K^0 and $\overline{K}^0$. They obtained [30]:

$$m_{K^0} - m_{\overline{K}^0} = (-1.5\pm2.0)\times10^{-18}\ \mathrm{GeV}$$

$$\Gamma_{K^0} - \Gamma_{\overline{K}^0} = (3.9\pm4.2)\times10^{-18}\ \mathrm{GeV}$$

These results were obtained with a coherent analysis without theoretical assumptions apart unitarity.

References

1.   J.H. Christenson et al., Phys. Rev. Lett. 13 (1964) 138
2.   L. Wolfenstein, Phys. Rev. 135 (1964) 562
3.   J.H Kobayashi and T. Maskawa, Prog. of Theor. Phys. 49 (1973) 652
4.   J. Schwinger, Phys. Rev. 82 (1951) 914.
     G. Lüders, Dansk. Mat. Fys. Medd. 28 (1954) 1
     W. Pauli, Niels Bohr and the development of Physics, Ed. W. Pauli, L. Rosenfeld and W. Weisskopf, Mc Graw Hill, New York, 1955
     R. Jost, Helv. Phys. Acta 31 (1958) 263
5.   S.W Hawking, Commun. Math. Phys. 43 (1975) 199, ibid. 87 (1982) 395

J. Ellis et al., Nucl. Phys. B241 (1984) 381

6. V.A. Kostelecky and R. Potting, Nucl. Phys. B359 (1991) 545
   V.A. Kostelecky and R. Potting, Phys. Lett. B381 (1996) 89

7. Sees for example P. Pavlopoulos, Proc. Work. on K physics, Orsay, 1996, 307

8. V.F Weisskopf and E.P. Wigner, Z. Phys. 63 (1930) 54, ibid. 65 (1930) 18

9. V.V. Barmin et al. Nucl. Phys. B247 (1984) 293, erratum ibid. B254 (1985) 747

10. C.D. Buchanan et al., Phys. Rev. D45 (1992) 4088

11. P.K. Kabir, Phys. Rev. D2 (1970) 54

12. A. Angelopoulos et al., CPLEAR Collaboration, Phys. Lett. B444 (1998) 43

13. A. Angelopoulos et al., CPLEAR Collaboration, Phys. Lett. B444 (1998) 38

14. A. Angelopoulos et al., CPLEAR Collaboration, Phys. Lett. B444 (1998) 52

15. P. Shanahan, for the KTeV Collaboration, in these proceedings
    E. Mazzucato, for the NA48 Collaboration, in these proceedings

16. E. Gabathuler and P. Pavlopoulos, Proc. Workshop on Physics at LEAR with
    Low Energy Cooled Antiproton, Erice, 1982, p. 747

17. R. Adler et al., Nucl. Instr. and Meth. in Phys. Res. A379 (1996) 76

18. KLOE, a general purpose detector for DAΦNE, The KLOE Collaboration
    LNF report LNF–92/019, 1992

19. A. Apostolakis et al., CPLEAR Collaboration, CERN–EP/99–05

20. Christenson et al., Phys. Rev. Lett. 43 (1979) 1209

21. Carosi et al., NA31 Collaboration, Phy. Lett. B237 (1990) 303

22. Matthews et al., E773 Collaboration, Phys. Rev. Lett. 75 (1995) 2803

23. R. Adler et al., CPLEAR Collaboration, Phys. Lett. B407 (1997) 193
    R. Adler et al., CPLEAR Collaboration, Eur. Phys. J C5 (1998) 389

24. R. Adler et al., CPLEAR Collaboration, Phys. Lett. B425 (1998) 391

25. A. Alavi-Harati et al., KTeV Collaboration, hep–ex/9908020
    B. Tshirhart, for the KTeV Collaboration, in these proceedings

26. L. Alvarez–Gaumé, C. Kounnas, S. Lola and P. Pavlopoulos, hep–ph/9903458

27. A. Apostolakis et al., CPLEAR Collaboration, Phys. Lett. B456 (1999) 297

28. V. Demidov et al., Phys. Atom. Nucl. 58 (1995) 968.

29. J.S Bell and J. Steinberger, Proc. of the Oxford Int. Conf. on Elementary Parti-
    cles, Oxford, 1965, p. 195

30. CPLEAR Collaboration, K^0 and $\overline{K}^0$ mass and width difference: CPLEAR evalu-
    ation, in preparation

The First $Re(\epsilon'/\epsilon)$ Result from the KTeV Experiment at Fermilab

Peter N. Shanahan
(for the Collaboration)
Fermilab - MS231
P.O. Box 500
Batavia, IL 60302 USA

By a precise comparison of $K_{L,S} \to \pi^+\pi^-, \pi^0\pi^0$ decay rates, KTeV has measured the direct CP violation parameter $Re(\epsilon'/\epsilon)$ to be $(28.0 \pm 4.1) \times 10^{-4}$. This preliminary result clearly establishes the existence of CP violation in a decay process.

1 Introduction

One of the most intriguing phenomena of particle physics is that of CP violation. Since its discovery in 1964[1], all observed CP violating effects have been found in the neutral kaon system, although the recent CDF measurement of $\sin(2\beta)$[2] is consistent with expectations for CP violation in the B sector. In the kaon system, the bulk of the CP violating effect is known to arise from an asymmetry in the $\Delta S = 2$ mixing transition $K^0 \to \bar{K}^0$. CP violation in the $\Delta S = 1$ decay transition $K \to \pi\pi$, or "direct" CP violation, is the focus of the KTeV E832 experiment.

1.1 ϵ and ϵ'

The CP eigenstates of the neutral kaon system, expressed in terms of its strong-force eigenstates $K^0, \bar{K}^0$, can be chosen to be

$$K_1 = \frac{1}{\sqrt{2}}(K^0 + \bar{K}^0) \qquad \mathrm{CP} = +1, \tag{1}$$

$$K_2 = \frac{1}{\sqrt{2}}(K^0 - \bar{K}^0) \qquad \mathrm{CP} = -1. \tag{2}$$

The parameter $\epsilon = 2.28 \times 10^{-3}$ measures the size of CP violation in the kaon system arising from the mixing asymmetry, so that the mass eigenstates of the systems are

$$K_S = \frac{1}{\sqrt{(1 + |\epsilon|^2)}}(K_1 + \epsilon K_2) \tag{3}$$

$$K_L = \frac{1}{\sqrt{(1 + |\epsilon|^2)}}(\epsilon K_1 + K_2) \tag{4}$$

The CP violating signal of $K_L \to \pi\pi$ (BR$= 3 \times 10^{-3}$) is known to be almost entirely an "indirect" CP violating effect, resulting from a CP conserving decay of the ϵK_1 component of the K_L[3].

The accommodation of this effect in the Standard Model via a non-trivial phase in the Cabbibo-Kobayashi-Maskawa quark mixing matrix allows for "direct" CP violation in the existence of a $\Delta S = 1$, $K_2 \to \pi\pi$ transition, connecting a CP-odd initial state to a CP-odd final state. A relative phase between the $\Delta I = 1/2$ and $\Delta I = 3/2$ isospin amplitudes of $K^0, \bar{K}^0 \to \pi\pi$ is needed for the direct CP violation to be detected. The degree of interference arising from the relative phase is parameterized by ϵ', giving a total CP violating amplitude (normalized to the CP conserving K_S decays) that depends on the final state:

$$\frac{A(K_L \to \pi^+\pi^-)}{A(K_S \to \pi^+\pi^-)} \equiv \eta_{+-} \approx \epsilon + \epsilon' \tag{5}$$

$$\frac{A(K_L \to \pi^0\pi^0)}{A(K_S \to \pi^0\pi^0)} \equiv \eta_{00} \approx \epsilon - 2\epsilon' \tag{6}$$

The magnitude of ϵ' is expected to be several orders of magnitude smaller than ϵ.

Alternatively, ϵ' is identically 0 in pure "Superweak" models, which posit a new CP violating $\Delta S = 2$ interaction, with no requirement of CP violation in the $\Delta S = 1$ transition[4].

As of the mid-1990's, the experimental status of ϵ' was ambiguous. The previous round of experiments had measured[5,6]

$$Re(\epsilon'/\epsilon) = (7.4 \pm 5.9) \times 10^{-4} \qquad \text{E731} - \text{FNAL}$$
$$Re(\epsilon'/\epsilon) = (23 \pm 6.5) \times 10^{-4} \qquad \text{NA31} - \text{CERN}.$$

Although the two experiments were in reasonable agreement with each other, their results yielded different conclusions on whether direct CP violation had been observed.

1.2 KTeV

The KTeV E832 experiment is one of 3 new experiments which aim to measure $Re(\epsilon'/\epsilon)$ to a precision of $\simeq 10^{-4}$, using simultaneously collected K_S and K_L decays. Each uses a form of the double ratio

$$R \equiv \frac{\Gamma(K_L \to \pi^+\pi^-)/\Gamma(K_S \to \pi^+\pi^-)}{\Gamma(K_L \to \pi^0\pi^0)/\Gamma(K_S \to \pi^0\pi^0)} = 1 + 6Re\left(\frac{\epsilon'}{\epsilon}\right), \tag{7}$$

which results in a cancellation of many of the uncertainties affecting absolute decay rate measurements.

The KTeV technique uses a comparison of decays from a K_L beam and a coherently regenerated beam of K_S. The $\pi^+\pi^-$ final states are reconstructed from charged tracks in a spectrometer; the $\pi^0\pi^0$ decays are reconstructed using the detection of the 4 photon final state in a high precision electromagnetic calorimeter. Backgrounds including non-$K \to \pi\pi$ decays and decays from non-coherently produced K_S are subtracted. Since the acceptance varies primarily with distance of the decay from the detector, and as a function of energy, the large $K_L - K_S$ lifetime difference necessitates a Monte Carlo-based acceptance correction before ϵ'/ϵ can be extracted from a fit to the event statistics. The coherent regeneration of K_S also allows the extraction of the $K_L - K_S$ mass difference, ΔM, as well as other kaon parameters.

1.3 Data Sample

The results presented here are obtained from 2.3×10^6 $K \to \pi^0\pi^0$ decays recorded in our 8 week 1996 run. Due to a bias arising from small but nontrivial drift chamber inefficiencies, coupled with an online cut applied to the 1996 data, we take the $K \to \pi^+\pi^+$ sample of 7.1×10^6 events from the first 18 days of our 1997 run. In addition to this signal sample, we collected 10's of millions of $K \to \pi e\nu$, $K \to \pi^+\pi^-\pi^0$ and $K \to 3\pi^0$ decays for calibration and acceptance studies. The $K \to \pi\pi$ data used for this result represents about 20% of the data that we acquired in our 1996 and 1997 runs.

2 The KTeV Detector

A schematic of the KTeV detector is shown in Fig. 1. The major components are: two neutral kaon beams, a 60m decay vacuum, a spectrometer for reconstructing charged tracks, a system of photon vetoes, a CsI calorimeter, and muon filter steel and vetoes.

Neutral Beams - Two neutral beams are created from the interactions of the 800 GeV Tevatron proton beam on a BeO target, 110 m upstream of the start of the fiducial decay region. A targeting angle of 4.8 $mrad$, combined with a Be absorber in the kaon beams, decreases the neutron contamination of the beams to about 65%. A system of sweeping magnets, including a magnetized beam dump, reduces the muon flux in the experiment to 10^{-7}/proton on target. K_S are regenerated in a 1.83 interaction length block of 85 plastic scintillator segments, which alternates between the 2 beams once per minute. Vetoing on the scintillation light

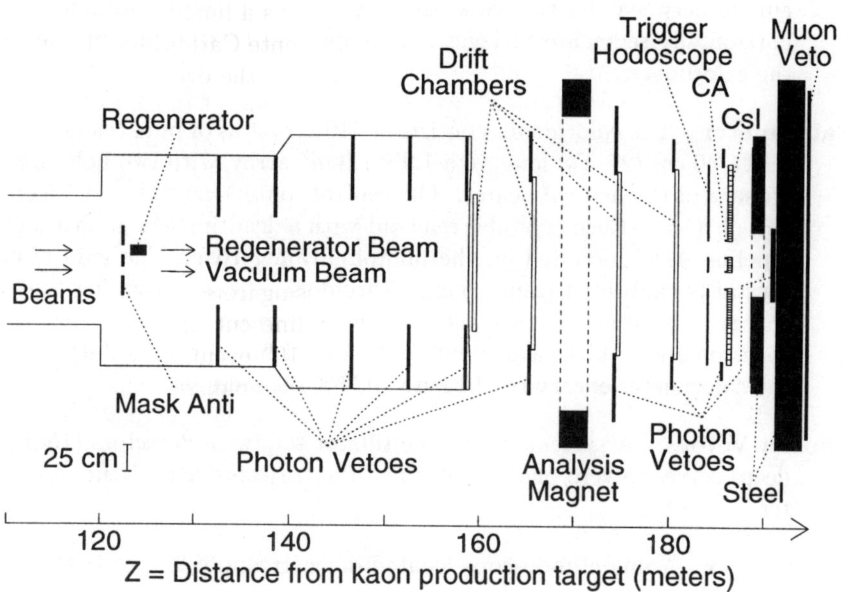

Figure 1: A schematic view of the KTeV detector, from above.

associated with the recoil of inelastic interactions reduces the background from non-coherent K_S production. The last segment contains two $6mm$ thick Pb slabs, to convert photons from upstream kaon decays.

Spectrometer - 4 Drift chambers, each with 2 staggered planes in both the vertical and horizontal views, and a high-uniformity magnet with a 0.412 GeV/c transverse kick are used to track charged particles. The single hit resolution of the 1.27 cm pitch chambers is 150 μm. Matching ambiguities between the two view are resolved using the calorimeter. Two horizontally segmented trigger hodoscopes are located just upstream of the calorimeter.

Calorimeter - The calorimeter consists of 3100 crystals of pure cesium-iodide (CsI), 50 cm (27 X_0) long, in a $1.9 \times 1.9\ m^2$ array, with two holes for the passage of the neutral beams. The central (outer) crystals are 2.5 cm (5 cm) on a side. Each crystal is read out with a deadtimeless, multiranging, pipeline ADC mounted on the photomultiplier base. The gain of each crystal is calibrated using electrons from $K \to \pi e \nu$, assuming that the energy deposited (E) is equal to the momentum (p) observed in the spectrometer. With nearly 198 million $e^\pm$ from our 1996 data sample, the calorimeter energy resolution is 0.75% on average.

Photon Vetoes - A system of lead-scintillator sandwich detectors (16-32 X_0 deep) serves to veto photons of which the trajectories miss the calorimeter.

Muon Veto - A muon hodoscope located downstream of 2 m of steel is used to distinguish muons from charged pions.

Trigger - The trigger system consists of 3 levels: Level-1 is a deadtimeless trigger with a typical rate of 45kHz. The detector is read-out at Level-2 (c. 10kHz). The Level-3 trigger is a full online reconstruction of each event.

3 Reconstruction and Analysis

3.1 $K \to \pi^+\pi^-$

$K \to \pi^+\pi^-$ decays are reconstructed from two tracks, forming a vertex in both the horizontal and vertical views. The mean $\pi^+\pi^-$ mass resolution is 1.6 MeV.

In addition to track quality and fiducial aperture cuts, the following cuts are applied:

Regenerator Veto Activity - reject inelasticly produced K_S;

Photon Veto Energy - reject events with extra photons from $K_L \to 3\pi^0$;

χ_V^2 - vertex matching χ^2;

Track Separation - events in which the 2 pion tracks pass within adjacent drift cells (or closer) are rejected because a hit from one of the tracks will be missing;

Muon Veto - no activity in the MU2 counter is permitted, to reject $K_L \to \pi^\pm \mu^\mp \nu$. Tracks are required to have a momentum of at least $8GeV$ to ensure that background muons do not range out in the muon filter steel;

E/p - the ratio of energy measured in the calorimeter to momentum measured in the spectrometer, for each track, is required to be less than 0.85, to reject electrons from $K_L \to \pi^\pm e^\mp \nu$;

$m_{p\pi}$ - the mass of the 2 track system is calculated assuming the higher mass track to be a proton, and rejected in the case it is consistent with the lambda mass;

p_T^2 - the transverse momentum squared of the $\pi^+\pi^-$ system; relative to the kaon trajectory at the downstream end of the regenerator, is required to be less than $250 MeV^2$. This cut rejects semileptonic kaon decays with a misidentified lepton, and non-coherently regenerated K_S decays;

Kaon Energy - required to be between 20 and 160 GeV;

Vertex Position - required to be between 110 and 158 m from the target;

$\pi^+\pi^-$ **mass** - required to be between 0.488 and 0.508 MeV/c^2.

After these cuts, the level of the remaining backgrounds to $K \to \pi^+\pi^-$ in the vacuum beam is 0.083%, which is dominated by $K_L \to \pi\mu\nu$ and $K_L \to \pi e\nu$ decays. The total background is visible in the P_T^2 spectra for the two beams (Fig. 2). The regenerator beam vacuum level, dominated by non-coherent K_S regeneration, is 0.089%. The background subtraction is understood to about 10% of the background level.

3.2 $K \to \pi^0\pi^0$

$K \to \pi^0\pi^0$ decays are reconstructed by assuming that each pair of photons resulted from the decay of a π^0. The distance from the decay vertex to the calorimeter is extracted from the positions and energies of the photon clusters in the calorimeter, under the constraint of the neutral pion mass. Incorrect photon pairings are rejected by requiring consistent vertex positions for the two photon pairs. The 4 photon invariant mass is calculated once the vertex position has been determined, with a kaon mass resolution of 1.5 MeV. The transverse "center-of-energy" of the 4 clusters gives the position at which the kaon would have impacted the calorimeter, had it not decayed.

Cuts are applied on the following variables:

Regenerator Veto Activity - reject inelasticly produced K_S;

Photon Veto Energy - reject events with extra photons from $K_L \to 3\pi^0$;

χ_p^2 - photon pairing χ^2 - to reject incorrectly paired events;

χ_S^2 - transverse shower shape χ^2 - to reject clusters of more than 1 photon, reducing $K_L \to 3\pi^0$ background;

Cluster Positions - clusters are required to be separated from each other and from the edges of the calorimeter enough to ensure an accurate reconstruction of their energies;

Charged Activity - cuts are made on the presence of charged tracks in the spectrometer, to reduce non-$K \to \pi^0\pi^0$ background;

Center of Energy - the center of energy is required to be within a square centered on either beam, of area $100 cm^2$, to reduce background from non-coherently produced K_S;

Kaon Energy - required to be between 40 and 160 GeV;

Vertex Position - required to be between 110 and 158 m from the target;

4γ mass - required to be between 0.490 and 0.505 MeV/c^2.

Following these cuts, the most important remaining backgrounds are $K_L \to 3\pi^0$ with fused or missing photons, and K_S produced non-coherently in the regenerator. The latter background is particularly important, since the lack of direction information allows the non-coherent K_S to be reconstructed in either beam.

These and other backgrounds are modeled in a detailed Monte Carlo simulation, and subtracted from the data by normalizing their contributions to the sidebands of the mass and center of energy distributions. Fig. 2 shows the backgrounds as a function of "ring number", which is the number of $1 cm^2$ square rings from the center of energy to the geometric center of the beam. The background level is $\simeq 1\%$, and is understood to 10% of itself.

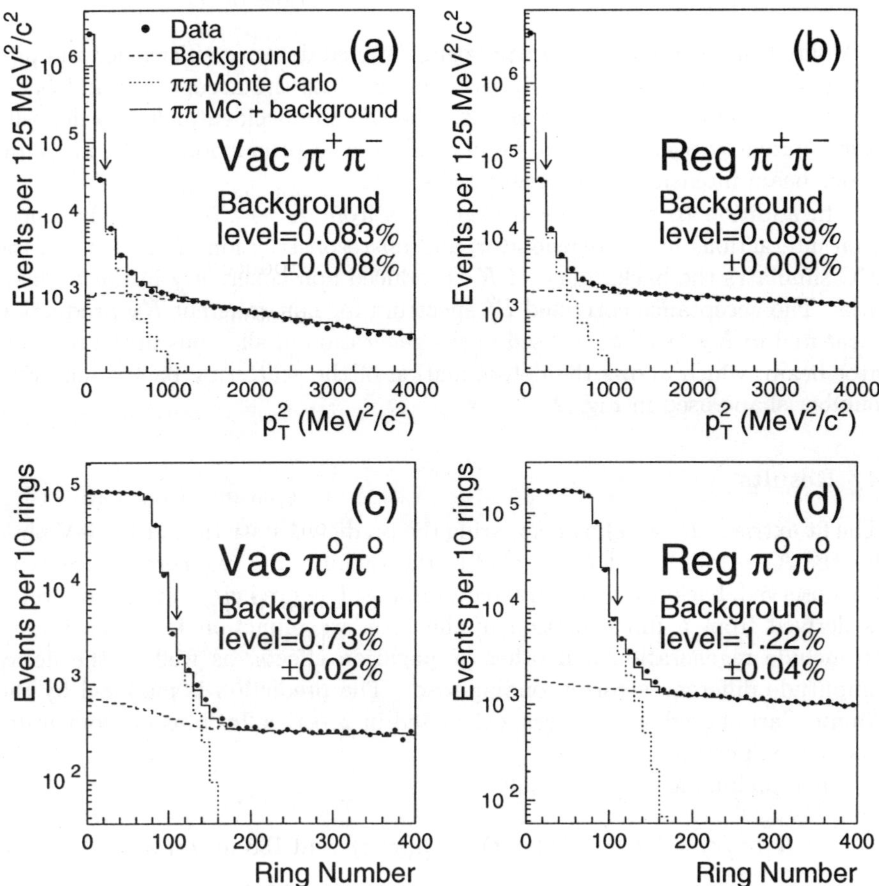

Figure 2: The data and MC distributions of P_T^2 for $K \to \pi^+\pi^-$ *(a,b)*, and Ring Number for $K \to \pi^0\pi^0$ *(c,d)*. The signal is in the peak at low values of the corresponding variables in each case, with the background dominant at higher values. The background distributions from Monte Carlo are scaled to match the data in the sidebands.

3.3 Monte Carlo

A Monte Carlo simulation, incorporating detailed detector efficiencies and precisely determined aperatures, is used to calculate the acceptance and backgrounds of the $K \to \pi\pi$ signal modes. The effect of accidental energy deposits is simulated by overlaying real events collected in proportion to the instantaneous beam intensity on the generated events.

In addition to simulating backgrounds from $K \to \pi l \nu$, $K \to 3\pi^0$, neutron interactions in the regenerator, and misreconstruction of $K \to \pi^0\pi^0$, the MC simulates the background of K_S produced non-coherently in the regenerator. The acceptance corrected P_T^2 spectrum for non-coherent K_S production measured in $K_S \to \pi^+\pi^-$ is used in the generation of all kaons in the regenerator beam, which allows the determination of the scattering background ring number shape used in Fig. 2.

4 Results

The fit extracts $Re(\epsilon'/\epsilon)$ by comparing the predicted statistics, in 10 GeV bins, for the $K \to \pi^+\pi^-$ and $K \to \pi^0\pi^0$ in the vacuum and regenerator beams to the observed, background subtracted samples. The predicted number of events is derived from a full evolution of the kaon wavefunction from the target, to include regeneration and other propagation effects, as well as the decay amplitude differences corresponding to ϵ'. The prediction is modified by the Monte Carlo-based acceptances, calculated in $2GeV \times 2m$ bins of momentum and vertex position.

The preliminary KTeV result [7] is

$$Re(\epsilon'/\epsilon) = (28.0 \pm 3.0(\text{stat}) \pm 2.6(\text{syst}) \pm 1.0(\text{MCsyst})) \times 10^{-4} \qquad (8)$$

Additional parameters are extracted in fits to the shape of the regenerator beam vertex distributions. The K_S lifetime τ_S and $K_L - K_S$ mass difference ΔM are obtained from a fit to the $K_S \to \pi^+\pi^-$ distributions assuming CPT invariance via the constraint that the phase of η_{+-} be equal to the superweak phase: $\phi_{+-} = tan^{-1}[2(\Delta m)/(\Gamma_S - \Gamma_L)]$ (Table 1).

The assumption of CPT conservation is abandoned in a fit for ϕ_{+-} and Δm, fixing τ_S to the value from the previous fit. Due to the strong correlations, we present a linear expansion of our ϕ_{+-} in Δm and τ_S. Finally, the phase difference $\Delta\phi \equiv \phi_{00} - \phi_{+-}$, for which an absolute value of more than $\approx 0.2^o$ would be a signal of CPT violation, is fit using both $K_S \to \pi^0\pi^0$ and $K_S \to \pi^+\pi^-$ distributions in the regenerator beam. The result (Table 2) is consistent with 0°.

Table 1: Preliminary kaon parameter results from vertex distribution fits, assuming CPT conservation.

Parameter	Result assuming CPT
	($\phi_{+-} = tan^{-1}[2(\Delta m)/(\Gamma_S - \Gamma_L)]$)
τ_S	$(0.8967 \pm 0.0007) \times 10^{-10}$ s)
Δm	$(0.5280 \pm 0.0013) \times 10^{-10}$ h/s
ϕ_{+-}	fixed

Table 2: Preliminary kaon parameter results from vertex distribution fits, not assuming CPT conservation.

Parameter	Result not assuming CPT
Δm	$(0.5286 \pm 0.0023) \times 10^{-10}$ h/s
ϕ_{+-}	$(43.66 \pm 0.30)^o + 0.23^o \times (\Delta m - 0.5286)/0.0010 - 0.26^o \times (\tau_S - 0.8967)/0.0010$
$\Delta\phi$	$(0.09 \pm 0.43 \pm 0.15)^o$

5   Systematic Uncertainties

The systematic uncertainties on $Re(\epsilon'/\epsilon)$ generally arise from a relative bias between K_S and K_L. Data collection inefficiencies, biases in reconstructing and defining the background subtracted event samples, misunderstanding of the detector acceptance, and uncertainties on external parameters used to extract $Re(\epsilon'/\epsilon)$ all contribute to the overall uncertainty.

A major uncertainty in defining the final event sample is due to the uncertainty on the overall scale of photon energies in the calorimeter, despite the calibration with electrons from $K \to \pi e\nu$. An energy scale different from 1 causes a loss or gain in $K_L \to \pi^0\pi^0$ statistics at the downstream end, due to the resulting migration of events across the fiducial vertex cut. The energy scale is determined by comparing the observed and predicted cutoffs in the vertex distribution of $K_S \to \pi^0\pi^0$ at the regenerator edge, resulting from photon conversions in the last Pb slab of the regenerator. The position of the edge depends linearly on the energy scale: the observed shift corresponds to an energy scale correction of -0.125% (Fig. 3). The uncertainty on this correction is determined by comparing it to 4 other measurements: the edge of $K_L \to 3\pi^0$ at the regenerator, the edge of $K_L \to 3\pi^0$ at a tungsten slab near the downstream end of the decay region (inserted for special calibration data taking), and the reconstructed position of $\pi^0\pi^0$ pairs produced in beam interactions in the regenerator Pb and the vacuum window at the downstream

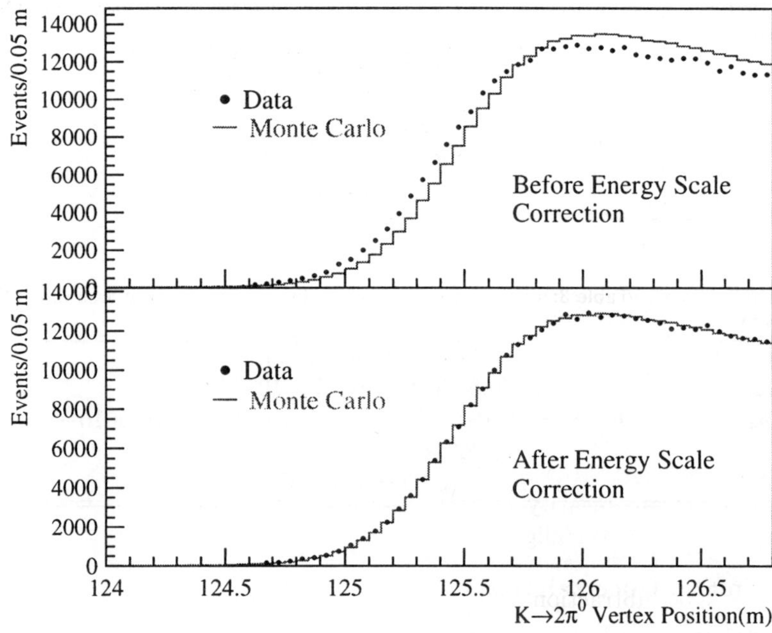

Figure 3: The $K_S \to \pi^0\pi^0$ vertex distributions near the regenerator edge for data and Monte Carlo, before (*top*), and after (*bottom*) the energy scale correction of -0.125%.

end of the decay region. The comparison of these energy scale measurements leads to an uncertainty of $\delta(Re(\epsilon'/\epsilon)) = 0.7 \times 10^{-4}$. An additional uncertainty $\delta(Re(\epsilon'/\epsilon)) = 0.6 \times 10^{-4}$ is assigned based on a residual nonlinearity in $m_{\pi^0\pi^0}$ as a function of kaon energy.

The uncertainties associated with background subtraction are dominated by the non-coherent background to $K_S \to \pi^0\pi^0$, due to the uncertainty in the P_T^2 spectrum input to the Monte Carlo. Based on the change in $Re(\epsilon'/\epsilon)$ observed when the P_T^2 spectrum is varied, we assign an uncertainty of $\delta(Re(\epsilon'/\epsilon)) = 0.8 \times 10^{-4}$.

The largest contribution to the total systematic error comes from the determination of the acceptance. One of the most direct tests of our understanding of the acceptance comes from comparing the shape of the vertex position

Table 3: Systematic uncertainties on $Re(\epsilon'/\epsilon)$.

Source of uncertainty	$Re(\epsilon'/\epsilon)$ uncertainty for:	
	$\pi^+\pi^-$ Analysis ($\times 10^{-4}$)	$\pi^0\pi^0$ Analysis ($\times 10^{-4}$)
Trigger and level-3 filter	0.5	0.3
Calorimeter energy scale	0.1	0.7
Calorimeter nonlinearity	-	0.6
Detector calibration/alignment	0.3	0.4
Cuts	0.6	0.8
Background subtraction	0.2	0.8
Aperatures	0.3	0.5
Detector resolution	0.4	< 0.1
Drift Chamber simulation	0.6	-
Acceptance vs. Z_{vtx}	1.6	0.7
MC statistics	0.5	0.9
Regenerator beam K_L transmission:		
1996 vs. 1997	0.2	
Energy dependence	0.2	
External Parameters	0.2	
TOTAL	2.8	

distributions for data and MC in the vacuum beam, where the only relevant physics parameter is the K_L lifetime. Although the overall agreement between data and MC is good in each case (Fig. 4), a small relative slope between data and MC can lead to a significant uncertainty on $Re(\epsilon'/\epsilon)$, due to the 6 m difference in the average vertex positions between K_L and K_S decays. Despite the indication from the $K_L \to \pi e \nu$ sample that the spectrometer acceptance is well understood, and the weak statistical significance of $(-1.6 \pm 0.6) \times 10^{-4}$ for relative slope of $K_L \to \pi^+\pi^-$, we assign an uncertainty corresponding to the full value of the slope - $\delta(Re(\epsilon'/\epsilon)) = 1.6 \times 10^{-4}$. Given the good data-MC agreement for the $K_L \to \pi^0\pi^0$ and $K_L \to 3\pi^0$ overlays, we set a limit of the possible $Re(\epsilon'/\epsilon)$ bias due to the neutral mode acceptance of 0.7×10^{-4}.

Another potential bias in $Re(\epsilon'/\epsilon)$ is related to the use of $\pi^0\pi^0$ data from 1996 and $\pi^+\pi^-$ from 1997. A change in the relative K_S to K_L beam flux could arise, in principle, from changes in an extra absorber upstream of the decay region, which is always in the same beam as the regenerator. Measurements of the K_L transmission through the regenerator beam, using $K_L \to 3\pi^0$ and $K_L \to \pi^+\pi^-\pi^0$ decays from both the 1996 and 1997 runs, show no evidence of such a change. Furthermore, estimates of an upper limit in the absorber attenuation due to thermal expansion sets an uncertainty on $Re(\epsilon'/\epsilon)$ of 0.2×10^{-4}. The total quadratic sum of all systematic uncertainties on $Re(\epsilon'/\epsilon)$, summarized in Table 3, is 2.8×10^{-4}.

Several cross-checks of our $Re(\epsilon'/\epsilon)$ result were performed, such as measuring $Re(\epsilon'/\epsilon)$ as a function of beam intensity and run period, all with no significant variation. In particular, an alternate technique of calculating $Re(\epsilon'/\epsilon)$ from a direct comparison of the regenerator and vacuum beam vertex distributions, which avoids the need to understand the variation of the acceptance along the beam axis at the cost of statistical power, gives a consistent result for $Re(\epsilon'/\epsilon)$. Finally, there is no significant variation of $Re(\epsilon'/\epsilon)$ as a function of kaon energy (Fig. 5).

6 Conclusion

We measure $Re(\epsilon'/\epsilon) = (28.0 \pm 3.0(\text{stat}) \pm 2.8(\text{syst})) \times 10^{-4}$, or $Re(\epsilon'/\epsilon) = (28.0 \pm 4.1) \times 10^{-4}$, adding the uncertainties in quadrature. This result taken by itself establishes the existence of CP violation in a decay process. Our result is in better agreement with the previous result from CERN (NA31) than with the previous Fermilab result (E731). The average of the E731, NA31, NA48, and KTeV measurements is $(21.2 \pm 2.8) \times 10^{-4}$, ruling out a superweak interaction as the sole source of CP violation in kaon decays.

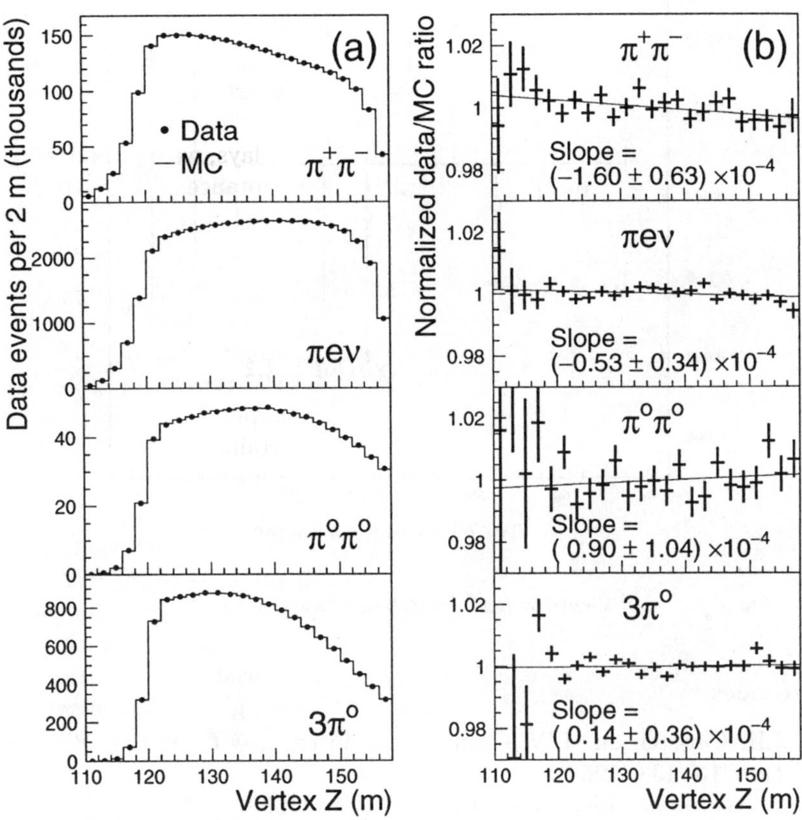

Figure 4: (a):The data and MC (normalized to data) distributions of the vacuum beam vertex position distributions for $K \to \pi^+\pi^-, \pi e\nu, \pi^0\pi^0$ and $3\pi^0$. (b): Linear fits to the ratio data/MC for each mode.

350

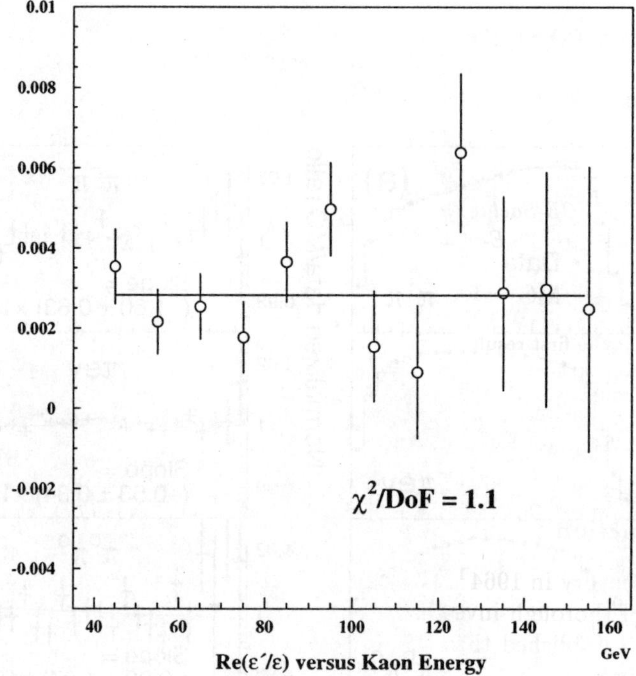

$\chi^2/\text{DoF} = 1.1$

Figure 5: $Re(\epsilon'/\epsilon)$ in bins of kaon energy.

References

1. J.H. Christenson, J.W. Cronin, V.L. Fitch, and R. Turlay, *Phys. Rev. Lett.* **13**, 138 (1964).
2. F. Abe, *et al.*, *Phys. Rev. Lett.* **81**, 5513 (1998).
3. B. Winstein and L. Wolfenstein, *Rev. Mod. Phys.* **65**, 1113 (1993).
4. L. Wolfenstein, *Phys. Rev. Lett.* **13**, 562 (1964).
5. L.K. Gibbons, *et al.* ,*Phys. Rev.* D **55**, 6625 (1997).
6. G.D. Barr, *et al.* , *Phys. Lett.* B **317**, 233 (1993).
7. A. Alavi-Harati, *et al.* , *Phys. Rev. Lett.* **83**, 22 (1999).

FIRST MEASUREMENT OF Re(ε'/ϵ) WITH THE NA48 DETECTOR AT CERN

E. MAZZUCATO

CEA, DSM/DAPNIA,
CE-Saclay, 91191 Gif-sur-Yvette Cedex, France
E-mail: edoardo@hep.saclay.cea.fr

for the NA48 Collaboration[a]

We present the first result on direct CP violation in the neutral kaon system from the NA48 experiment at CERN. A preliminary measurement of Re(ε'/ϵ) based on the data collected in 1997 has been performed by studying decays of K_L and K_S into $\pi^+\pi^-$ and $\pi^0\pi^0$ collected simultaneously in the NA48 detector. The value obtained is Re(ε'/ϵ) = $(18.5 \pm 4.5(stat) \pm 5.8(syst)) \times 10^{-4}$.

1 Introduction

Since its discovery in 1964[1], CP violation in the neutral kaon system has been the object of thorough investigations both theoretically and experimentally[2]. It was soon established that the dominant component of the violation occurs in the mixing between K^0 and $\bar{K}^0$ eigenstates[3]. This main component is characterized by the parameter ε which gives the size of the deviation from pure CP=± 1 states of the K_S and K_L physical states. Whether CP violation can also take place in the $K \to 2\pi$ decay process itself is a question of fundamental importance for the understanding of the origin of the effect. Direct CP violation in neutral kaon decays into two pions would imply different decay rates for K^0 and $\bar{K}^0$ states. It is parametrized by ε' and can be observable as a non-zero value of the double ratio of decay widths:

$$R = \frac{\Gamma(K_L \to \pi^0\pi^0)}{\Gamma(K_S \to \pi^0\pi^0)} \Big/ \frac{\Gamma(K_L \to \pi^+\pi^-)}{\Gamma(K_S \to \pi^+\pi^-)} \cong 1 - 6Re(\varepsilon'/\varepsilon). \tag{1}$$

CP violation can be naturally accomodated within the Standard Model by the introduction of a complex phase in the CKM matrix[4]. The direct contribution arises from the phase difference in ΔS=1 diagrams between ΔI=1/2 and ΔI=3/2 $K \to 2\pi$ transitions[5]. Recent calculations[6] of Re(ε'/ϵ) give values typically in the $0 - 10^{-3}$ range, but values consistent with 3×10^{-3} have also been reported. The computation of Re(ε'/ϵ) suffers from large theoretical

---

[a] Cagliari-Cambridge-CERN-Dubna-Edinburgh-Ferrara-Firenze-Mainz-Orsay-Perugia-Pisa-Saclay-Siegen-Torino-Vienna-Warsaw Collaboration

uncertainties as it is rather sensitive to input parameters and to the method used to evaluate the hadronix matrix elements.

Historically, the NA31 collaboration [7] at CERN performed the first precision experiment which claimed evidence for direct CP violation in neutral kaon decays. They reported a final value [8] of $\mathrm{Re}(\varepsilon'/\epsilon) = (23. \pm 6.5) \times 10^{-4}$. However, their result was only marginally in agreement with the Fermilab E731 measurement [9] of $(7.4 \pm 5.9) \times 10^{-4}$, consistent with no effect. Very recently, the KTeV collaboration [10] has obtained $\mathrm{Re}(\varepsilon'/\epsilon) = (28.0 \pm 4.1) \times 10^{-4}$ relying on a technique similar to the one used by E731.

In this paper, we present a preliminary measurement of $\mathrm{Re}(\varepsilon'/\epsilon)$ based on the first data collected in 1997 by the NA48 experiment.

2 Experimental method

The NA48 experiment [11] has been designed to measure the $\mathrm{Re}(\varepsilon'/\epsilon)$ parameter with an accuracy of 2×10^{-4}. In order to reach that level of precision, 4 to 5 million events from the limiting $K_L \to \pi^0\pi^0$ mode have to be collected and systematic uncertainties kept at very small levels.

The principle of the NA48 experiment consists in detecting the four relevant decay modes concurrently in the same decay region, using almost collinear K_L and K_S beams converging at the centre of the NA48 detector and with similar momentum spectra. In this way, differences and variations in detection efficiencies for short- and long-lived neutral kaons become unimportant, as do rate dependent effects introduced by accidental activity in the detector elements. K_S decays are distinguished from the K_L decays by tagging the protons which produce the K_S particle.

High-rate capability and high-resolution detectors for the identification of the charged and neutral pion modes and for the proton tagging system are necessary to provide excellent background rejection and to keep the systematic uncertainties at the desired level. Moreover, to minimize as much as possible acceptance differences between K_L and K_S decays, a lifetime weighthing procedure is applied to the K_L events used in the computation of the double ratio: reconstructed K_L events are weighted with a factor $e^{z \cdot \left(\frac{1}{\Lambda_L} - \frac{1}{\Lambda_S} \right)}$ where z is the vertex position along the beam axis and $\Lambda_{L,S}$ are the decay lengths of the $K_{L,S}$. Such a procedure has the advantage of reducing heavily the dependence of the systematic accuracy of the result on detailed Monte Carlo simulations, at the cost, however, of some loss in statistical precision.

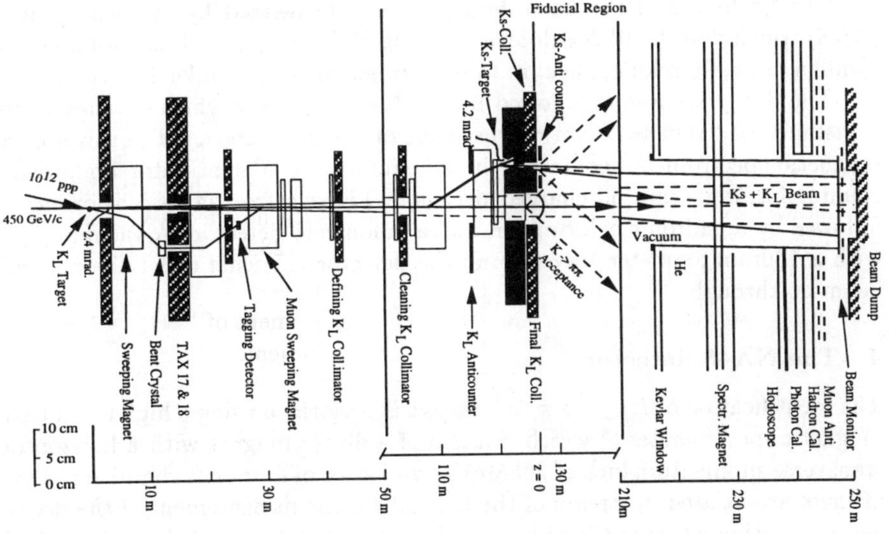

Figure 1: Layout of the NA48 experiment.

3 The neutral kaon beams

A schematic view of the kaon beam lines[12] and of the detector layout is shown in Figure 1. A 450 GeV/c primary proton beam with a nominal flux of 1.5×10^{12} particles per SPS pulse (ppp) impinges on a Be target at a downward angle of 2.4mrad to produce the K_L beam. A dipole magnet located downstream of the target station sweeps away the emitted particles and deflects the remaining primary protons towards a Si bent crystal[13]. A small fraction ($\approx 5 \times 10^{-5}$) of these protons are channeled by the crystal and deflected back onto the the K_L beam line after traversing a tagging detector. All other charged particles are absorbed in a beam dump.

The resulting low intensity ($\approx 3 \times 10^7$ ppp) proton beam is then transported towards a second Be target for the production of the short-lived kaons. The K_S target is positioned 72mm above the K_L beam axis and 120m downstream of the K_L target. The proton beam strikes the K_S target at an angle of 4.2mrad. This production angle is chosen in order to render the momentum spectrum of the detected K_S as similar as possible to the K_L one over the 70-170 GeV/c studied momentum range.

The beginning of the kaon decay is precisely defined by an anti-counter (AKS) which detects all K_S decays occuring upstream. The decay volume lies inside a large, 90m long, vacuum tank terminated by a thin kevlar window.

The decay volume is followed by the NA48 detector which extends from a distance of 216m to 255m downstream of the K_L target. It consists of a magnetic spectrometer contained in a helium tank, a scintillator hodoscope counter for triggering on charged decays, a LKr electromagnetic calorimeter followed by a hadronic calorimeter, and a muon veto detector. A vacuum pipe with a 160mm diameter traverses all the detector elements to let the neutral beam go through.

4   The NA48 detector

The identification of $K_{L,S} \to \pi^+\pi^-$ decays is performed using a high resolution magnetic spectrometer [14] which consists of a dipole magnet with a horizontal transverse momentum kick of 265MeV/c and a set of four drift chambers. Two of them are located upstream of the magnet for the measurement of the decay vertex position whereas the other two, located downstream of the magnet, are used for the bending angle determination of the tracks. Each drift chamber has an octagonal shape with a transverse dimension of 2.9m and contains eight sense wire planes oriented in four different directions, orthogonal to the beam axis: $0°$ (X,X′), $90°$ (Y,Y′), $-45°$ (U,U′) and $+45°$ (V,V′). The space resolution in each projection is $90\mu m$ and the average efficiency is 99.5% per plane. The geometrical accuracy of the chambers is better than $100\mu m/m$. The spectrometer provides a measurement of the decay position of the $K_{L,S}$ in the charged mode with longitudinal and transverse resolutions of about 50cm and 2mm respectively. Since the two beams are separated vertically by about 6cm, a clean identification of $K_S \to \pi^+\pi^-$ and $K_L \to \pi^+\pi^-$ is possible. The momentum determination of a track is achieved with a resolution given by:

$$\frac{\sigma_p}{p}(\%) \approx 0.5 \oplus 0.009p[GeV/c]. \tag{2}$$

The mass resolution (σ_M) for recontructed $K_{L,S} \to \pi^+\pi^-$ decays is better than 2.5MeV/c^2 (Figure 2). The precise time reference of tracks is provided by the two scintillator planes of the hodoscope, one with vertical slabs and the other with horizontal ones. The time resolution is about 200ps per track.

The signature of the $K_{L,S} \to \pi^0\pi^0$ mode is obtained by detecting the decay photons of the neutral pions in a quasi-homogeneous liquid krypton (LKr) calorimeter. It has a 125cm long projective tower structure which is made of copper-beryllium ribbons extending between the front and the back of the detector with a ±48mrad accordion geometry. The 13212 readout cells each

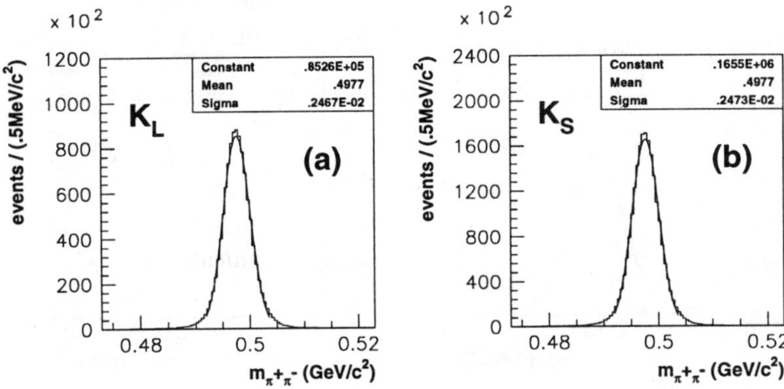

Figure 2: Invariant $\pi^+\pi^-$ mass distributions for (a) K_L and (b) K_S decays.

have a cross section of $2\times2\mathrm{cm}^2$ at the back of the active region. The projective towers point to the average K_S decay position located about 110m upstream of the detector so that the measurement of the angles between photons is almost insensitive to the initial conversion depth. The transverse scale of the calorimeter is measured to be better than $200\mu\mathrm{m/m}$. The initial current induced on the electrodes by the drift of the ionisation is measured using 80ns FWHM pulse shapers digitized with 40MHz FADCs[15]. The energy resolution of the calorimeter is:

$$\frac{\sigma_E}{E}(\%) \approx 3.2/\sqrt{E} \oplus 12.5/E \oplus 0.5 \quad (E \ in \ GeV). \tag{3}$$

The achieved time and spatial resolutions for 20GeV photons are better than 300ps and 1.3mm respectively. The mass resolution on reconstructed neutral pions is about 1 MeV/c^2 (Figure 3).

A 4mm thick detector consisting of scintillation fibres has been inserted inside the electromagnetic calorimeter to provide an independent time measurement as well as minimum bias neutral triggers. This device is located near the maximum of the shower development.

Behind the LKr calorimeter, a 6.7 nuclear interaction lengths thick calorimeter made of iron and scintillator is used to measure the energy of hadrons in the trigger. The muon identification for the suppression of the semileptonic $K_{\mu3}$ background is performed by a set of three planes of 25cm wide scintillation counters shielded by 80cm thick iron walls.

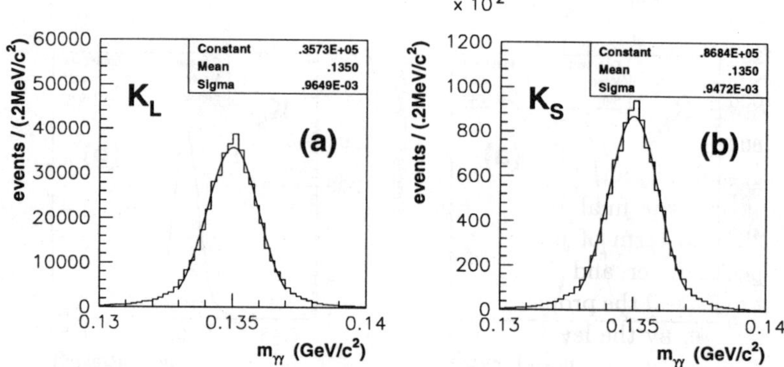

Figure 3: Reconstructed neutral pion mass in K_L (a) and K_S (b) decays into $2\pi^0$.

5 The 1997 run

The first $\mathrm{Re}(\varepsilon'/\epsilon)$ data taking period of the NA48 experiment started in September 1997 and lasted 42 days. The data was collected at a proton beam intensity of about 10^{12} ppp corresponding to nearly 500kHz of K_L decays in the detector. About 12k events were collected per 2.4s SPS spill and a total amount of 25Tbytes of data was recorded on tape. The readout capability of the DAQ system reached 80Mbytes/s. The on-line selection of the $K_{L,S} \to 2\pi$ events was performed by a level 2 charged trigger and a fully pipelined neutral trigger system.

In 1997, the LKr calorimeter was operated with a reduced value of 1.5kV of the high voltage on the anodes due to some faulty blocking capacitors. As a consequence, the electronic noise was increased by about 20% and small space charge effects ($< 0.5\%$) from the accumulation of positive ions were observed. In addition, a vertical 4cm wide strip remained inactive due to a bad high voltage connection in the calorimeter. This caused a reduction of about 15% in the acceptance for $\pi^0\pi^0$ events. These two problems were solved in 1998 and since then, the detector has been operating in a very stable manner at the nominal 3kV drift voltage.

5.1 Trigger system

The $\pi^+\pi^-$ trigger [16] consists of two levels. The pre-selection of events having a two-body decay topology is performed at the level 1 (L1) stage by using informations from the two planes of the scintillator hodoscope and by requiring a summed energy above 30 GeV in the calorimeters. The L1 trigger output rate, downscaled by a factor of 2 to limit dead time in the subsequent stage, was about 70kHz. The final selection of $\pi^+\pi^-$ candidates is made at the second level (L2) by a farm of processors which performs the track reconstruction in the spectrometer and computes the vertex position as well as the $\pi^+\pi^-$ invariant mass and the proper time-of-flight of the decaying kaon. The rejection factor obtained by the level 2 charged trigger was typically 60 and the overall L1·L2 efficiency was measured to be $(91.68 \pm 0.09)\%$. In 1998, this efficiency was substantially improved reaching a value above 97%. The maximum latency of the L2 trigger was $100\mu s$ and its dead time was kept at the 0.3% level.

The neutral trigger [17] is a 40MHz pipelined system based on the LKr information from 64 vertical and 64 horizontal projections. The rejection of the dominant $K_L \to \pi^0\pi^0\pi^0$ background is obtained by computing and cutting on the number of in-time clusters per projection, the total energy in the LKr calorimeter and the reconstructed decay vertex position. The neutral trigger has negligible dead time and a $3\mu s$ latency. Its ouput rate was about 2kHz and the efficiency for selecting good $\pi^0\pi^0$ events was measured to be $(99.88 \pm 0.04)\%$, independent of the decay mode, K_L or K_S.

5.2 Event selection and background rejection

The identification of the $\pi^+\pi^-$ events is obtained by requiring two in-time (± 6ns) reconstructed tracks of oppostive charge forming a vertex located in the K_S or K_L decay regions. The energy of the event is measured using the angle of the two tracks before the magnet and the ratio of the track momenta, assuming the two tracks originate from $K_{L,S} \to \pi^+\pi^-$ decays. A cut on the asymmetry variable $A_p = (p_1 - p_2)/(p_1 + p_2)$, where p_1 and p_2 are the two track momenta, is also made in order to avoid events with tracks near the beam hole for which a precise Monte Carlo modelling is required. Since A_p is related to the decay orientation in the kaon rest frame, the rejection obtained is $K_L - K_S$ symmetric. Furthermore, an adequate cut on A_p allows to remove entirely Λ and $\bar{\Lambda}$ decays which are present in the K_S beam. $\pi^+\pi^-$ candidates were required to satisfy the energy dependent condition $|A_p| < 1.08 - 0.0052$E (E in GeV) with $|A_p| < 0.62$.

In order to obtain clean samples of $\pi^+\pi^-$ events, cuts are applied on the two-pion invariant mass ($\pm 3\sigma_M$) around the kaon mass value and also on

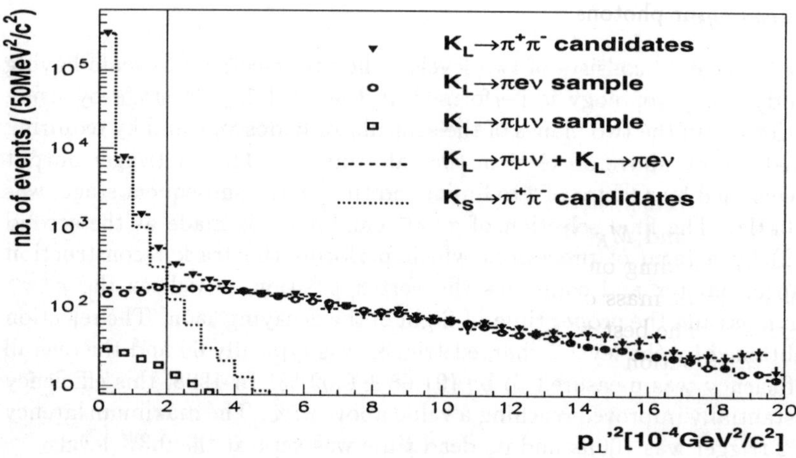

Figure 4: $p_\perp'^2$ distribution for K_L candidates (triangles) compared to the K_S ones (dotted histogram), normalized in the first bin. The open circles and open squares are respectively the K_{e3} and the $K_{\mu3}$ background contributions in the K_L beam.

the square of the momentum transfer ($p_\perp'^2 < 2 \times 10^{-4}$ GeV^2/c^2). The large background from semileptonic K_L decays is rejected using electron (E/p < 0.8) and muon identification. These requirements reduce both the K_{e3} and the $K_{\mu3}$ contributions by about a factor 500 while removing 8% of the good $\pi^+\pi^-$ events. The remaining K_L background in the signal region was found to be $(2.3\pm0.2\pm0.4)\times10^{-3}$. Figure 4 shows a comparison of the $p_\perp'^2$ distributions for K_L and K_S events after applying all the other analysis cuts.

Other charged background sources, like high-$p_\perp'$ events due to scattered kaons in the collimators or $K_{L,S} \to \pi^+\pi^-\gamma$, have also been investigated and their contribution to the correction on the double ratio measurement has been taken into account.

$K_{L,S} \to \pi^0\pi^0$ decays are selected by demanding four reconstructed in-time ($\pm$ 5ns) clusters in the LKr calorimeter with energy between 3GeV and 100GeV. The minimum distance required between photon candidates is 10cm. In order to reduce the background from $K_L \to 3\pi^0$ decays, events with an additional cluster of energy above 1.5GeV and within $\pm$ 3ns of the $\pi^0\pi^0$ candidate are rejected.

The kaon energy is obtained with a resolution of 0.6% from the summed energy of the four photons while the longitudinal position of the decay vertex relative to the front face of the calorimeter D is given by:

$$D = \frac{\sqrt{\sum_i \sum_{j>i} E_i E_j r_{ij}^2}}{M_K} \tag{4}$$

where E_i is the reconstructed energy of cluster i, r_{ij} the distance between two clusters i and j and M_K the kaon mass. The resolution of D varies between 50 to 70cm depending on energy.

The invariant mass of each of the two photon pairs m_1, m_2 is then computed using D. The best π^0 combination is found by pairing the photons which minimize the function

$$\chi^2 = \left\{ \left(\frac{\frac{m_1+m_2}{2} - m_{\pi^0}}{\sigma_{\frac{m_1+m_2}{2}}} \right)^2 + \left(\frac{\frac{m_1-m_2}{2}}{\sigma_{\frac{m_1-m_2}{2}}} \right)^2 \right\} \tag{5}$$

where $\sigma_{\pm}$ are the measured resolutions of $m_1 \pm m_2$. Typically, $\sigma_+ = 0.45 \text{MeV}/\text{c}^2$ and $\sigma_- = 1.1 \text{MeV}/\text{c}^2$.

Good $\pi^0\pi^0$ events are selected by requiring $\chi^2 < 13.5$. This condition rejects about 7% of the $K_S \to \pi^0\pi^0$ signal in a pure K_S beam where there is no background. These are due to photon conversions in the spectrometer or in the kevlar window, Dalitz decays and photon-nucleon interactions in the LKr calorimeter. Figure 5 shows the χ^2 distributions for $K_L \to \pi^0\pi^0$ and $K_S \to \pi^0\pi^0$ candidates after lifetime weighting. The excess of events in the K_L sample for $\chi^2 > 13.5$ is due to $K_L \to 3\pi^0$ decays and is used to determine the background level in the signal region. This was estimated to be $(8 \pm 2) \times 10^{-4}$ averaged over energy.

To obtain symmetric samples of charged and neutral events in the K_L and K_S modes, few additional cuts (detector activity, acceptance) were applied in the analysis. For instance, events which contained an accidental activity with a large number of hits in the drift chambers (> 7 per plane) were rejected if it occured within ± 312ns around the event time. This requirement was applied to both $\pi^+\pi^-$ and $\pi^0\pi^0$ modes in order to avoid possible biases related to different average beam intensity. The probability for such high multiplicity accidental events was found to be about 20%. Clusters in the vicinity of dead cells or the inactive vertical strip in the Lkr calorimeter were removed as well as events with at least one track pointing to these dead areas.

All events were also required to satisfy a symmetric cut on the position of the centre of gravity defined as the energy- or momentum-weighted average transverse position of the showers or extrapolated tracks at the front face of

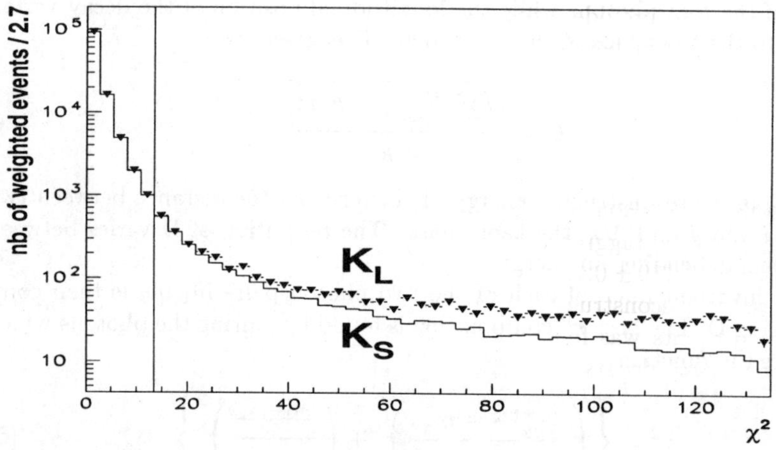

Figure 5: χ^2 distribution for weighted K_L events (triangles) compared to the shape for pure K_S events (histogram), normalized in the first bin.

the LKr calorimeter. Events with a radial position of the centre of gravity within 10cm of the beam axes were removed. By applying the same cuts to both modes, the sensitivity to differences in beam halo and to K_S scattering in the collimators or the AKS anti-counter is minimized. This centre of gravity is relatively wide compared to both the K_L and the K_S beam spot radii which are 3.6cm and 4.6cm respectively.

5.3 K_L versus K_S identification: proton tagging

The knowledge of the origin of a decay in the K_L or K_S beam relies on the information provided by the tagging counter located upstream of the K_S target. The K_L or K_S assignment is obtained by measuring the time difference between the passage of the proton in the tagging counter and the event time in the detector. Events with a time difference inside an interval of $\pm$2ns are called K_S, any other events are called K_L. Inefficiencies in the tagging counter (α_{SL}) such as misalignment and deadtime cause K_S decays to be identified as K_L. On the other hand, any accidental hit in the tagging counter produces a K_L to K_S transition (α_{LS}). Since these misidentifications are nearly decay mode independent, they lead primarily to a dilution of $\mathrm{Re}(\varepsilon'/\epsilon)$.

The proton tagging detector [18] has been designed to cope with high beam intensities and to provide a detection efficiency close to 100% combined with a time resolution of few hundred ps. It consists of two sets of twelve staggered thin scintillator foils arranged alternately in the horizontal and vertical directions, so that each counter sees only a fraction of the beam.

During the 1997 run, the tagging system was operated at a proton rate of 28MHz. An average time resolution of about 200ps per counter and a 4ns double pulse separation were achieved.

The accidental tagging probability was carefully measured for $\pi^+\pi^-$ events to be $\alpha_{LS}^{+-} = 0.1119 \pm 0.0003$ by making use of the good vertical separation of the K_L and K_S reconstructed decay vertices. Similarly, the tagging inefficiency for charged events was found to be $\alpha_{SL}^{+-} = (1.5 \pm 0.1) \times 10^{-4}$ by looking at vertex signed K_S decays. Detailed studies indicate that this is mainly due to inefficiencies in the tagger which is intrinsically symmetric between $\pi^+\pi^-$ and $\pi^0\pi^0$ events. Using a sample of neutral events containing π^0 Dalitz decays and gamma conversions, the comparison of the event time obtained from the LKr calorimeter with the time of the tracks measured with the hodoscope allowed us to extract the difference between tagging efficiencies for decays into $\pi^+\pi^-$ and $\pi^0\pi^0$. The tagging inefficiencies α_{SL}^{+-} and α_{SL}^{00} were found to be equal with an accuracy of $\pm 1 \times 10^{-4}$ leading to an uncertainty on R of $\pm 6 \times 10^{-4}$.

The charged-neutral accidental tagging probability $\alpha_{LS}^{+-} - \alpha_{LS}^{00}$ is obtained from the study of $K_L \to \pi^+\pi^-$ and $K_L \to \pi^0\pi^0$ events with out-of-time proton tags. The measured value $\alpha_{LS}^{+-} - \alpha_{LS}^{00} = (-0.10 \pm 0.05)\%$ yields a correction on R of $(18 \pm 9) \times 10^{-4}$.

Figure 6 shows the time correlation between signals observed in the tagging system and the NA48 detector for charged and neutral events.

6   The $\mathrm{Re}(\varepsilon'/\epsilon)$ value

The $\pi^+\pi^+$ and $\pi^0\pi^0$ events considered in the double ratio measurement are chosen to have reconstructed kaon energies in the 70-170GeV interval and lifetime τ less than $3.5\tau_S$. In the case of K_S decays, the beginning of the decay region is sharply defined by the AKS anti-counter, while in the K_L beam, events with a negative reconstructed τ value are rejected. Table 1 gives the number of events collected in each channel (no lifetime weighting) after removing background and correcting for accidental mistagged events.

The computation of R was done in 20 energy bins in order to decrease the sensitivity of the measurement to the difference between the beam spectra (about $\pm 10\%$ in the 70-170GeV range). In addition to the corrections on the R value due to tagging efficiencies and accidental tagging, the numbers of K_S and weighted K_L candidates were corrected in each bin for trigger inefficiency, background subtraction and acceptance.

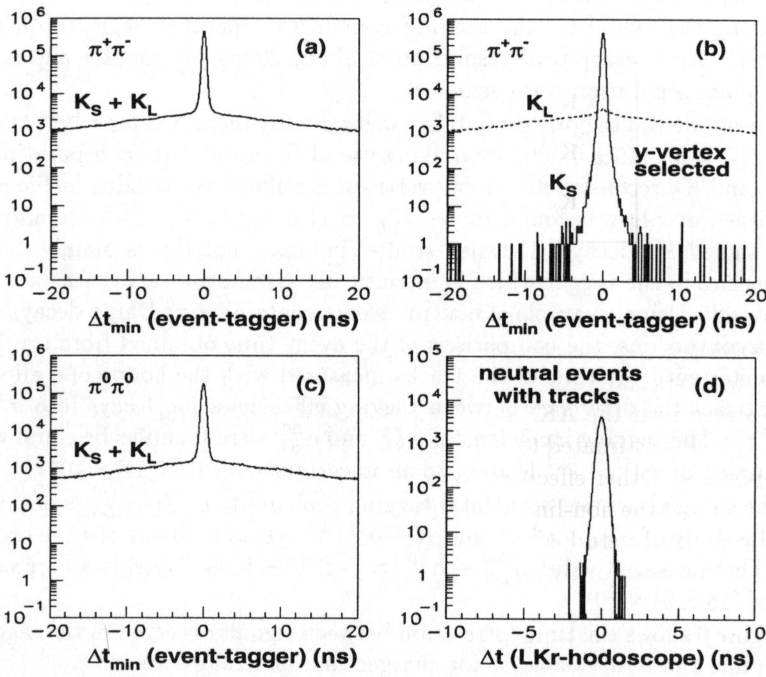

Figure 6: (a) Distribution of the minimum difference between tagger and event times for decays into $\pi^+\pi^-$. The peak corresponds to in-time K_S events. (b) Same distribution, separated into K_L and K_S using vertex information. (c) Same distribution as in (a) for the $\pi^0\pi^0$ mode. (d) Time correlation between the LKr and the hodoscope detectors for neutral events with tracks (see text).

The precise knowledge of the absolute energy scale in the experiment is crucial since a small error on it directly affects the estimated number of neutral kaons as a function of energy or lifetime. For charged $\pi^+\pi^-$ events, the vertex position and the kaon energy are completely determined by the detector geometry. Uncertainties in the geometry and on residual effects due to the magnetic field in the decay region (e.g. earth's magnetic field) lead to an estimate of the uncertainty on R of $\pm 5 \times 10^{-4}$. This is in agreement with the measurement of the AKS anti-counter longitudinal position obtained from the z-vertex distribution for good $\pi^+\pi^-$ events (Figure 7(a)).

Table 1: Statistical samples.

Mode	Events (10^6)
$K_L \to \pi^0\pi^0$	0.489
$K_S \to \pi^0\pi^0$	0.975
$K_L \to \pi^+\pi^-$	1.071
$K_S \to \pi^+\pi^-$	2.087

For $\pi^0\pi^0$ events, the neutral scale was adjusted by fitting the sharp edge of the reconstructed z-vertex (Figure 7(b)). The energy scale was also cross-checked with special runs performed with a π^- beam striking a thin CH_2 target located near the AKS position. The residual uncertainty on the overall energy scale was estimated to be $\pm 5 \times 10^{-4}$ which leads to a similar uncertainty on the R value. Other effects like the transverse scale of the calorimeter, non-uniformities and the non-linearity (0.3%) of the response of the LKr calorimeter were also carefully studied. They lead to a total systematic uncertainty of $\pm 12 \times 10^{-4}$ on the double ratio.

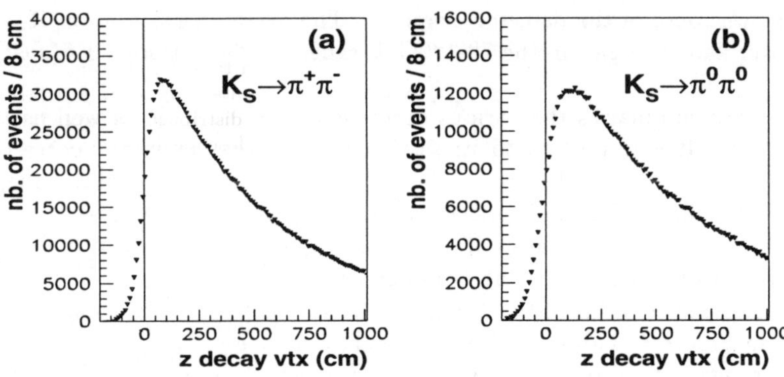

Figure 7: Distribution of the reconstructed decay vertex in K_S events for (a) the $\pi^+\pi^-$ mode and (b) the $\pi^0\pi^0$ mode. The rising edge corresponds to the position of the AKS counter.

Table 2: Corrections and systematic uncertainties on R.

Source	Correction (10^{-4})	Uncertainty (10^{-4})
Tagging $(\Delta\alpha_{LS}, \Delta\alpha_{SL})$	+18	11
Charged Trigger	+9	23
Charged Background	+23	4
Neutral Background	-8	2
Energy Scale/Linearity		12
Charged Vertex		5
Acceptance	+29	12
Accidental Activity	-2	14
Beam Scattering	-12	3
All	+57	35

Corrections due to accidental effects were determined by overlaying events with random triggers which represent the random accidental activity in the beam. Since all four decay modes are detected simultaneously, accidental effects are expected to cancel in the double ratio. The correction on the double ratio was estimated to be $(-2 \pm 14) \times 10^{-4}$.

Finally, the acceptance correction was measured using a detailed Monte-Carlo simulation taking into account the beam characteristics, the geometry and the response of the detector elements. The overall correction on R, averaged over kaon energies in the 70-170GeV range is $(29 \pm 11(stat) \pm 6(syst)) \times 10^{-4}$.

Table 2 summaries the various corrections applied to R as well as their uncertainty. It is important to stress that most of the systematic corrections for the 1997 run, namely those related to acceptance, accidentals, tagging and trigger efficiencies, have uncertainties which are dominated by statistics. These systematic uncertainties are therefore expected to significantly decrease as more data is collected by the NA48 experiment.

The corrected value of the double ratio as a function of the kaon energy is shown in Figure 8. Values of R in regions beyond the investigated 70-170GeV energy range have also been measured for cross-check purposes and are shown in the same figure.

The average of R over the bins in the 70-170GeV interval was done with an unbiased estimator of log(R) and was found to be $0.9889 \pm 0.0027(stat) \pm$

0.0035($syst$). The corresponding value for the direct CP violation parameter is:

$$Re(\varepsilon'/\epsilon) = (18.5 \pm 4.5(stat) \pm 5.8(syst)) \times 10^{-4} \qquad (6)$$

This result confirms that direct CP violation occurs in neutral kaon decays into two pions.

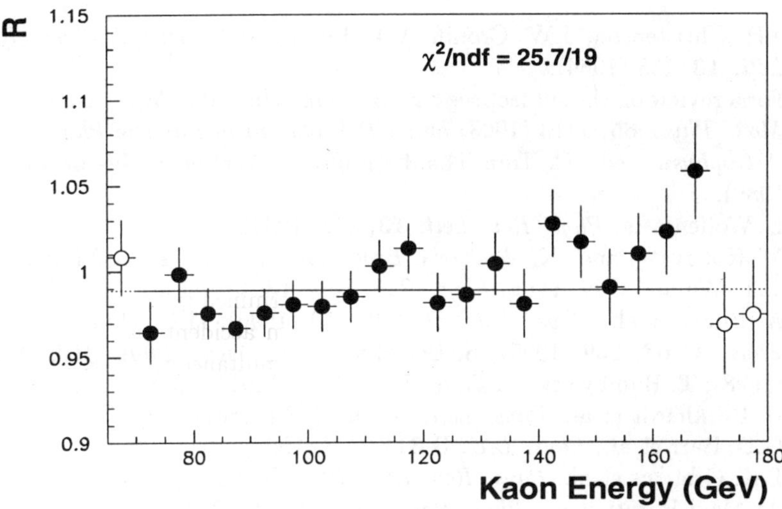

Figure 8: Measured double ratio R in energy bins. The points used for the determination of $Re(\varepsilon'/\epsilon)$ are shown in black. The open circles are used for cross-check purposes only.

7   Prospects and conclusion

Using a novel technique, the NA48 experiment has obtained a preliminary measurement of $Re(\varepsilon'/\epsilon) = (18.5 \pm 7.3)$ based on about 10% of the total expected amount of data. This result favours a non-zero value of $Re(\varepsilon'/\epsilon)$ at a 2.5σ level with an uncertainty that is dominated by statistics.

In 1998, the experiment accumulated respectively about 2.5 and 5 times more $\pi^0\pi^0$ and $\pi^+\pi^-$ statistics than in 1997 with improved detector and data acquisition system. The NA48 experiment is scheduled to take data until year 2000 in order to reach the proposed accuracy of 2×10^{-4} on $Re(\varepsilon'/\epsilon)$.

The average of the four most recent experimental results is $\text{Re}(\varepsilon'/\epsilon) = (21.2 \pm 2.8) \times 10^{-4}$ $(\chi^2/ndf = 8.4/3)$ which strongly supports the existence of direct CP violation in the neutral kaon system. The ultimate precision on $\text{Re}(\varepsilon'/\epsilon)$ that will be reached in the next few years by KTeV and NA48 will provide tight constraints on theoretical models. The scientific community is also awaiting results from the KLOE experiment in Frascati which started running in 1999.

References

1. J.H. Christenson, J.W. Cronin, V.L. Fitch and R. Turlay, *Phys. Rev. Lett.* **13**, 138 (1964).
2. For a review on the subject see e.g. B. Weinstein and L. Wolfenstein, *Rev. Mod. Phys.* **65**, 1113 (1993) and *CP Violation in Particle Physics and Astrophysics*, ed. C. Tran Thanh (Editions Frontières, Gif-sur-Yvette, 1989).
3. L. Wolfenstein, *Phys. Rev. Lett.* **13**, 562 (1964).
4. M. Kobayashi and. K. Maskawa, *Prog. Theor. Phys.* **49**, 652 (1973).
5. T.T. Wu and C.N. Yang, *Phys. Rev. Lett.* **13**, 380 (1964).
6. A. Buras et al., *Phys. Lett.* B **389**, 749 (1996); M.Ciuchini et al., *Z. Phys.* C **68**, 239 (1995); S. Bertolini et al., *Nucl. Phys.* B **514**, 93 (1998); T. Hambye et al., *Phys. Rev.* D **58**, 14017 (1998).
7. H. Burkhardt et al., *Phys. Lett.* B **206**, 169 (1988).
8. G.D. Barr et al., *Phys. Lett.* B **317**, 233 (1993).
9. L.K. Gibbons et al., *Phys. Rev. Lett.* **70**, 1203 (1993).
10. A. Alavi-Harati et al., *Phys. Rev. Lett.* **83**, 22 (1999).
11. G.D. Barr et al., *CERN/SPSC/90-22/P253*.
12. C. Biino et al., *CERN-SL-98-033 (EA)*.
13. N. Doble, L. Gatignon and P. Grafström, *Nucl. Instrum. Methods* B **119**, 181 (1996).
14. D. Béderède et al., *Nucl. Instrum. Methods* A **367**, 88 (1995); I. Augustin et al., *Nucl. Instrum. Methods* A **403**, 472 (1998);
15. B. Hallgren et al., *Nucl. Instrum. Methods* A **419**, 680 (1998);
16. S. Anvar et al., *Nucl. Instrum. Methods* A **419**, 686 (1998);
17. B. Gorini et al., *IEEE Trans. Nucl. Sci.* **45**, 1771 (1998);
18. P. Grafström et al., *Nucl. Instrum. Methods* A **344**, 487 (1994); H. Bergauer et al., *Nucl. Instrum. Methods* A **419**, 623 (1998).

Participant List

Jeffrey Appel
Fermilab
MS 122
PO Box 500
Batavia, IL 60510
USA
(630) 840-3922
appel@fnal.gov

Bernard Aubert
LAPP
Chemin de Bellevue
Annecy le Vieux, 74940
FRANCE
(334) 5009-1500
aubert@lapp.in2p3.fr

Cesare Bacci
INFN
Roma 3
V.d. Vasca Navale 84
Roma, 146
ITALY
(39) 655 177 234
bacci@aroma3.roma3.infn.it

Gaspar Barreira
LIP
Av. Elias Garcia 14-1.o
Lisbon, 1000
PORTUGAL
(35) 117 973 880
gaspar@lip.pt

Stan Bentvelsen
CERN
CH-1211
Geneva, 23
SWITZERLAND
(41) 22 767-4236
stan.bentvelsen@cern.ch

Dave Besson
University of Kansas
Department of Physics & Astronomy
1082 Malott Hall
Lawrence, KS 66045-2151
USA
(785) 864-4741
dbesson@ukans.edu

Stefano Bianco
Laboratori Nazionali di Frascati
via E. Fermi 40
Frascati, (Roma) 00044
ITALY
(39) 694 032 793
bianco@lnf.infn.it

Kenneth Bloom
University of Michigan
2477 Randall Laboratory
500 East University
Ann Arbor, MI 48109-1120
USA
(734) 763-2329
kenbloom@umich.edu

368

Walter Blum
MPI Munich and CERN
Div. EP, CERN
Ch 1211
Geneva, 23
SWITZERLAND
(41) 22 767-3943
walter.blum@cern.ch

Maurizio Bonesini
Sezione INFN, Milano
via Celoria 16
Milano, 20133
ITALY
(39) 022 392-388
bonesini@mi.infn.it

Jean-Marie Brom
IReS
Institute de Recherches Subatomiques
BP 28 Cro
Strasbourg Cedex 02, 67037
FRANCE
(33) 388 1062-72
jean-marie.brom@ires.in2p3.fr

Peter Bussey
University of Glasgow
University Avenue
Glasgow, G12 8QQ
UNITED KINGDOM
(14) 1 330-6417
p.bussey@physics.gla.ac.uk

Myron Campbell
University of Michigan
Physics Department
2477 Randall Laboratory
Ann Arbor, MI 48109-1120
USA
(734) 764-2192
myron@umich.edu

John Carlson
University of Michigan
Physics Department
2477 Randall Laboratory
Ann Arbor, MI 48109-1120
USA
(734) 936-6648
jmcarlso@umich.edu

John Carlson
University of Michigan
Physics Department
2477 Randall Laboratory
Ann Arbor, MI 48109-1120
USA
(734) 936-6688
jmcarlso@umich.edu

Carlo Caso
Genova University and INFN
Via Dodecaneso, 33
Genova, 16146
ITALY
(39) 010 353-6206
carlo.caso@ge.infn.it

Heriberto Castilla-Valdez
CINVESTAV-IPN
Av. IPN 2508
Col.. San Pedro Zacatenco
Mexico City, CP 07300 D.F.
MEXICO
(011) 525-747-7098
castilla@fis.cinvestav.mx

Giovanni Chiefari
INFN Sezione Di Napoli
Complesso Universitario di
Monte S. Angelo
Napoli, 80126
ITALY
(39) 81 676-181
chiefari@na.infn.it

Roberto Contri
University & INFN - Genova
Dip. Di Fisica - Via Dodecaneso 33
Genova, I 16146
ITALY
(39) 010 353-6465
contri@ge.infn.it

Anna Dabrowska
Institute of Nuclear Physics
Kawiory 26A
Krakow, 30-055
POLAND
(48) 12 633-3366
anna.dabrowska@ifj.edu.pl

Simone Dell'Agnello
INFN-LNF
PO Box 13
via E. Fermi, 40
Frascati Rome, 00044
ITALY
(39) 06 9403-2730
simone.dellagnello@lnf.infn.it

Mikhail Dubinin
Institute of Nuclear Physics
Moscow State University
Leninskie Gory
Moscow, 119899
RUSSIA
(095) 939-2393
dubinin@theory.npi.msu.su

Franco Fabbri
INFN
Laboratori Naz. Di Frascati
via E. Fermi 40 - CP 13
Frascati, Roma 00044
ITALY
(39) 06 9403-2913
flf@lnf.infn.it

James Fast
Purdue University
1396 Physics Building
West Lafayette, IN 47907
USA
(765) 494-5193
jfast@purpcj.physics.purd

Pauline Gagnon
CPP
1125 Colonel by Drive
Ottawa, ON, K1S 5B6
CANADA
(41) 22 767-6810
pauline.gagnon@cern.ch

David Gerdes
University of Michigan
Physics Department
2477 Randall Laboratory
Ann Arbor, MI 48109-1120
USA
(734) 647-3807
gerdes@umich.edu

Al Goshaw
Fermilab (Duke University)
CDF, Mail Station 318
PO Box 500
Batavia, IL 60510
USA
(630) 840-8612
goshaw@fnal.gov

Herbert Greenlee
Fermilab
MS 357
PO Box 500
Batavia, IL 60510
USA
(630) 840-3618
greenlee@fnal.gov

John Hauptman
Iowa State University
Physics Department
Ames, IA 50011
USA
(515) 294-8572
hauptman@iastate.edu

Yimei Huang
University of Michigan
Physics Department
2477 Randall Laboratory
Ann Arbor, MI 48109-1120
USA
(734) 763-7209
yimeih@umich.edu

Joey Huston
Michigan State University
Physics and Astronomy Dept.
East Lansing, MI 48824
USA
(517) 353-8783
huston@pa.msu.edu

Gianluca Introzzi
Univ. and INFN of Pavia
Via Bassi, 6
Pavia, 27100
ITLAY
(630) 840-8426
introzzi@fnal.gov

Gordon Kane
University of Michigan
2477 Randall Laboratory
500 East University
Ann Arbor, MI 48109-1120
USA
(734) 764-4451
gkane@umich.edu

Kenji Kaneyuki
Tokyo Institute of Technology
Ookayama, 1-12-1
Meguro
Tokyo, 152-8551
JAPAN
(81) 35 734-2388
kaneyuki@hp.phys.titech.ac.jp

Michael Kelly
University of Michigan
Physics Department
2477 Randall Laboratory
Ann Arbor, MI 48109-1120
USA
(734) 763-2323
cdfkelly@umich.edu

Soo-Bong Kim
Seoul National University
Department of Physics
Shilim-dong, Kwanak-ku
Seoul, 151-742
KOREA
(82) 2 880-5755
sbkim@phya.snu.ac.kr

Oleg Kouznetsov
DAPNIA/SPP
CEA-Saclay
Gif-sur-Yvette
Paris, FR-91191
FRANCE
(16) 90 885-33
oleg.kouznetsov@cern.ch

Mark Lancaster
University College

Renaud LeGac
CPPM
163, avenue de Luminy
Case 907
Marseille, 13288
FRANCE
(33) 49 182-7264
legac@in2p3.fr

Serguei Levonian
DESY
Notkestrasse, 85
Hamburg, 22607
GERMANY
(49) 408 998-3767
levonian@mail.desy.de

Ivan Logashenko
Budker Institute of Nuclear Physics
11 Lavrentiev Prospect
Novosibirsk, 630090
RUSSIA
(73) 832 394-366
logash@inp.nsk.su

Wolfgang Lorenzon
University of Michigan
Physics Department
2477 Randall Laboratory
Ann Arbor, MI 48109-1120
USA
(734) 647-6825
lorenzon@umich.edu

Abdenour Lounis
Institut de Recherches Subatomiques
Rue du Loess, 23
Strasbourg Cedex 02, 67037
FRANCE
(33) 38 810-6622
abl@sbghp3.in2p3.fr

Naomi Makins
University of Illinois
Loomis Laboratory of Physics
1110 West Green Street
Urbana, IL 61801
USA
(49) 408 998-4614
makins@hermes.desy.de

Reinhard Maschuw
University of Bonn
Nussalle 14-16
Bonn, D-53115
GERMANY
(228) 732-201
maschuw@iskp.uni-bonn.de

Edoardo Mazzucato
CEA DAPNIA/SPP
CEA/Saclay
gif sur Yvette, 91191
FRANCE
(33) 16 908-4476
edoardo@hep.saclay.cea.fr

Timothy McKay
University of Michigan
Physics Department
2477 Randall Laboratory
Ann Arbor, MI 48109-1120
USA
(734) 763-1462
tamckay@umich.edu

Ronald Moore
University of Michigan
Physics Department
2477 Randall Laboratory
Ann Arbor, MI 48109-1120
USA
(734) 763-2323
ronmoore@umich.edu

William Murray
Rutherford Appleton Laboratory
Chilton
Didcot, Oxon. OX11 0QX
UNITED KINGDOM
(44) 12 354-46256
w.murray@rl.ac.uk

Pavel Nadolsky
Michigan State University
Department of Physics & Astronomy
East Lansing, MI 48824-1116
USA
(517) 355-3519
madolsky@pa.msu.edu

Aleksei Pavlinov
Wayne State University
Department of Physics & Astronomy
6666 W. Hancock
Detroit, MI 48201
USA
(313) 577-1203
rykov@physics.wayne.edu

Emmanuelle Perez
CEA-Saclay
DAPNIA/SPP
Gif-sur-Yvette
Paris, 91191
FRANCE
(33) 16 908-5228
eperez@hep.saclay.cea.fr

Roberto Perrino
INFN Sezione di Lecce
via per Arnesano
Lecce, 73100
ITALY
(39) 083 2320-460
roberto.perrino@le.infn.it

Jianming Qian
University of Michigan
Physics Department
2477 Randall Laboratory
Ann Arbor, MI 48109-1120
USA
(734) 936-1033
qianj@umich.edu

Heather Ray
University of Michigan
Physics Department
2477 Randall Laboratory
Ann Arbor, MI 48109-1120
USA
hray@tquark.physics.lsa.umich.edu

Keith Riles
University of Michigan
Physics Department
2477 Randall Laboratory
Ann Arbor, MI 48109-1120
USA
(734) 764-4652
keithr@umalp1.physics.lsa.umich.edu

Krzysztof Rybicki
Institute of Nuclear Physics
Cracow
ul. Kawiory 26 A
Krakow, 30-055
POLAND
(48) 12 633-3366
krzysztof.rybicki@ifj.edu.pl

Vladimir Rykov
Wayne State University
Department of Physics & Astronomy
6666 W. Hancock
Detroit, MI 48201
USA
(313) 577-2781
rykov@physics.wayne.edu

Yves Sacquin
CEA DSM/DAPNIA
CEA/Saclay
gif sur Yvette, 91191
FRANCE
(33) 16 908-6081
sacquin@dapnia.cea.fr

Alexander Savin
DESY/F1
Notkestr. 85
Hamburg, 22603
GERMANY
(49) 40 899-83601
savin@mail.desy.de

Bruce Schumm
University of California-Santa Cruz
SCIPP
Natural Sciences II
Santa Cruz, CA 95064
USA
(831) 459-3034
schumm@scipp.ucsc.edu

Peter Shanahan
Fermilab
PO Box 500
MS 231
Batavia, IL 60510
USA
(630) 840-8378
shanahan@fnal.gov

Monika Szarska
Institute of Nuclear Physics
Kawiory 26A
Krakow, 30-055
POLAND
(48) 12 633-3366
monika.szarska@ifj.edu.pl

Monica Tecchio
University of Michigan
Physics Department
2477 Randall Laboratory
Ann Arbor, MI 48109-1120
USA
(734) 763-2329
tecchio@umich.edu

Roberto Tenchini
INFN - Pisa
Via Livornese 1291
San Peiro a Grado
Pisa, I-56010
ITALY
(39) 050 880-296
roberto.tenchini@cern.ch

Bob Tschirhart
Fermilab
PO Box 500
Batavia, IL 60510
USA
(630) 840-4100
tsch@fnal.gov

Alexei Varganov
University of Michigan
Physics Department
2477 Randall Laboratory
Ann Arbor, MI 48109-1120
USA
(734) 936-6648
varganov@umich.edu

Horst Wahl
Florida State University
Physics Department
MS 4350 KEN 512
Tallahasse, FL 32306-4350
USA
(850) 644-3509
wahl@hep.fsu.edu

Rudolf Thun
University of Michigan
Physics Department
2477 Randall Laboratory
Ann Arbor, MI 48109-1120
USA
(734) 936-0792
rthun@umich.edu

Mike Turner
University of Chicago
Department of Astronomy and Astrophysics
5640 S. Ellis Avenue
Chicago, IL 60637
USA
(773) 702-7974
mturner@oddjob.umich.edu

Maneesh Wadhwa
University of Basel
Klingelbergstrasse 82
Basel, CH-4056
SWITZERLAND
(41) 22 767-4250
maneesh.wadhwa@cern.ch

Arthur B. Wicklund
Argonne National Laboratory
3511 Fremont Street
Chicago, IL 60657
USA
(630) 252-6215
abw@hep.anl.gov

John Wilkerson
University of Washington
Nuclear Physics Laboratory
Box 354290
Seattle, WA 98195
USA
(206) 685-9061
jfw@u.washington.edu

David Wolinski
University of Michigan
Physics Department
2477 Randall Laboratory
Ann Arbor, MI 48109-1120
USA
(734) 936-6648
wolinski@umich.edu

Sarah Wolinski
University of Michigan
Physics Department
2477 Randall Laboratory
Ann Arbor, MI 48109-1120
USA
(734) 936-6688
truitt@umich.edu

Qichun Xu
University of Michigan
Physics Department
2477 Randall Laboratory
Ann Arbor, MI 48109-1120
USA
(734) 763-7209
xuq@umich.edu

Bing Zhou
University of Michigan
Physics Department
2477 Randall Laboratory
Ann Arbor, MI 48109-1120
USA
(734) 647-3760
bzhou@umich.edu

Andrzej Zieminski
Indiana University
Physics Department
Bloomington, IN 47408
USA
(812) 855-4089
zieminsk@indiana.edu

Participating Vendor List

ADCO Circuits, Inc.   (248) 853-5525
1900 Northfield Drive   FAX: (248) 348-4819
Rochester Hills, MI 48309
USA

Exabyte   (330) 626-8000
1286 Vantage Way
Streetsboro, OH 44241
USA

Hewlett Packard   (248) 380-2545
29550 Orchard Hills Place Drive
PO Box 8024
Novi, MI 48376-8024
USA

Micron Semiconductor   (407) 421-5977
478 E. Altamonte Drive, Suite 108-570   FAX: (407) 360-1732
Altamonte Springs, FL 32701
USA

Rittal Corporation   (800) 477-4220
28800-A Beck Road
Wixom, MI 48393
USA

Tektronix   (248) 305-6429
29555 Orchard Hills Place Drive, Suite 525
Novi, MI 48375
USA